Robin Marc Plum

Fatigue crack detection on structural steel members by using ultrasound excited thermography

BAND 3

Versuchsanstalt für Stahl, Holz und Steine
Berichte zum Stahl- und Leichtbau

Fatigue crack detection on structural steel members by using ultrasound excited thermography

by
Robin Marc Plum

KIT Scientific Publishing

Dissertation, Karlsruher Institut für Technologie (KIT)
Fakultät für Bauingenieur-, Geo- und Umweltwissenschaften, 2015
Tag der mündlichen Prüfung: 13. Februar 2015
Referent: Prof. Dr.-Ing. Thomas Ummenhofer
Korreferent: Prof. Dr.-Ing. Martin Mensinger

Impressum

KIT Scientific Publishing

Karlsruher Institut für Technologie (KIT)
KIT Scientific Publishing
Straße am Forum 2
D-76131 Karlsruhe

KIT Scientific Publishing is a registered trademark of Karlsruhe Institute of Technology. Reprint using the book cover is not allowed.

www.ksp.kit.edu

Print on Demand 2015

ISSN 2198-7912
ISBN 978-3-7315-0417-7
DOI: 10.5445/KSP/1000048191

“Vibrothermography [...] has transitioned from a laboratory curiosity to one of the most exciting new NDE methods.”

Stephen D. Holland, Iowa State University

Vorwort

Die vorliegende Arbeit entstand im Rahmen meiner Tätigkeit als wissenschaftlicher Mitarbeiter an der Versuchsanstalt für Stahl, Holz und Steine des Karlsruher Instituts für Technologie (KIT). Grundlage der Ausarbeitung ist ein am KIT Stahl- und Leichtbau durchgeführtes Forschungsprojekt, welches von der Deutschen Forschungsgemeinschaft (DFG) finanziert wurde. Wesentliche Voraussetzungen für die erfolgreiche Bearbeitung des Themas wurden jedoch schon Jahre zuvor an der Technischen Universität Braunschweig geschaffen, an der ich sämtliche experimentellen Untersuchungen durchführen durfte. Zum Gelingen dieser Arbeit haben schließlich zahlreiche Personen beigetragen, denen ich an dieser Stelle gerne danken möchte.

Besonderer Dank gebührt meinem Doktorvater, Prof. Dr.-Ing. Thomas Ummenhofer, der diese Dissertation angeregt und fachlich begleitet hat. Sowohl als Braunschweiger als auch als Karlsruher Mitarbeiter konnte ich stets auf seine volle Unterstützung zählen. Der mir gewährte Freiraum bei der Ausgestaltung meiner Forschungsarbeit und sein großes Vertrauen in meine Person haben mich stark motiviert. Sehr abwechslungsreiche und herausfordernde Tätigkeiten unter seiner Leitung lassen mich auf eine fachlich sowie persönlich prägende und erfüllte Zeit zurückblicken.

Für das Interesse an meiner Arbeit und die freundliche Übernahme des Korreferats bedanke ich mich bei Herrn Prof. Dr.-Ing. Martin Mensinger. Ich danke Herrn Prof. Dr.-Ing. Hans Joachim Blaß für die Übernahme des Prüfungsvorsitzes sowie Herrn Prof. Dr.-Ing. Karl Schweizerhof für die Mitgestaltung der Prüfung.

Ausdrücklich bedanken möchte ich mich bei Herrn Dr.-Ing. Justus Medgenberg, der durch seine Arbeit am Braunschweiger Institut für Bauwerkserhaltung und Tragwerk mein Interesse an der Infrarotthermografie und der zerstörungsfreien Prüfung geweckt hat und mir in vielen Diskussionen Anregungen zur Untersuchung weitergehender Fragestellungen gab.

Herr Ronny Wegener hat einen erheblichen Anteil zur effizienten Durchführung meiner Laborversuche beigetragen, indem er mich über Jahre hinweg mit maßgeschneiderten, professionell programmierten Software-Lösungen unterstützt hat. Hierfür danke ich ihm vielmals.

Für die wissenschaftliche Kooperation mit der Siemens AG und insbesondere für die freundliche Geräteleihgabe der Abteilung Corporate Technology in München danke ich stellvertretend Herrn Dr.-Ing. Christian Homma.

Herrn Prof. Dr.-Ing. Udo Peil und Herrn Dr.-Ing. Matthias Reininghaus danke ich herzlich für das Ermöglichen meines Gastaufenthalts am Institut für Stahlbau der TU Braunschweig. Die Herren Olaf Einsiedler, Stephan Amelung und Andreas Limbach haben mir nicht nur vollen Zugriff auf die Laborausstattung gewährt, sondern mich tatkräftig bei der Realisierung zahlreicher Versuchsaufbauten unterstützt. Alle technischen Mitarbeiter sowie die Auszubildenden haben ihr Können bei der Anfertigung komplexer Bauteile unter Beweis gestellt. Sie verdienen großes Lob und hohe Anerkennung für diese Leistung.

Den vielen Kollegen, die ich in Braunschweig am Institut für Bauwerkserhaltung und Tragwerk und am Institut für Stahlbau sowie am KIT Stahl- und Leichtbau kennengelernt habe, danke ich für die gute Zusammenarbeit und die angenehme Arbeitsatmosphäre an beiden Standorten.

Mein ganz besonderer Dank gebührt meinen Eltern Evelin und Werner Plum für ihre immerwährende Unterstützung und ihren Rückhalt. Schließlich danke ich meiner Freundin Carolin Sass für ihre Motivation und ihren Beistand aber auch für ihre große Geduld und Nachsicht.

Braunschweig, im Februar 2015 Robin Marc Plum

Abstract

In the field of non-destructive testing (NDT), ultrasound excited thermography has been recognised as a promising technique that was successfully applied to metals, fibre composites and many more engineering materials in order to detect cracks, delaminations and other types of internal flaws. Dating back to the late 1970s, the idea of high-frequency vibration excitation of structural members combined with monitoring the surface temperature by means of infrared thermography aims at the localised energy dissipation at defect regions and its thermal detection. The purpose of this thesis is to investigate the potential use of ultrasound excited thermography for detecting surface breaking fatigue cracks in thick-walled components relevant to steel construction. The presented research is motivated by a lack of fast and imaging crack detection methods in the field and the growing acceptance and technological progress of active thermography techniques.

After introducing the concept of ultrasound excited thermography or vibrothermography, its current state of the art is described by means of a comprehensive literature review focusing on research activities towards crack detection on metals. Owing to the interdisciplinarity of the test method, all relevant technical subdisciplines from the excitation of plate vibrations via potential heat generation mechanisms and heat transfer to infrared thermography are outlined.

The experimental work starts with the manufacture and fatigue loading of suitable plate specimens made from low-carbon steel S355, mostly in the high cycle fatigue regime, to generate through-thickness cracks with specified depths. Using a modified high-power ultrasonic welding generator, basic dependencies of the defect heating on frequency, coupling location and excitation duration are clarified at first. Besides of an estimation of realistic detection limits depending on the plate thickness, main issues such as the relation between vibration intensity and defect heating, wear and damage due to repeated ultrasound insonification and the influence of a static preload are addressed in detail. A procedure for 3D scanning of vibration mode shapes using a single-point laser vibrometer is developed in order to predict the frictional power dissipation based on crack face displacements. In-situ crack detection is simulated on a hot-rolled girder featuring multiple cracks. The efficiency of several prototypes of a portable excitation device is evaluated.

Finally, a numerical analysis using commercial finite element codes is proposed to simulate the high-frequency mechanical excitation, the frictional heat generation and the transient heat diffusion of a cracked plate specimen. The random distribution of crack face asperities is taken into account.

The performed vibrothermographic experiments combined with laser vibrometry reveal that significant defect heating can only be achieved at resonant excitation. Depending on the plate thickness, corner flaws smaller than 1 mm are securely detectable by means of a tunable ultrasound generator. The thermal response is found to be highly sensitive to small variations of the local gap between the crack edges. Therefore, a static preload acting on the structure but also residual compressive stresses generally have a strong impact on the crack detection capability and affect the exact location of the thermal hot spot along the crack path. Excessively repeated ultrasound excitation leads to a decrease in defect heating and localised wear of the crack faces. Further damage to the component in the form of surface cracking is possible. The simulated in-situ testing reveals an encouraging ratio of detected to total present cracks and proves that portable excitation devices with a magnetic coupling can be effectively used. Instead of a universally valid relationship between vibration intensity and defect heating, a variety of reasonable functional relations are found, which can be explained theoretically. The analysis of vibration mode shapes indicates that the relative heating at different frequencies can be estimated on the basis of 3D displacement components when considering basic friction laws together with a normal contact modulation. The numerical simulation results in a good agreement with experiments and can be enhanced to predict the detectability of arbitrarily shaped cracks in complex steel structures.

The present work contributes to the implementation of vibrothermographic crack detection on large-scale steel structures. Experimental investigations prove the technique's potential to be further developed towards a fail-safe, imaging NDT method with a significant range of operation and intuitive result interpretation. Future efforts are required in order to design optimised portable excitation devices and to overcome the dependencies of defect heating on the crack mouth opening and the coupling position.

Kurzfassung

Eine vielversprechende Methode im Bereich der zerstörungsfreien Prüfung metallischer und faserverstärkter Bauteile ist die ultraschallangeregte Thermografie, mit der Risse, Delaminationen und andere innere Fehlstellen detektiert werden können. Die Idee der hochfrequenten Schwingungsanregung von tragenden Komponenten bei gleichzeitiger Verfolgung der Oberflächentemperatur ist seit den späten 1970er Jahren bekannt. Sie zielt darauf ab, dass an Fehlstellen lokal Energie dissipiert wird und diese somit thermisch nachweisbar werden. Gegenstand dieser Arbeit ist die Klärung des Potentials der ultraschallangeregten Thermografie, offene Ermüdungsrisse in dickwandigen Stahlbauteilen zu detektieren. Die Untersuchungen sind zum einen durch den Mangel an schnellen, bildgebenden Prüfverfahren im konstruktiven Stahlbau und zum anderen durch die wachsende Akzeptanz und die technologischen Fortschritte aktiver thermografischer Methoden motiviert.

Nach einer Darstellung des Prinzips der „Vibro-Thermografie" wird ihr Stand der Technik anhand einer Literaturrecherche zur Risserkennung an Metallen beschrieben. Aufgrund der starken Interdisziplinarität des Verfahrens folgt ein Abriss aller relevanten Teildisziplinen, von Plattenschwingungen über die Mechanismen der Wärmeerzeugung und des -transports bis hin zur Infrarotthermografie.

Die experimentelle Arbeit beginnt bei der Herstellung und Ermüdung geeigneter Blechprobekörper aus Baustahl S355, um vornehmlich im high cycle fatigue Bereich Seitenanrisse vorgegebener Tiefe zu erzeugen. Mit Hilfe eines leistungsfähigen Ultraschallschweißgenerators werden zunächst grundlegende Abhängigkeiten der Risserwärmung von der Anregungsfrequenz und -dauer sowie der Ankopplungsstelle erörtert. Neben einer Abschätzung der blechdickenabhängigen Nachweisgrenze werden das Verhältnis zwischen Vibrationsstärke und Erwärmung, das Verschleißverhalten bei wiederholter Anregung sowie der Einfluss statischer Vorspannung untersucht. Die dreidimensional scannende Laservibrometrie hat zum Ziel, basierend auf punktuell gemessenen Verschiebungen, die Dissipationsleistung in Folge Rissuferreibung zu prognostizieren. Ein 3,2 m langer, mehrfach gerissener Walzträger erlaubt die Simulation einer In-situ-Rissprüfung, wobei verschiedene Prototypen einer mobilen Anregungskonstruktion erprobt und hinsichtlich Risserkennung bewertet werden.

Schließlich erfolgt eine numerische Simulation des gerissenen Blechprobekörpers mit kommerzieller Finite-Elemente-Software, welche die hochfrequente mechanische Anregung, die Erzeugung von Reibungswärme und die transiente Wärmeausbreitung wiedergibt. Die zufällige Verteilung der Rauigkeitsspitzen auf den Rissflächen findet dabei explizit Berücksichtigung.

Thermografische Untersuchungen mit begleitenden vibrometrischen Messungen zeigen deutlich, dass sich nur bei resonanter Anregung eine signifikante Risserwärmung erzielen lässt. Mittels abstimmbarem Ultraschallgenerator sind, abhängig von der Plattendicke, Eckanrisse kleiner als 1 mm nachweisbar. Hierbei haben bereits kleinste Veränderungen der lokalen Rissklaffung große Auswirkungen auf den Temperaturanstieg. Äußere statische Lasten, aber auch Druckeigenspannungen, welche auf den Rissbereich einwirken, beeinflussen daher generell das Potential einer Risserkennung und bei näherer Betrachtung die genaue Position der thermischen Anzeige entlang des Risspfades. Die wiederholte Beaufschlagung mit Ultraschall führt zu verminderter Risserwärmung und lokalem Verschleiß der Rissflächen. Eine zusätzliche Oberflächenrissbildung kann dabei nicht ausgeschlossen werden. In einer nachgestellten realen Prüfsituation wurden vielversprechende Ergebnisse verzeichnet, da sechs von acht Rissen im Walzträger erkannt wurden und sich mobile Anregungskonstruktionen mit magnetischer Ankopplung als effektiv herausstellten. Statt einer allgemeingültigen Beziehung zwischen Vibrationsstärke und Risserwärmung wurden verschiedene, theoretisch nachvollziehbare funktionale Zusammenhänge festgestellt. Die Analyse der rekonstruierten Schwingmoden deutet darauf hin, dass sich die relative Dissipationsleistung bei verschiedenen Frequenzen auf Basis der räumlichen Verschiebungskomponenten abschätzen lässt. Der Ansatz Coulombscher Reibung und modulierter Kontaktpressung ist dabei ausreichend. Die numerische Simulation zeigt eine gute Übereinstimmung mit experimentellen Ergebnissen und kann erweitert werden, um die Detektierbarkeit beliebiger Risse in komplexen Stahlbauteilen zu prognostizieren.

Die vorliegende Arbeit stellt einen Beitrag zur Übertragung der ultraschallangeregten Thermografie auf die Prüfung von Stahlbauten dar. Experimentelle Untersuchungen zeigen das Potential dieser einzigartigen Methode auf, zu einem zuverlässigen bildgebenden Prüfverfahren weiterentwickelt zu werden, welches sich durch eine große Reichweite und einfache Ergebnisinterpretation auszeichnet. Zukünftiger Forschungsbedarf wird insbesondere in der Optimierung der Schalleinleitung mit Hilfe mobiler Anregungsapparaturen und in der Kompensation starker Abhängigkeiten der Risserwärmung, z.B. vom Ankopplungsort und der Rissklaffung, gesehen.

Contents

Notation

Abbreviations

1D	one-dimensional
2D	two-dimensional
3D	three-dimensional
avg.	average
A/D	analogue/digital
CFRP	carbon-fibre-reinforced plastic
CMOD	crack mouth opening displacement
CNC	computer numerical control
CV	coefficient of variation
diff.	differential
DAQ	data acquisition
DFT	discrete Fourier transform
ECT	eddy current testing
FAA	(U.S.) Federal Aviation Administration
FEA	finite element analysis
FE(M)	finite element (method)
FFT	fast Fourier transform
FPA	focal plane array
FPI	fluorescent penetrant inspection
HCF	high cycle fatigue
HgCdTe	mercury cadmium telluride
HP	high-pass
InSb	indium antimonide
IR	infrared

LCF	low cycle fatigue
LP	low-pass
LWIR	long wave infrared
max.	maximum
min.	minimum
MT	magnetic particle inspection
MWIR	mid-wave infrared
NDT	non-destructive testing
NETD	noise equivalent temperature difference
POD	probability of detection
PT	(dye) penetrant testing/inspection
PTFE	polytetrafluoroethylene
PVC	polyvinyl chloride
RMS	root mean square
SEM	scanning electron microscopy
SIF	stress intensity factor
SNR	signal-to-noise ratio
STD	standard deviation
SWIR	short wave infrared
THD	total harmonic distortion
US	ultrasound, ultrasonic
UT	ultrasonic testing
var.	variable

Prefixes and units

μ...	micro...
m...	milli...
k...	kilo...
M...	mega...
G...	giga...
°C	degree Celsius
gv	grey value

Hz	Hertz
K	Kelvin
m	metre
N	Newton
Pa	Pascal (N/m^2)
px	pixel
s	second
sr	steradian
S	sample
W	Watt

Indices

$(...)_0$	reference value, normalised
$(...)_{\mathrm{a}}$	amplitude
$(...)_{\mathrm{act}}$	actual
$(...)_{\mathrm{atm}}$	atmosphere
$(...)_A$	plate side A
$(...)_{\mathrm{b}}$	bending
$(...)_{\mathrm{bb}}$	blackbody
$(...)_B$	plate side B
$(...)_{\mathrm{cum}}$	cumulated
$(...)_{\mathrm{d}}$	damping
$(...)_{\mathrm{e}}$	elastic
$(...)_{\mathrm{eff}}$	effective
$(...)_{\mathrm{env}}$	environment
$(...)_{\mathrm{exc}}$	excitation
$(...)_\lambda$	spectral
$(...)_{\mathrm{m}}$	mean
$(...)_{\mathrm{nom}}$	nominal
$(...)_{\mathrm{obj}}$	object
$(...)_{\mathrm{p}}$	plastic
$(...)_{\mathrm{tot}}$	total

Operators and diacritical marks

y, Y	scalar or real function
$\underline{y}, \underline{Y}$	complex number or function
$\vec{y}, \vec{Y}$	vector
$\mathbf{Y}$	matrix
$\mathrm{Im}\{\underline{y}\}$	imaginary part of y
$\mathrm{Re}\{\underline{y}\}$	real part of y
$\|y\|$	absolute value
$\hat{y}$	amplitude
$\bar{y}$	average
$\mathrm{Med}\{y\}$	median value of y
Δ	Laplace operator
Δy	difference, increment or peak to peak variation of y
∇	nabla operator
∂	partial derivative operator
d	absolute derivative operator
$*$	convolution operator

Greek and latin letters

α	thermal diffusivity, coefficient of thermal expansion
α_{c}	convective heat transfer coefficient
α_{r}	radiative heat transfer coefficient
α_{T}	linear thermal expansion coefficient
a	crack depth
A	area
A_5	failure strain based on a reference length $L_0 = 5 \cdot \varnothing_0$
β	volume expansion coefficient
B	(real) plate bending stiffness
c	wave propagation velocity, crack length
c_{l}	longitudinal wave propagation velocity
c_{t}	transversal wave propagation velocity

c_b	bending wave propagation velocity
C_p	specific heat capacity at constant pressure
d	duration of (pulse or sweep) excitation
ϵ	emissivity
ε	strain
E	(real) modulus of elasticity
E^*	composite modulus of elasticity
ζ	Lehr's damping ratio
η	mechanical loss factor
f	frequency
f_mod	modulation frequency
f_s	sampling frequency
F	force
F_n	normal contact force
G	shear modulus
G^*	composite shear modulus
Gr	Grashof number
h	plate thickness
i	imaginary unit with $\mathrm{i}^2 = -1$
κ	circular wavenumber
k	coefficient of thermal conductivity
K	stress intensity factor (SIF)
K_I, K_II, K_III	mode I, II and III stress intensity factor (SIF)
λ	wavelength
μ	coefficient of friction (general)
μ_s	static coefficient of friction
μ_d	dynamic coefficient of friction
μ_th	thermal diffusion length
M	specific radiance
ν	Poisson's ratio
$\vec{n}$	normal vector
N	number of load cycles
N_a	number of load cycles to nominal crack depth
N_c	number of load cycles to crack initiation

N_f	number of load cycles to failure
Nu	Nusselt number
ω	angular (eigen)frequency
Ω	angular excitation frequency
P	relative excitation level, power
P_v	volumetric heat source power
Pr	Prandtl number
q	heat flux (with unspecified direction)
$\vec{q}$	heat flux vector
q_c	convective heat flux
q_r	radiative heat flux
ρ	mass density
r	empirical correlation coefficient
R	load ratio $R = F_\mathrm{min}/F_\mathrm{max} = \sigma_\mathrm{min}/\sigma_\mathrm{max}$, reflection coefficient, roughness profile
Ra	Rayleigh number
R_eH	upper yield strength
R_m	ultimate strength
$R_\mathrm{p0.2}$	proof strength
σ	stress, Stefan-Boltzmann constant, standard deviation
σ_1	first principle stress
σ_I	first stress invariant
σ_y	yield stress (axial)
τ	shear stress, friction stress, transmissivity
τ_y	yield stress (shear)
t	time
T	absolute temperature, transmission coefficient
υ	kinematic viscosity
v	velocity
φ	phase shift, eigenfunction
W	work
z	approach of mean planes

1 Introduction

1.1 Problem statement and motivation

In the past, undetected flaws in steel members caused catastrophic damage events in nautical, aerospace, rail vehicle and civil engineering. Thus, one of the main objectives in non-destructive testing/inspection (NDT/NDI) of steel constructions is to detect and localise surface breaking and near-surface cracks. A main reason for cracking of steel members is the accumulation of fatigue damage due to cyclic loading. Here, microstructural material changes set in well below the static proof stress. Growth and coalescence of the yet undetectable defects yield the formation of one dominant flaw. The macroscopic crack initiation is followed by the stage of crack growth, which will finally lead to a complete fracture, unless a load redistribution shields the damaged component. Sharp notches, rapid stiffness variations and weld seams are the most critical points concerning fatigue design. In zinc-coated steel structures potential damages due to liquid metal embrittlement

Zinc-coated steel girder with attached head plate in a multi-storey car park

Crack caused by liquid metal embrittlement, found at the end of the half-plate weld

Figure 1.1: Magnetic particle inspection (MT) of a welded steel structure.

(LME) are to be detected as well (Fig. 1.1). These cracks are filled with zinc, which might impede the detection using established NDT methods such as magnetic particle inspection (MT) or liquid penetrant testing (PT). Both techniques represent the most widely used standard methods, whose flaw indications are very easy to interpret. As a drawback, manual operation is demanding and time-consuming if many details are to be examined. In-situ use of radiographic methods (RT) is not as flexible as MT and PT but very expensive and thus considered only for safety-relevant components like welded pipelines, etc. Ultrasonic inspection (UT) and eddy-current testing (ECT) are advanced but non-imaging techniques. Highly qualified personnel are required to perform the test and judge on the achieved output signals. A major disadvantage shared by all common NDT methods is the limited range of operation. Thus, in the field of steel construction there is a need for innovative and imaging NDT methods, which are fast, fail-safe, and easy to handle.

1.2 Strategy and objectives of this work

Due to the rapid technical progress of staring array infrared (IR) cameras thermal methods recently gained strong attention among the NDT research community. A promising approach based on active thermography that was never transferred and applied to civil engineering structures might be further developed towards a powerful test method for the detection of cracks in massive steel components: Ultrasound excited thermography, which is also known as vibrothermography, thermosonic activation or sonic IR could not only fulfil the above stated requirements put on NDT methods but would also have a strongly enlarged range of operation. The idea of high-power ultrasound excitation is to generate elastic waves that propagate far inside the investigated structure. In case of internal flaws, such as cracks, the boundary faces move relatively to each other. The resulting rubbing and clapping of crack faces generates frictional heat, which is detected by means of an infrared camera. Cyclic plastic deformation at the crack tip potentially causes a localised energy dissipation that heats up the crack, but could also induce further damage to the structure.

Within the last four decades numerous research groups dealt with active thermography approaches and successfully broadened the field of applications. In contrast to the well established thermal excitation methods, which seem to be most developed, there is no unified theory that explains all observed phenomena involved with vibrothermographic crack detection. This lack of understanding is mainly due to the complicated interaction of opposing crack faces. Dry friction, as one of the main heating mechanisms, becomes extremely complicated if one switches from a macro to a meso or microscopic point of view. Considering the typically achieved displacement amplitudes in the order of a few micrometres this change of perspective seems to be fully justified and necessary to better understand the underlying physical effects. Similar difficulties regarding measurement and identification apply to the subject areas including cyclic plasticity, material degradation and fracture. As a result, the following issues of ultrasound excited thermography are currently addressed to evaluate the method's potential and to extend its range of application:

- Limit flaw size being securely detectable
- Geometric and morphologic conditions met by detectable flaws
- Relationship between vibration amplitude and defect heating
- Interference due to material damping
- Wear and fatigue damage due to intense ultrasonic vibration
- Reproducibility and probability of detection
- Efficiency of portable exciters

The purpose of this thesis is to investigate the potential use of ultrasound excited thermography as a non-destructive test method for detecting surface breaking fatigue cracks in thick-walled steel components relevant to steel construction. In contrast to many other studies a systematic procedure is chosen regarding the generation of low-carbon steel samples and the execution of main test series. All of the above stated aspects are covered by various investigations on realistic defects, especially against the background of a harsh testing environment. The number of considered defect types is intentionally limited to two: Only corner flaws and through-edge cracks in plate-like steel members are examined in this work.

To provide a comprehensive state of the art, important results of existing works in the field but also theoretical approaches and measurement strategies are outlined first. Owing to the pronounced interdisciplinarity of vibrothermography various principles are reviewed following the logical succession: Energy input, energy conversion, energy detection. The main aspects involved are the vibration of plates, potential heat generation mechanisms, heat transfer and infrared thermography. Special attention is given to the analysis of dissipative effects such as dry friction including normal force modulation and micro-slip behaviour, but also material damping and cyclic plasticity concerning the theoretical relation between vibration amplitude and defect heating. Finally, the commonly used ultrasound excitation methods and the corresponding data processing techniques are briefly discussed.

Experimental investigations of the crack detection capabilities of vibrothermography are based on cracked plate specimens that have been subjected to controlled fatigue loading. One of the primary goals is to identify the flaw detection limit depending on the plate thickness and the static preload using a commercially available high-power ultrasound generator and a highly sensitive focal plane array infrared camera. Further effects on the thermal response due to the variation of ultrasound frequency, bulk temperature and excitation position are separately analysed. Additionally, the occurrence and intensity of standing wave patterns, which also affect the temperature distribution, is examined and compared to theoretical predictions. The question whether vibrothermography is a truly non-destructive method represents another important issue to be addressed. Therefore, excessively repeated testing is carried out by means of a fully automated setup so that long-term effects with respect to morphologic alteration of fracture surfaces and crack growth can be assessed. In-situ testing of a load-bearing member is simulated by coupling portable excitation devices to a hot-rolled girder featuring several fatigue cracks of different depths. Realistic testing conditions are met by an artificial increase of the structural damping. The defect detectability is estimated regarding different excitation locations and coupling devices.

In order to identify the relation between vibration amplitude and defect heating, vibrometric measurements of the crack edges are conducted during vibrothermographic testing. Single-point and two-point differential laser vibrometry enables a phase sensitive measurement of the displacement components of both crack sides, which are related to the thermal response. Finally, a semi-automatic scanning procedure is developed that allows for the three-dimensional measurement of vibration mode shapes. Based on a frictional normal contact model the achieved crack edge displacements are evaluated and correlated to the extent of heat generation found for different vibration modes.

Quantitative simulations of ultrasound excited thermography can be exploited to predict the detectability of a given defect in arbitrarily shaped components. The experimental work is therefore complemented by numerical analyses. A highly adjustable finite element model of a cracked plate specimen is built in order to reproduce the mechanical excitation, heat generation and heat transfer observed experimentally.

2 Vibrothermography research and applications – State of the art

After a brief introduction to the use of infrared technology in the field of non-destructive testing, this chapter presents the general concept and interdisciplinarity of ultrasound excited thermography. Subsequently, it is attempted to give a comprehensive outline of research activities, in which the emphasis is clearly on the inspection of metal components.

2.1 Infrared thermography in non-destructive testing

Since the first evidence of heat rays beyond the red portion of the spectrum was provided by Sir William Herschel in 1800, there is an ever increasing interest in detecting and quantifying thermal or infrared radiation. Unfortunately, after the first photon detectors have been produced in 1920, the driving force of further developments was mainly attributed to military aims. In 1939 infrared night vision devices, as for instance the "Snooperscope", have been invented in the United States [HW90]. Infrared-homing missiles established since 1950. The first non-military applications using scanning cameras with cooled quantum detector materials, such as indium antimonide (InSb) and mercury cadmium telluride (HgCdTe), have been delayed and came up in the mid 1960s [FR06]. Military approved technology finally penetrated civil engineering areas [MC04b].

Currently, full-field temperature measurements are mostly based on two types of infrared cameras, both equipped with focal plane array (FPA) detectors. These are the low-cost microbolometer cameras at the one hand and state-of-the-art quantum detector cameras on the other hand. Photon detector based thermal imagers are generally faster and provide a higher thermal sensitivity as well as a greater array size (number of pixels) compared to bolometer cameras. Drawbacks can be seen in the much higher pricing and the need for detector cooling, which involves maintenance of the coolant system after a certain number of operating hours. Uncooled infrared cameras are becoming more and more affordable, which leads to a growing group of potential users across the industries. Although the thermal and spatial resolution of bolometer cameras has been significantly improved in the last years, researchers usually rely on photonic cooled detectors with a typical noise level (NETD) below 20 mK and a minimum detector size of 320×256 to 640×512 points to obtain high-quality thermal images [Mal02].

In recent years, permanently increasing requirements on the quality and reliability of components have led to a growing demand for fast and imaging non-destructive test methods. Ever since in the 1980s the understanding of heat transfer has been consequently combined with the monitoring of surface temperatures, the thermographic inspection of parts (material acceptance/rejection) and production processes (machine performance) was proven extremely useful in many fields [WKN$^+$99, MC04b]. Since process integrated and fully automated testing has been successfully realised, today, infrared camera manufacturers and test system suppliers appreciate the growing acceptance of innovative thermographic methods throughout the industries.

At this point it appears meaningful to distinguish between active and passive thermography approaches. The passive thermographer is observing an object without changing its "natural" conditions regarding heating/cooling processes and the involved heat transfer mechanisms. Here, the temperature itself represents the quantity that is actually of interest [WKN$^+$99]. A well-known example related to constructional engineering is the passive thermographic inspection of a building's façade, which allows for the detection of poor insulation and thermal bridges [FR06]. The resulting thermograms visualise the stationary temperature distribution caused by a constant heat flow from the inside of the building through its envelope.

In contrast to the passive approach active thermography implies an artificial disturbance of the object's energetic equilibrium either in the fashion of a pulsed, stepped or periodic stimulation. This can be achieved in various ways. Optical/thermal, inductive and mechanical excitations represent only the main disciplines of thermal NDT. Based on the spatially and temporally resolved thermal response of the investigated body, information on the structural integrity regarding defects like cracks, delaminations or cavities can be derived, whereas the absolute value of the temperature is of minor importance. Instead, the common objective of active thermographic NDT methods is to generate and to detect a locally deviating thermal response at flaws compared to that of sound regions [WKN$^+$99]. More precisely, the transient dynamic or stationary 3D heat conduction that sets in with the energetic stimulation is assessed by monitoring the 2D surface temperature over time. Choosing the most appropriate excitation method strongly depends on the material to be investigated and the type of defect to be detected. Table A.1 in Appendix A.1 exemplarily summarises the scope of different excitation methods regarding non-destructive testing of solid components.

The optical/thermal excitation of a plate-like structure by means of a flash generator, a halogen lamp or an infrared heat source will lead to an areal heating of the exposed surface. Neglecting any lateral heat flow, a one-dimensional thermal wave will travel orthogonally to the surface while being absorbed by the material. A planar defect, such as a delamination in a plate made from fibre reinforced plastic (FRP), represents an inner boundary face that reflects the incoming thermal wave earlier compared to sound regions, where the rear side of the plate acts as a boundary. Both the locally increased temperature response as well as the different time history can be exploited for flaw detection [BWK92, WKN+99, Mal01]. Making use of the thermal wave theory, analysis of the infrared record additionally allows for retrieval of the defect depth [IC05]. Here, knowledge of the thermal material properties is explicitly required. In case of a tightly closed fatigue crack in a steel plate or coating, the above mentioned approach will most likely fail to detect this flaw, especially if the crack faces are directed perpendicular to the plate surface (see for example [PBML08]). The crack faces oriented parallel to the propagation direction of the thermal wave will not significantly affect the conductive heat transfer so that there is no contrast available for defect detection.

Inductive and mechanical excitations are usually better suited to detect surface cracks in metal structures. The non-contact, electromagnetic excitation via an inductor coil induces an alternating electric current in the sample, which is limited to a rather small depth below the surface, known as the skin depth [Rie06]. Depending on the material (e.g. ferromagnetic or austenitic steel) a perpendicular surface crack affects the distribution of the current density and leads either to a positive or a negative temperature contrast, which allows for the thermographic flaw detection in both cases [WVN+09]. For larger structures induction thermography requires to be used in the fashion of a scanning method, moving the inductor along the surface [BGHR13]. A locally applied high-power ultrasound excitation can generate sufficient vibration of the structure, so that crack faces rub and clap against each other. The involved heat dissipation is primarily restricted to the flaw region. Compared to induction thermography, ultrasound excited thermography as a (partly) contacting method has a greatly increased coverage. Due to the wave guide characteristics of the investigated structure, cracks can be detected even if located in a distance of several metres from the coupling location [PU09]. The defect-selective heating and the large operating range render vibrothermographic testing very intuitive and promising with regard to the application to large-scale, heavy load-bearing steel members.

Progress in the widespread use of infrared thermography can be found in the proceedings of two specific conferences, namely the "Quantitative Infrared Thermography (QIRT)" and the "SPIE Thermosense", as well as in the international QIRT Journal. The deployment of active thermographic methods towards non-destructive testing represents a significant portion of the total number of publications in the field. Advanced excitation methods and post-processing approaches are continuously being developed, while known methods are still being optimised to further push the limits of thermographic testing.

2.2 Concept and interdisciplinarity of ultrasound excited thermography

The idea of subjecting structural members to a cyclic mechanical loading in order to generate a thermal contrast at existing ("developed") damages was realised in the late 1970s by the pioneers of vibrothermography Edmund G. Henneke, Kenneth L. Reifsnider and Wayne W. Stinchcomb at the Virginia Polytechnic Institute and State University in Blacksburg, USA. In a paper from 1979 [HRS79] the research group coined the term vibrothermography to emphasise the fact that the novel approach relies on actively introducing steady-state mechanical energy into the test piece. A high potential of vibrothermography was revealed by successfully detecting fatigue-related delaminations at the edge of boron-epoxy coupons of varying thickness. The test pieces were inertially loaded by three different exciters, a shaker, a transducer from a glassware cleaner and an ultrasonic horn. Covering a frequency range from 1 to 20 kHz, all configurations had equal success. The authors intentionally applied a non-damaging, low-amplitude mechanical load and explicitly mentioned the possible use in the field of non-destructive testing [HRS79]. Later, Mignogna et al. [MGD+81] investigated the bulk heating of steel, brass and copper specimens when subjected to resonant high-power ultrasonic loading. The non-resonant testing of a cracked aluminium bar finally revealed a pronounced heating at the defect site providing further evidence of the method's crack detection capability.

The principle of ultrasound excited thermography as used in today's applications is illustrated in Figure 2.1. The test equipment usually comprises a high-power ultrasonic generator working at a constant or slightly variable frequency in the order of 20 or 40 kHz. More recently, the use of tunable broadband exciters established. An ultrasound converter, which is brought into mechanical contact with the structure under investigation, induces elastic waves that propagate inside the structure.

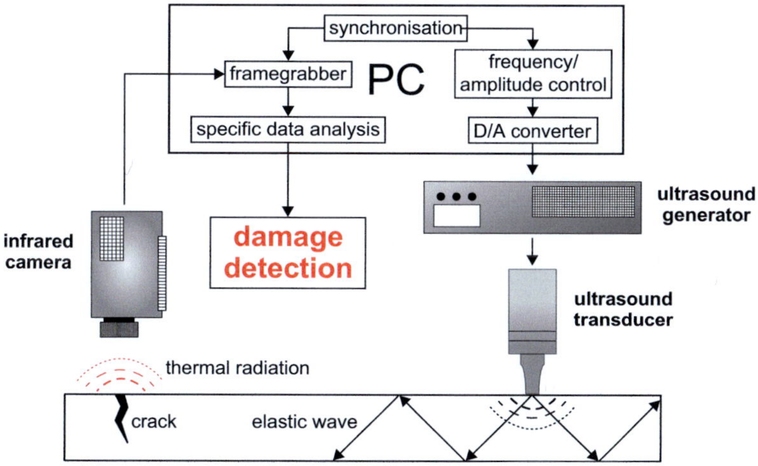

Figure 2.1: Principle of ultrasound excited thermography.

The continuously excited component tends to vibrate at the transducer's frequency, whereas the oscillation of the structure is permanently damped. Attenuation is mainly caused by the inherent material damping and friction at every support or contact interface, such as bolt connections, where parts of the structure slide on each other. In case of internal flaws like cracks or delaminations, their boundary faces possibly move relatively to each other in a shearing or clapping mode. Kinetic energy is locally dissipated as heat and the resulting temperature increase at the flaw region is detected and visualised by means of an infrared camera.

A standard PC is sufficient for controlling the generator, grabbing the frames acquired by the camera and post-processing the data stream. Typically, a synchronisation between generator and camera operation is realised to allow for reproducible test procedures. Since both parameters, the excitation level and the ultrasound frequency are continuously adjustable, different excitation methods and corresponding data processing techniques have been developed in the past. The most accepted procedures are described in Section 3.5.

The concept of ultrasound excited thermography is apparently straightforward and easily comprehensible. But actually, the testing procedure described above involves a broad spectrum of dependent and independent technical disciplines. Vibrothermography can be logically separated into three physical domains, namely the mechanical energy input, the energy conversion and the thermal energy detection. A further subdivision into six thematic groups allows to illustrate the functional interactions in Figure 2.2.

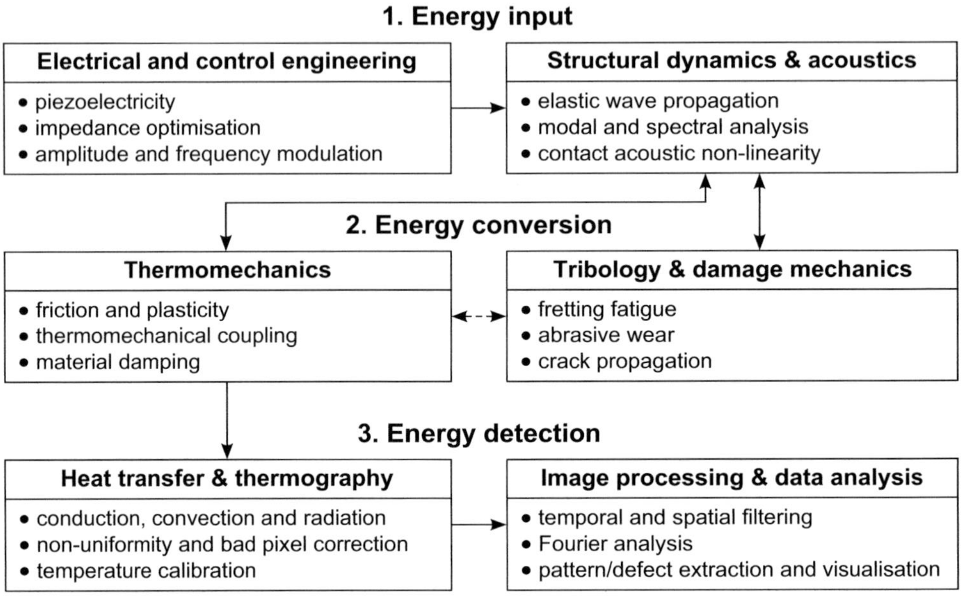

Figure 2.2: Interdisciplinarity of ultrasound excited thermography.

The need for high-intensity ultrasound generation including widely adjustable amplitude and frequency modulation capabilities imposes high requirements on the generator control. In the past, hardware and software modifications of generators with up to 4 kW electric power, which are originally used for ultrasonic welding of plastics, have overcome the strongly limited frequency response of standard units. The field of structural dynamics and acoustics is of great importance to vibrothermography, since the mechanical excitation of elastic waves and vibrations provides the basis for all potential heating mechanisms. The non-linear interaction of boundary faces represents a challenge in the dynamic analysis of cracked components.

The mechanism of heat generation constitutes a main issue in vibrothermography. At least the effects of dry friction, plastic dissipation and material damping have to be distinguished and assessed individually. Furthermore, tribological damages related to the interaction of crack faces as well as vibration induced crack growth are essential aspects that have been rarely investigated, yet.

The spatio-temporal evolution of a body's temperature field is fully determined by the heat transfer mechanisms once a heat source has been activated. An accurate infrared camera calibration together with an equalised detector response becomes important if smallest surface temperature variations are to be discerned. Finally, the full range of image processing techniques is available for the thermographic data analysis and defect visualisation. Sophisticated procedures, often based on filtering in the spatial, time and frequency domains, allow for case-specific post-processing methodologies. The efficient implementation of digital signal processing towards defect characterisation clearly represents an own discipline.

In a strict sense, vibrothermographic testing involves two or even three (rather weakly) coupled problems. Clearly, the mechanical vibration is understood as the driving force for all heat generation mechanisms and the evolution of potential damages. However, heat generated at defects or inside the bulk material affects the thermomechanical material properties and results in a non-uniform thermal expansion and a change of the residual stress state. This in turn influences the dynamic properties of the structure and therefore the mechanical domain. Similarly, severe tribological damages or crack growth can manipulate the contact behaviour at the crack faces and thus the dynamic response of the structure. A third coupling may be seen in the interaction of wear and frictional heating.

Ultrasound excited thermography proves to be a multidisciplinary, application-related research topic that quickly reveals its complexity when examined in more detail. The following sections provide an overview of published applications and research on the subject with particular emphasis on the testing of metal structures.

2.3 Range of application

The potential use of vibrothermography as a non-destructive test method has been demonstrated by numerous research groups. It has been successfully applied to components made of fibre-reinforced plastics (FRP), metals, ceramics, ceramic-matrix composites, wood-based materials and other engineering materials, mainly in order to detect cracks, delaminations and disbonds [RWSB96, SDW+98, DBW99, GSB07]. Furthermore, the quality of bondlines with regard to local regions of missing adhesion or non-cured adhesive was assessed [DZRB03, HDZ03, Has06]. More exotic case studies such as the detection of cracks in human teeth [HFT01] clarify that the field of possible applications is quite extensive.

A considerable number of researchers detected typical flaws in plate and shell structures made from carbon fibre reinforced plastics (CFRP), namely the areal delamination of individual composite layers and localised impact damages featuring complex matrix/fibre fractures [RWB96, WB98, FHO+00, HFOT01, FHO+01, ZDR+03, MHIN04, BACM06].

Regarding the maintenance of aircraft structures, vibrothermography has been proven useful to detect the disbonding and cracking of stringer stiffened CFRP structures, such as a landing flap [ZDB01, ZDR+02] or a side panel of the fuselage [ZDB00, BDZ01, Dil01]. Moreover, small areas affected by hidden corrosion could be identified between two plates of an aluminium component [WB98, SDW+98]. Zweschper and colleagues located loose rivets in an aluminium assembly after being exposed to a shear force overload and found cracks along a row of rivets [ZDB00, ZDB01]. A similar work demonstrated a good correlation between the interference fit level of a fastener in an aluminium plate and the thermal signature produced during ultrasound excitation [TR07]. Artificial defects intended to simulate delaminations in GLARE (glass-fibre reinforced aluminium) and CFRP face sheets of honeycomb structures were found by Ibarra-Castanedo et al. [ICGS+07, ICSK+08].

Based on a literature review, the automated detection of cracks in turbine blades and vanes is probably the most advanced industrial application of ultrasound excited thermography. Since the premature failure of a single blade, which is exposed to extreme thermomechanical loads, can lead to a total loss of the gas turbine, a great effort is expended to maintain those highly critical parts. Several authors reported a successful vibrothermographic crack detection on aero-engine blades [HFOT02, HHW+07, HRBS06, BGPH10, THFN10] and disks [HZL+04b, Bod11]. Delaminations and hairline cracks of the thermal barrier coating have been detected by Holland [Hol07] and Dillenz [Dil01]. A related study revealed that ultrasound excited thermography is particularly suited to detect tight flaws in thin plasma-sprayed coatings [PBML08].

In the field of fossil power generation turbine manufacturer Siemens recently achieved a notable progress in the routine testing of blades towards a fully automated procedure. The implementation of the SIEMAT® or Smart SIEMAT® [HHW+07] acoustic thermography into the inspection process aims at the replacement of dye penetrant testing [ZS06].

Crack detection by means of vibrothermography is not restricted to plate-like and thin-walled structures. Favro et al. [FHO+01] successfully tested a massive aluminium cylinder head, which has been thermally cycled to provoke cracking. The authors finally detected several induced fatigue cracks as well as a sub-surface crack. Zweschper et al. [ZRDB03] also investigated automotive components made from die-cast aluminium and found cracks in a ribbed part and in a motor-cycle cylinder head. Another example for the testing of complex shaped bodies is given by Montanini et al. [MRF10], who identified inner flaws in cast iron turbocharger housings. Regarding the industrial series production, Dillenz et al. [DSZ05] reported on a fully automated and integrated test stand for the vibrothermographic inspection of cast metal components to provide a one hundred percent inspection.

Comparatively few tests of weighty, massive metal components have been published so far. Han et al. [HFOT02] investigated an 84 kg monolithic aluminium structure that was accidentally involved in a high-speed collision and detected several 5 cm long cracks. Focusing on civil engineering structures, He [He10] found cracks in a C-section steel beam after intentionally breaking off a small part to produce some fracture faces spreading into the specimen. Own experiments revealed that the thermographic detection of tight cracks in heavy load-bearing steel components is also possible. Figure 2.3a depicts a 50 kg T-joint, which was extracted from a thick-walled welded girder with damages due to liquid zinc assisted cracking while being hot-dip galvanised. Several cracks have been identified along the vertical fillet weld between web and end plate.[1] Applying an intense, mono-frequency ultrasound excitation to the extracted T-joint results in a crack indication, which is characteristic for the vibrothermographic testing of metal components. The four thermograms I to IV in Figure 2.3b illustrate the typical sudden response in the form of a strongly localised temperature increase along the weld toe, a non-uniformly distributed heating and the subsequent heat diffusion after the excitation source has already been switched off.

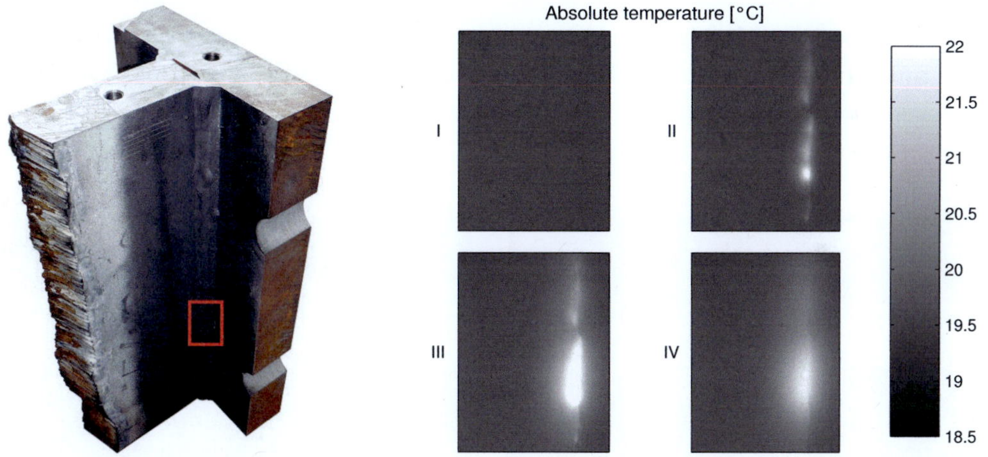

(a) 50 kg massive T-joint, extracted from a thick-walled welded girder [PU09]

(b) Temperature distribution along the weld toe; initial state (I), shortly after the excitation is switched on (II), after 2 s of insonification (III) and 200 ms after switching off (IV)

Figure 2.3: Thermographically detected crack in a massive welded T-joint.

[1]The girder, made from low-carbon steel (European grade S235), was already assembled and mounted in the "Fritz-Walter-Stadion", a football arena in Kaiserslautern, Germany. During renovation many hidden cracks of the same type were found in newly manufactured components, causing an enormous economic loss.

Vibrothermography has been mostly performed on relatively small parts that can be mounted in a stationary test setup. Here, the ultrasound transducer is either spring-loaded (e.g. [BGPH10, MC09b]) or pneumatically pressed against the specimen (e.g. [DR03, CKLR07, Hom07]) prior to testing. So far, very few examples of an in-situ inspection were published. Han et al. [HZL$^+$04a, HIN$^+$06] developed prototypes of a 20 kHz and a 40 kHz hand-held excitation device, which are manually pressed against the structure. The detection of cracks in the skin of a Boeing 737, operated by the FAA Airworthiness Assurance NDI Validation Center (AANC), demonstrated the potential use for the inspection of an aircraft fuselage. Later, Polimeno and Barden [PA09] also developed a portable inspection system, consisting of a more complex hand-held unit, a spring-loaded 35 kHz horn-transducer assembly and a very compact microbolometer camera. The authors successfully tested their system on aerospace components with impact damages and proved that vibrothermography can be packaged to provide a practical and rapid means of inspection for composite parts.

Figures 2.4 and 2.5 show the test setup and results of a preliminary experiment conducted on a heavy load-bearing steel girder with double T-section and welded head plate. The zinc-coated girder features a crack at the end of the half-plate weld caused by liquid metal embrittlement[2], very similar to the girder shown in Figure 1.1a. A massive steel clamp served to couple a high-power 20 kHz ultrasound transducer to the sample. Ultrasound excited thermography clearly enabled the crack detection as illustrated in the difference image (Fig. 2.5), where the initial thermogram acquired

(a) Test setup (US generator and PC not shown)

(b) Clamped transducer and marked crack location

Figure 2.4: Portable setup for simulating the in-situ use of vibrothermographic testing.

[2]The presented specimen was kindly provided by the Institute for Steel Structures at RWTH Aachen University, Germany.

before testing has been subtracted from the highest contrast image achieved during excitation. Judging by the remaining crack indication left from a standard black on white magnetic particle inspection, the vibrothermographic testing did not unveil the whole crack path but rather 50 % of it. The region next to the crack tip could not be identified in the course of this ad-hoc experiment, giving rise to further investigation of the physics involved in ultrasound excited thermography.

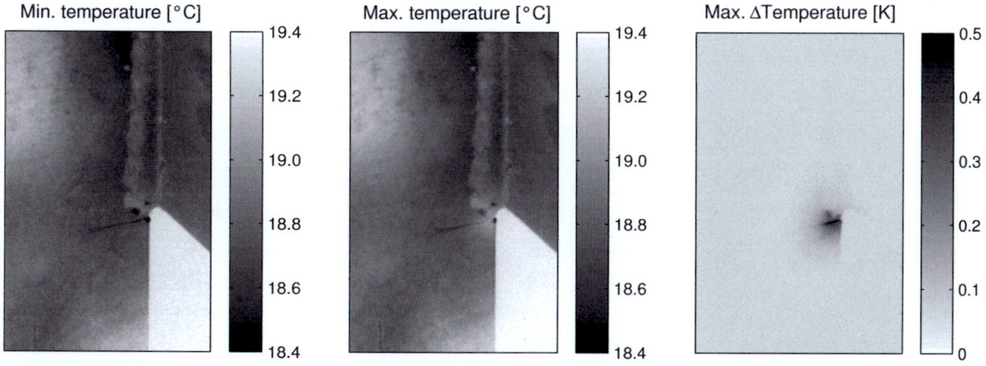

(a) Initial thermal state prior to ultrasound excitation (b) Maximum defect temperature achieved during stimulation (c) Computed difference image indicating the crack location

Figure 2.5: Vibrothermographic crack detection on a zinc-coated steel girder.

2.4 Research activities

Despite of the fact that vibrothermography has been recently introduced in the industry as a potential alternative to standard NDT methods, substantial questions remain about its reliability and the underlying physics of the heat-generation process [Iow13, SAL04]. Since the complex phenomena of localised heat generation at crack regions are not yet completely understood, quite different mechanisms and interdependencies have been considered in the past [RCHT11]. This section outlines the most relevant research activities towards the reliable application to metal components. Here, friction and plasticity are regarded as the most relevant dissipative effects provoking heat generation. Besides of the basic influences on defect heating, such as the excitation frequency, vibration modes and the residual stress state at the crack, more advanced aspects are introduced. These are the prediction of the thermal response, the relation between vibration intensity and defect heating, wear of fracture surfaces, crack growth, repeatability and probability of crack detection.

2.4.1 Heating mechanisms and basic influences on crack detection

From a macroscopic point of view, frictional energy dissipation at the crack faces arises depending on the interfacial shear stress and the tangential relative velocity of the crack faces. The complex morphology of metallic fracture surfaces can only be described by roughness measures based on the statistical asperity distribution. Considering very small relative displacements, it is questionable whether the definition of macroscopic friction coefficients according to the classical Amonton-Coulomb model of dry friction is possible and meaningful at all. However, following a friction theory seems to be justified, since many published experimental results can be adequately explained based on this model concept. In the past, mostly surface cracks, corner flaws and through-thickness cracks according to Figure 2.6 have been vibrothermographically inspected.

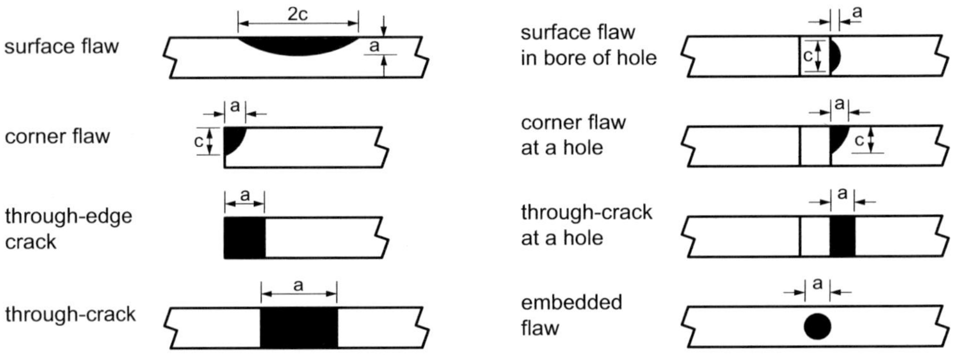

Figure 2.6: Crack type definitions and geometric parameters [Wan86].

In order to investigate the dependence of crack heating on contact stress Lu et al. [LHN+07] examined a through-thickness crack (compare Fig. 2.6) in an aluminium plate specimen, which allows for applying a static compression load acting on the crack faces. A 40 kHz ultrasonic weld gun was then utilised to excite the specimen and heat up the crack region. It has been shown very clearly that the maximum thermal response is continuously lowered when increasing the applied compression force. Obviously, a relative motion of the crack faces, which is substantial for the generation of frictional heat, is progressively prevented by the normal force. Further evidence is provided by out-of-plane differential laser vibrometry measurements of both crack faces. The correlation between the vibration energy derived from relative velocity and the temperature increase indicates that compressive stresses may fully prevent relative displacements and thus defect heating. Similar results have been achieved on a connecting rod, which was cut into two parts and pressed together by means of screws [RH05]. This pseudo-crack has also been subjected to a compressive prestress controlled by the screw torque. Plotting the achieved IR signal against the torque yields a bell-shaped curve, so the authors found that there is an optimum tightening torque at which the strongest interface heating is achieved. Both lower and higher preloads resulted in a greatly reduced heating. Such basic investigations clarify that the actual stress state and the possible gap between crack faces strongly influence the vibrothermographic crack detection.

Tsoi and Rajic [TR04] tested cracked aluminium plates and reported a non-uniform heating along the crack path, where the strongest heating occurred at the first half. The crack tip region was not detected at all. Possibly, residual compressive stresses locally prohibit the generation of frictional heat. This hypothesis is strongly supported by Renshaw et al. [RHT08], who examined a titanium beam with rectangular cross-section and a semi-elliptical surface crack centred on one face (compare Fig. 2.6). A bending moment was applied at the ends of the test beam so that the cracked side is subjected to flexural tension in order to compensate for the residual closure stress and to open the crack. Variation of the bending stress clearly revealed that only certain regions along the crack path heat up during ultrasound excitation, namely those with an adequate contact stress state or gap between the crack faces, respectively. As long as the closure stress is not compensated by the loading there is no frictional heating. On the other hand, a fully opened crack does not heat up either. Instead, thermal hot spots reveal the transition zones between the closed and opened regions along the crack path.

Several research groups experimentally revealed that in contrast to fibre reinforced plastics the heating of cracks in metal components strongly depends on the actual working frequency of the excitation device [GSB07, PU08]. As a consequence, the usage of tunable exciters and associated measurements of the specimen's vibration established in the research field of ultrasound excited thermography. Based on detailed investigations of cracked turbine vanes, Homma [HRBS06, Hom07] concluded that a strong thermal response of defects is only achieved at resonance frequencies. He further demonstrates that even a resonant excitation does not necessarily lead to heat dissipation at all (known) crack locations. While one crack is detectable at all considered natural frequencies, a second flaw can only be identified at some of them. Similar ultrasound sweep experiments on steel samples revealed the same typical behaviour [GSB07, PU09].

At this point it should be noted that even a fixed frequency ultrasound system is generally able to excite a rich frequency spectrum of vibrations. Han and colleagues utilised 20 and 40 kHz ultrasonic welders to detect cracks in various metal structures and verified by means of laser vibrometry measurements that the unloaded tip of a 20 kHz ultrasonic weld gun indeed oscillates purely harmonically [HZL+04b]. The corresponding amplitude spectrum of the acquired velocity data does not reveal any distortions of the signal. In contrast, the analysis of vibrational waveforms measured on the specimen's surface proved the existence of numerous frequency components besides of the driving frequency. According to the authors, the non-linearity in the coupling between transducer and sample appears to be the source of the "acoustic chaos", which typically contains harmonics and many multiples of a certain subharmonic frequency. This way, a nominal 20 kHz excitation can result in a broad spectrum, starting well below the driving frequency, e.g. at 1.7 kHz, and extending beyond 200 kHz [HFNT06].

Han et al. [HLZT04] proposed a reduced mechanical model of an ultrasonic horn bouncing against a rigid post to simulate the generation of acoustic chaos. The achieved transducer waveforms were not stable and extremely sensitive to a change of the initial conditions. A Fourier analysis revealed the existence of subharmonics, similar to those fractions of the driving frequency, which have been experimentally observed on the sample. Later, an improved dynamical model including the elasticity, damping and motion of the sample was analysed by Zheng et al. [ZZZF06]. The authors found that for adequate initial conditions the sample vibrates with stable subharmonics, different from the theoretical prediction by Han et al. Chaotic and aperiodic states could only be provoked by inadequate initial conditions, such as an insufficient restoring force applied to the transducer.

Despite of the controversial discussion about the stability of sample vibrations, Han and colleagues emphasised that a chaotic insonification of metal structures leads to a greatly increased thermal response at defect locations in contrast to a purely sinusoidal excitation with very little harmonic distortion. In some cases only the chaotic excitation enabled a sufficient defect heating [HFNT06, HZL+04a] while the non-chaotic sound left cracks undetected. Furthermore, a low-power pulse of chaotic sound could be equally efficient for crack detection compared to a high-power pulse of non-chaotic sound [THFN10, HZL+04b]. Although conceivable, it is not suggested by Thomas et al. [THFN10] that a broader spectrum significantly increases the probability of matching an efficient eigenfrequency. According to Thomas and colleagues, two effects contribute to an enhanced ability of chaotic sound to heat cracks even if arbitrarily located in an object: Firstly, it appears virtually impossible for a crack to be located at the nodal point of an acoustic mode pattern. Secondly, it seems chaotic sound can guarantee a certain gradient of vibration amplitude across the crack, which is essential in order to have differential motion between the crack faces. Holland et al. [HUO+11] generally agreed to the improved thermosonic efficiency of a broad range of frequencies but state that some of those harmonics and subharmonics inevitably excite specimen resonances.

The pronounced frequency dependence of crack heating gives rise to the question to what extent different vibration mode shapes promote or constrain the involved heating mechanisms. Rothenfusser and Homma used a 3D scanning laser vibrometer to measure the surface displacement components near the crack in a turbine blade. An equidistantly spaced 9×9 point measurement grid was defined to cover a 10×10 mm region of interest including the crack, which represents a discontinuity of the mesh. The authors were able to distinguish three different crack vibration modes, namely a clapping mode (in-plane movement, crack mode I), a rubbing mode (out-of-plane shear, crack mode III) and a third mode in which both crack faces move uniformly with little relative motion. It is pointed out that mixed modes, combinations of the described ones, occur very often. Although the cited paper does not contain corresponding temperature measurements, the authors presume the rubbing mode being mostly responsible for crack heating. Later, Homma [Hom07] found that the mode I relative motion can also be decisive for the vibrothermographic detection of tight cracks. Holland et al. [HUO$^+$11] pointed out that due to the irregular interlocking of asperities even a clapping mode will give rise to significant shearing motion and thus substantial friction.

The occurrence of different vibration mode shapes does not only depend on the excitation frequency. Instead, the coupling location of the exciter and the direction of applied forces influence the dynamic response of the investigated structure. Hassan et al. [HHW$^+$07] tested aero-engine fan blades and varied the horn contact position incrementally along the bottom face of the blade in 12 steps covering a distance of about 60 mm. Mono-frequency excitation of the blade revealed a dramatic scatter of the achieved crack heating. Here, the ratio between highest and lowest infrared signal is estimated to be about 12 [Hom07]. This finding clarifies the need for tunable exciters in order to (partly) compensate for such a strong dependency.

Besides friction between opposing crack faces, plastic deformation at the crack front is considered as a second distinct mechanism of heat generation in vibrothermography. The potential occurrence of cyclic plasticity in the vicinity of crack tips is beyond debate, since crack propagation was documented a number of times, mostly in thin aluminium plates but also in stainless steel and titanium [CKR06]. Kephart et al. [KCZ05] applied a cyclic loading to notched aluminium single edge tension specimens (ESET, according to ASTM standard E647) with a thickness of 2 mm in order to generate small through-cracks. After ten ultrasound pulses of 250 ms duration each, the initial crack depth has grown from 1.6 mm to 2.5 mm, yielding a crack growth rate of 85 μm per pulse or 0.017 μm per vibration cycle. Excessive crack propagation was demonstrated by Homma [Hom07]. An aluminium plate, weakened by a milled slot on one side, was cyclically bended to produce a small crack. During a 200 ms ultrasonic excitation most of the defect heating appeared to be generated directly at the crack tip. The experiment resulted in a crack growth of about 4.8 mm, which corresponds to a rate of 1 μm per vibration cycle. A recent publication by Renshaw et. al [RCHT11] gives further evidence of plasticity-induced heat generation. They reported on a growing crack in an aluminium sample and observed significant heating more than 0.5 mm past the crack tip, where plastic deformations are expected to occur.

Possible influences on the ultrasound induced crack propagation have been studied by Chen et al. [CKLR07]. Based on 3.2 mm thick aluminium ESET specimens with fatigue cracks generated at a constant load ratio $R = 0.1$ but with varying load amplitudes, Chen and colleagues found a decrease in ultrasound induced crack growth with increasing stress intensity range ΔK. It is believed that higher stress intensities create a larger plastic zone and therefore a blunted crack tip, which results in a retardation of further crack growth. Sectioning of the specimens cycled at lower stress intensity revealed significant fretting patterns, which are largely confined to the outer regions of the crack face. In case of higher ΔK values, the signs of fretting were also identified on the interior of the fracture surface. Compared to the low amplitude specimens the crack topography of high amplitude samples was characterised by highly curved or slanted profiles. According to the authors, this feature leads to higher energy dissipation at the crack interface, thus reducing the energy available for crack propagation. The hypothesis that a more pronounced interlocking of asperities shields the crack front from being loaded is supported by additional experiments. Specimens, which have been milled off half the crack depth exhibited an average crack propagation rate increased by a factor of six compared to reference samples. Another set of specimens was subjected to a heat treatment (400 °C for 2 h) intended to eliminate the wake and plastic zone around the fatigue crack. This procedure resulted in a dramatic increase of the crack growth rate achieved by nominally the same ultrasound excitation. Further opening of the crack mouth after annealing finally caused a vibrothermographic defect heating, which is isolated at the crack tip. Here, the average crack growth rate increased again [CKLR07].

Tsoi and Rajic, who found the strongest crack heating in the middle of the crack path, far from the crack tip, did not observe any crack growth in aluminium samples. Furthermore, they determined that a 10 s ultrasound excitation, performed at three different crack depths, did not affect the crack growth rate under regular fatigue loading when compared to a set of reference samples [TR04].

Currently, it seems beyond the scope of most vibrothermography research activities to further investigate the impact of high-frequency excitation on the bulk material. Possible interactions of intense ultrasound with dislocations in various metals and the resulting changes of macroscopic material properties have been described by Abramov [Abr98]. Mignogna et al. observed a softening effect on specimens under tensile load when these are additionally subjected to ultrasound [TR04].

Bolu and colleagues [BGPH10] analysed the shift in resonance frequencies of cracked and uncracked turbine blades using a process compensated resonance testing (PCRT) device. After 400 repetitions of a 1 s ultrasound excitation the authors found a decrease in 15 out of 17 analysed eigenfrequencies. Although crack propagation was not investigated, obviously a change of the structural properties occurred, which was attributed to the insonifications. Interestingly, the authors also found small changes in the resonance behaviour for an uncracked blade subjected to the same number of excitations.

It is widely acknowledged that besides of friction and plasticity the material damping represents a third mechanism of heat generation in vibrothermographic experiments. The material inherent energy dissipation, which is activated by dynamic loading, results in a bulk heating, which might interfere with a local defect heating as shown by Zweschper et al. [ZDR$^+$03, ZRDB04]. Especially in case of a resonant excitation, where standing wave patterns are provoked, the structure is heated according to the positions of vibration antinodes. Hosten et al. [HBB08] modelled the temperature rise due to the propagation of ultrasonic waves in viscoelastic materials. The authors simulated the resonant in-plane excitation of a PVC plate and computed the temperature distribution over time based on the mechanical energy losses. A comparison of numerical and experimental temperature fields, showing a perfect agreement of the fixed pattern heating, validates the finite element model.

More examples of spatially resolved correlations between strain amplitude and thermal response of PVC, CFRP and metals, regarding different natural frequencies, have been reported by Homma [Hom07]. Renshaw et al. [RHB09] visualised standing wave patterns on a 7.3 mm thick titanium plate strip featuring a grid of blind holes filled with honey. Flexural resonance modes of the specimen could be identified due to the viscoelastic damping of the honey, which caused different heating levels depending on the local surface strain amplitude. The comparison of the heating profile across the specimen length with the measured displacement field clearly revealed the spatial coincidence.

There is a consensus that the energy dissipation caused by material damping is proportional to the square of the applied stress amplitude. Homma [Hom07] proved the quadratic dependence on a flawless PVC plate for three different frequencies. Renshaw and colleagues [RHB09] likewise achieved a quadratic relation between vibrational stress amplitude of the examined titanium plate and heating of the integrated viscous material-filled pockets.[3]

[3]The achieved data has been approximated by a second order polynomial with a non-zero coefficient of the linear term. However, omitting this linear term according to the theory of viscous damping has very little influence on the fit curve within the relevant range of the data presented.

Many experiments in the past were performed using a costly, commercial ultrasonic weld system with up to several kilowatts of electrical power. However, quantitative numerical studies showed that typically only a very small fraction of the available electrical power is released as heat at defect locations, see for example [BAM$^+$04]. Focusing on metals, Homma [Hom07] also noticed a rather inefficient operation of high-power but fixed-frequency devices unless a resonant excitation is either achieved accidentally or due to a non-linear transducer coupling that leads to multi-frequency excitation. In case of CFRP plates, where the driving frequency is less critical, Barden et al. [BACM06] demonstrated the detection of impact damages with only 1 W of electrical power.

Owing to the pronounced frequency dependence of crack heating in metals, alternative, tunable excitation sources have been investigated and found to have similar crack detection capabilities compared to fixed-frequency systems. The Smart SIEMAT® system, for example, has significantly less electric power (approximately 300 W) than the standard SIEMAT® setup (3000 W), but it can compete with the high-power system regarding crack detectability [HHW$^+$07]. Homma [HRBS06] earlier proved the tunable piezo actuator driven at only 50 W to be a viable ultrasound source for detecting cracks in turbine blades. Clearly, the excitation frequency is not restricted to the lower ultrasound range. Genest et al. [GMFM09] successfully used a tunable electromagnetic/piezoelectric dual shaker system providing an extended frequency range of 10 Hz to 20 kHz. The authors found 1, 2 and 4 mm deep single edge cracks in 3.2 mm thick aluminium sheets, whereas crack detection was only enabled in a low bending resonance mode within the audible range, namely at a frequency of about 212 Hz.

Numerous researchers inserted a thin sheet of a softer coupling material between horn tip and structure, either to prevent surface damages due to the transducer hammering or to achieve a certain acoustic impedance match between structure and horn material. However, there has been very little research on couplants particularly suited for vibrothermographic testing. Zweschper et al. [ZDR$^+$03] increased the coupling efficiency by 35 % when placing an adhesive aluminium tape between CFRP specimens and the employed titanium sonotrode. Others reported a successful use of metal foils, PTFE (Teflon), leather, paper, cork or felt, whereas no material has proven to be a universally good couplant. The selection of an appropriate material rather seems to be an empirical trial and error process [SAL04].

Besides of experimental investigations, numerical studies have been performed by several research groups in order to better understand the mechanisms of heat generation. The conducted finite element (FE) analyses mainly aim at the realistic description and, ideally, the prediction of the mechanical energy dissipation, which is released as heat at a modelled defect interface. In a first study Han et al. [HIN+05] simulated a flawless aluminium plate being excited by a simplified transducer, which was able to lift off from the sample and hammer against it. The result of the non-linear coupling was clearly revealed in the form of a complex vibration spectrum achieved in the plate including higher harmonics and subharmonics with respect to the fixed driving frequency. In a following paper the same authors introduced a rectangular crack-like discontinuity and adjusted the available friction model provided by the commercial FE code LS-DYNA. Once, the clapping effect of the transducer was incorporated and another time it was omitted by directly applying forced sinusoidal displacements of the coupling region instead. A comparison of the cumulated energy dissipation achieved with and without non-linearity of the excitation revealed that the induced chaotic vibration leads to a greater frictional power and would therefore generate more heat at the crack faces than a non-chaotic vibration [HIN+06]. Mian and colleagues [MHIN04, MNH+04] simulated the frictional heating of a graphite/epoxy composite sample with a centre hole and two semi-elliptical delaminations, which originated from the hole during fatigue loading. Basically, a very similar computational approach was adopted. The explicit FE code LS-DYNA was used to solve the mechanical problem and compute the sliding energy at the interfaces. In a next step the achieved heat load was applied to a separate thermal model in order to simulate the transient heat conduction using the FE software Abaqus. The thermal computation was based on the same solid element mesh that was used for the mechanical simulation. Minor adjustments of friction coefficients resulted in a very good (qualitative) agreement of experimentally and numerically achieved time histories of the temperature at both delaminations.

Contrary to the decoupling of the mechanical and the thermal problem, Mabrouki et al. [MTGF09, MTGF10] presented some coupled thermomechanical analyses of cracked steel compact tension specimens and aluminium bars using the Marc FE software package. Both effects, frictional heating at the sliding interfaces and volumetric dissipation due to cyclic plasticity have been considered. Pieczonka and colleagues [PSA+10] successfully simulated the thermal response of a 2 mm thick carbon/epoxy composite plate that was subjected to multiple impact tests using a drop-weight. The complex shape of the resulting delamination zones and tensile matrix cracks were assessed by conventional ultrasonic scans and radiography and then adopted in the FE model by realising multiple contact interfaces. The outcome of a transient coupled field thermomechanical analysis based on the energy input from a hammering excitation is promising since the different damage zones match both the relative extent of heating and the temporal occurrence when compared to the experimental results.

2.4.2 Prediction, repeatability, limits and probability of crack detection

Providing that a constant vibration amplitude is achieved, the crack heating is generally assumed proportional to the excitation frequency, thus to the number of cycles per time interval. However, the experimental verification is difficult because a change in frequency is very likely to affect the vibration mode so most tests on realistic samples lack comparability [RH05].

So far, comparatively few experimental works focusing on the relation between vibration intensity and defect heating have been published. Davis and Rankin [DR03] investigated a cracked aluminium compact tension specimen with a thickness of 2.4 mm that has been enlarged to have a defined area beyond the crack region, which is pneumatically clamped between the support and the used ultrasound weld gun. Based on the Coulomb friction law a linear relationship between crack heating and tangential relative displacement of the crack edges was expected. During the vibrothermographic tests the authors performed out-of-plane laser vibrometer measurements near the crack tip. The comparison of the velocity amplitude with the achieved thermal response at the crack resulted in a nearly linear relation with very little scatter. Although an extrapolation towards zero heating would leave a positive axis intercept, a potential threshold behaviour was not discussed.[4]

Barden et al. [BAM$^+$04] reported test results on a 3 mm thick plate made from nickel base superalloy featuring a 20 mm long surface crack of unknown depth. The surface temperature increase due to a 0.5 s pulse excitation at varying power was compared to the strain amplitude measured at the back side of the specimen. A chart presenting temperature increase over strain amplitude revealed a linear relation, which appears to be horizontally offset quite clearly. Although data points at lower strain amplitudes are not provided, the authors suggest there is a strain threshold, below which the crack does not heat up at all. It is further believed that the threshold behaviour is caused by friction between crack faces, which needs to be overcome before heat is generated. Rothenfusser [RH05] examined cracked gas turbine blades and found a quadratic relation between the crack heating and the velocity amplitude measured at one point. Differential vibrometry measurements of both crack faces were contributed by Homma [Hom07], who confirmed the type of functional relation. The dissipation power is presumed to be proportional to the square of the particle velocity in either case, clapping and rubbing modes of the crack faces. Some test results also indicated a significant threshold behaviour of the temperature increase. No crack heating was found unless a certain vibration amplitude was achieved. Similar to the argumentation of Barden et al. the authors suggest a strong interlocking of asperities. Hence, a certain activation energy is required to enable

[4]In the opinion of the author of this work, the presented data could also be interpreted as a truly non-linear relation when assuming another data point at the origin.

slippage. An alternative explanation could be that an initial gap between the opposing crack faces needs to be exceeded first by the normal relative displacement in order to generate frictional heat [RH05]. Later, the same authors found that the proposed quadratic dependence of the IR signal on the particle velocity is only satisfied at low vibration amplitudes. The relation seems to turn linear, if the measurement range is strongly extended towards higher amplitudes [HRBS06]. The same phenomenon was observed by Renshaw et al. [RHTA11], who presented similar charts relating crack heating and vibrational stress. Homma [Hom07] suggested a fundamental change of the heating mechanism and constructs a new hypothesis according to which the involved deformation is purely elastic at low vibration levels, whereas at higher levels plastic, irreversible material changes occur, which lead to the deviation from the quadratic dependence. He further discussed the frictional heating mechanism and points out that a static normal load during shearing motion corresponds to a linear relation between heating and relative velocity amplitude. Once the contact pressure is modulated due to the vibration, a quadratic relationship could be expected [Hom07]. Holland et al. [HUR08] exemplarily presented the entire spectrum of heating versus excitation intensity data. The diagrams include a perfectly straight line nearly through the origin, linear curves with a clear offset as well as an initially quadratic curve showing a decreasing slope at high vibration levels.

Apparently in contradiction to the theory of dry friction, Holland and colleagues [HUO$^+$11] assume the classical Amonton-Coulomb model being also suitable to predict the frictional heating of crack faces in case of a pure opening/closing mode (crack mode I) vibration. The authors argue that due to the inherent complexity of asperities local shearing motion can still be expected, although from a macroscopic view only normal, clapping relative displacements of the crack faces occur. The decomposition of the normal pressure into a static and a superimposed dynamic fraction leads to the following prognosis: When static normal pressure dominates dynamic pressure, a linear relation between vibration level and crack heating is brought about, whereas a quadratic relation results from a more significant dynamic normal pressure that dominates the static pressure. Thus, power law exponents between one and two are expected. The authors utilised a laser vibrometer during vibrothermographic testing to quantify the resonant bending motion of titanium and Inconel bars with a surface crack on one side. The specimens have been tuned to resonate in their 3rd order flexural mode approximately at 20 kHz, the working frequency of the deployed fixed-frequency welder. Knowledge of the mode shape allows to derive the longitudinal stress amplitude $\hat{\sigma}$ acting perpendicular to the half-penny shaped surface cracks, wholly based on a single-point vibration measurement. The authors fitted a power law to the achieved crack heating and found a temperature increase for the two sets of samples proportional to $\hat{\sigma}^{1.73}$ and $\hat{\sigma}^{1.49}$, respectively, which is in agreement with the proposed modelling.

Due to the limited number of systematic experiments, the different types of specimens used and also because of different excitation devices and measurement strategies, there is no conclusive theory available yet, which allows to estimate the thermal response of a flaw when exciting the structure. However, theoretical and experimental approaches have been developed towards the quantitative prediction of crack heating.

Miller et al. [MDBR03] proposed a predictive methodology based on numerical and analytical computations that finally aims at the definition of detection limits. The procedure requires a solid finite element model of the flawless structure, which is excited by harmonic displacements at the hypothetical transducer coupling location. It is assumed that the flaw is small compared to the part and does not affect the homogeneous displacement and stress fields. A periodic steady-state solution for the stress fields serves as input data for a simplified Coulomb model, which yields the frictional and thus the thermal power generated by a crack of a certain size. The heat generation depends on the flaw area, two crack dimension parameters, the elastic material properties, the friction coefficient, local shear and normal stress and the driving frequency. A major advantage of this approach is that only one FE solution is required to assess the friction power generated from flaws at any point and orientation within the structure. The required stress values are available at every single element. Now, many different orientations of a given crack at each solid element can be statistically evaluated. Providing a specific crack size, excitation amplitude and boundary conditions, the process gives the probability distribution (e.g. mapped onto the FE model) for this crack producing more than a certain heating power. If the heat generation probability is low in a critical region, variation of the constraints or the excitation position may yield a more favourable outcome [MDBR03].

Miller and colleagues further proposed a simplified conductive heat transfer model, which considers the defect as a planar, circular heat source, aligned parallel to the surface. The model defect is embedded in a semi-infinite, cylindrical domain and buried in a certain depth below the surface. For this configuration analytical solutions of the heat equation exist.[5] Based on the heat generation predicted by the friction model and a given duration of excitation, the maximum surface temperature over time can be derived. Finally, comparing the surface temperature rise with a specified minimum detectable temperature variation allows to define the crack detection limits. The theoretical envelope of detection limits includes the range of flaw parameters size and depth but also the required duration of forced vibration that can produce a sufficiently strong thermal response for detection [MDBR03].

[5] Solutions for upright and inclined heat sources are presented in [MAS$^+$09] and [OFTH02], respectively.

Morbidini and Cawley developed a sophisticated experimental procedure that aims at the quantitative prediction of the thermosonic signal based on the excited vibrations. The authors performed modal damping measurements on cracked and pristine beams made from steel and nickel-based superalloys. It is found that surface cracks on one side of the beam across the whole width cause a significant extra damping with respect to the flawless reference sample when exciting a low order flexural mode shape. Intuitively, the additional damping, as expressed by the loss factor η, increases with the crack size, specifically with the percentage of cracked cross-section [MCB$^+$06, MC09b]. Later, frequency and amplitude dependent damping measurements were reported, where the identified crack damping was only slightly affected by the two parameters [MCB$^+$07]. According to the authors, the energy released by the crack in one bending cycle is proportional to the extra loss factor caused by the crack and the involved modal strain energy. Since the strain energy grows with the strain amplitude squared, Morbidini expected the heating power and thus the temperature rise to be proportional to the square of the achieved longitudinal strain amplitude, which was monitored by a strain gauge on the uncracked back face. Strain amplitudes and derived power signals are averaged over a few milliseconds in the time domain considering the non-stationary vibration caused by the chattering between transducer and sample. The heating power distribution along the crack depth was estimated based on the infrared signature observed from the side of the specimen shortly after excitation when heat diffusion just set in (compare Fig. 2.3b, II). Finally, the derived heat load distribution was applied to a thermal finite element model in order to predict the temperature rise over time. The general approach was validated since numerous tests were conducted and a linear relation between the (averaged) measured and predicted values of the temperature rise was found [MCB$^+$06].

Further evidence that the excitation of numerous frequency components besides of the driving frequency produces a higher temperature rise was provided by Morbidini and Cawley [MC09b]. They inverted the described approach and estimated the required strain amplitude to achieve a certain crack heating, based only on the fundamental frequency. Exciting those predicted minimum amplitude levels often revealed temperature increases well above the target heating. This indicates an underestimation that was attributed to the actual non-linear excitation causing a rich frequency spectrum.

Encouraged by the question whether a vibrothermography test run that did not reveal any defect heating can be trusted, Morbidini and Cawley proposed a sophisticated calibration procedure in order to improve the reliability of vibrothermography applied to metals. Therefore, the above mentioned approach was further developed by introducing a single, time dependent parameter termed “heating index”, which has been proven to correlate well with the temperature rise of cracked beam specimens [Mor07, MC09a]. The multi-frequency excitation is now explicitly taken into account to evaluate each of the excited vibration modes. Thus, the instantaneous heating power depends on the squared strain amplitude, frequency and extra loss factor of all activated modes, whereas the loss factor can be assumed constant for small cracks [MC09a]. The proposed "heating index" (HI) additionally accounts for the heat generated earlier in time, which already has diffused in space but also contributes to the current temperature rise measured at the surface. This step incorporates a time depending weighting of the derived heating power. For that reason Morbidini employs a time constant that describes the exponential decay of the achieved thermal signal when switching off the excitation [Mor07, MC09a]. A detailed analysis of all conducted tests with different vibration levels revealed that the maximum temperature rise increases linearly with $\mathrm{HI}_{\mathrm{max}}$, the maximum value of HI achieved during the test [MC08, MC09a]. The presented charts containing all data for each specimen do not indicate a physical threshold behaviour since the linear fit curves are not shifted but reasonably match the coordinate origin. Morbidini and Cawley only found a threshold behaviour regarding defect detection, whereas the limited thermal sensitivity of the camera seems to be the only reason for that. The maximum HI achieved in those tests, which did not enable the detection of cracks were all lower than the smallest values of $\mathrm{HI}_{\mathrm{max}}$ at which a temperature rise was just perceptible [MC08, MC09a]. Therefore, the correlation between HI and the crack heating appears to be consistent.

The proposed calibration procedure includes testing of several cracked specimens with different flaw sizes [MC08, MC09a]. After estimation of the time constant, the "heating index" HI is calculated for each sample. Breaking the specimens to pieces allows to find the exact crack sizes and the calibration curve $\Delta T_{\mathrm{max}}/\mathrm{HI}_{\mathrm{max}}$ over crack size. The resulting curve corresponds to the thermosonic efficiency and was approximated by a quadratic relation between the normalised defect heating and the flaw dimension. Depending on the specific crack size to be detected and the given thermal sensitivity of the infrared camera a threshold value $\mathrm{HI}_{\mathrm{thr}}$ can be estimated from the curve fit. In a real testing situation the excitation is rated again by means of the "heating index", which is calculated using a conservative time constant from the calibration. If the threshold level $\mathrm{HI}_{\mathrm{thr}}$ is exceeded by the achieved $\mathrm{HI}_{\mathrm{max}}$, the test is considered to be reliable. The specimen will be judged as uncracked if the post-processing of the thermal record does not reveal any temperature rise [Mor07, MC09a]. Polimeno and Almond [PA09] installed a high-frequency microphone to their hand-held inspection device in order to evaluate the success of insonification by means of the described methodology proposed by Morbidini et al. The concept of assessing the "heating index" was also adopted by Bolu et al. [BGPT10], who compared the accuracy of HI derived from laser vibrometry, strain gauge measurements and microphone records.

In the past, researchers did not only question the reproducibility of vibrothermographic testing regarding the excitation and the heat generation. There is another issue that needs special attention if dry friction is presumed to be the prevailing effect: The tribological behaviour of contacting crack face asperities. Homma et al. [HRBS06] have shown a two-euro coin, which has been excited so strongly it separated into two pieces, the outer rim made from copper-nickel and the inner brass core. Significant wear was found at the same regions along the edge of the core, which clearly heated up during ultrasound excitation. In between, areas indicating only little wear perfectly correlate to the regions of no or little heat generation as seen in the thermograms. The connecting rod consisting of two parts, which was examined by Rothenfusser [RH05], revealed microscopic surface changes of the mating faces after ten ultrasonic pulses with a duration of 1 s each [Hom07].

Kato summarised the wear behaviour of different engineering materials. In case of metallic (multi-phase) alloys the "hard oxide film effect" prevails and the following responses of the tribo-system are typical results of repeated stress cycle and frictional heat cycle [Kat00]:

- Growth of oxide film on the surface and its fracture
- Severe plastic deformation in the surface layer and restructuring of its microstructure
- Adhesive material transfer to the counter surface and retransfer between mating surfaces
- Generation of wear particles
- Detachment, agglomeration, compaction and/or bedding of wear particles
- Conforming of interface morphology after roughening and/or smoothing of wear surfaces

An in-depth analysis of the tribological damage due to repeated vibrothermography testing of titanium and brass samples was conducted by Renshaw et al. [RHTA11]. They repeated a resonant flexural excitation of titanium beams with half-penny shaped surface cracks up to 1000 times, where different longitudinal stress amplitudes were realised. In case of high vibration amplitudes fretting damages were found on the fracture surface after the specimens have been sectioned. The fretting band with its perfectly semi-elliptical edges has left a region showing no tribological damage towards the crack initiation site (centred at the specimen surface) and towards the crack front. The temperature distribution achieved during the tests revealed two hot spots, which exactly match the intersection points of the fretting band with the surface plane. Unless the applied peak to peak stress exceeded about 30 % of the endurance limit, no clear evidence of surface alterations were observed when compared to pristine cracked, not vibrated specimens. Profilometry of the fretting band yielded about 60 % greater roughness measures with respect to the regions outside of it.

Renshaw et al. effectively increased the maximum vibration level by cutting some specimens into two halves along their longitudinal axis. In this way, new samples were produced, which have a quarter-elliptical edge crack instead of a semi-elliptical surface crack (compare Fig. 2.6) and a 50 % lower bending stiffness compared to the original specimen. Due to the resulting increase of achieved vibrational stress, additional changes to the fracture surface have been discovered by means of scanning electron microscopy (SEM). Firstly, significant oxidation was found in the form of small nucleation sites of oxide scale, clustered near the specimen surface. The oxide level quickly decreased towards greater crack depth, which is consistent with a lower availability of oxygen. Furthermore, a lower crack heating towards the neutral axis of the beam was attributed to the decrease in longitudinal stress and strain. Secondly, conspicuously smooth, featureless areas on the fracture surface have been identified. Although the measured bulk temperature increase did not exceed 7 K, Renshaw and colleagues [RHTA11] believe that extremely large oscillating stresses on contacting asperities caused a local material melting. Phase transformations such as melting are known to occur in metal-to-metal contacts undergoing relative motion, especially due to extreme flash temperatures at the microscopic level, which exceed by far the bulk temperature.

Vibrothermographic tests on extruded brass samples clearly revealed macroscopic signs of material deterioration after only 100 repetitions of the ultrasound excitation. During insonification wear debris were ejected from the crack, which gives evidence for the occurrence of adhesive wear according to Renshaw et al. [RHTA11]. Asperities that break off from a surface form loose particles, which either adhere to the opposing crack face or leave the vibrating interface. SEM imaging of the fracture surfaces proved not only the existence of loosely adhering material flakes but also manifested local melting and plastic deformations of asperities.

So far, very little research is available on the effect of excessively repeated ultrasound excitations with regards to the crack heating. Homma et al. [HRBS06] observed a decrease in the IR signal achieved from three consecutive test series on a single turbine blade with varying vibration levels. Interestingly, the non-linear curve described by the IR versus velocity data at low vibrations seemed to be exactly repeatable. Deviations with respect to the first test series appeared only at high vibration levels as if a threshold level was exceeded above which the crack faces are being modified in such a way that the heating mechanism gets less effective. Renshaw et al. [RHTA11] observed very similar phenomena when performing several series of vibrothermographic measurements on surface cracked titanium bars with increasing longitudinal stress amplitude. In total, a single specimen was subjected to about 1000 mono-frequency tone-bursts. While keeping the stress level below 20 % of the material's endurance limit, it is found that the overall slope of achieved crack heating versus stress amplitude data continuously decreased with each subsequent test series.

Here, the thermal responses achieved for each series follow a consistent trend with very little scatter. A contrary situation occurred when the vibrational stress range was greatly increased. The first 400 excitations were repeatable and showed little scatter but during the final 600 runs heat generation became non-repeatable. The crack heating tended to increase at a constant stress amplitude with a significantly enlarged scatter. A subsequent SEM analysis of the fracture surfaces revealed that severe tribological damage accumulated during the test series. Consequently, the authors recommend to limit the applied vibrational stress to ensure a truly non-destructive testing and to increase the test-retest repeatability as well as the overall inspection reliability.

The defect detection limit of vibrothermography clearly depends on a variety of parameters. Besides of the technical equipment like the ultrasound system, the type of infrared camera and the used camera optics, case-specific parameters will have a major impact on the detection capability. These are the material(s) or more precisely the tribological system, the static crack closure conditions and of course the crack dimensions. As stated above, the local vibration field, which depends on the dynamic system, the transducer coupling and the excitation frequency spectrum, represents the main variable in every single inspection. Small changes in geometry and mass of the structure under investigation or slightly different coupling conditions can result in divergent vibration spectra and thus defect heating despite conducting nominally equivalent test runs (compare [Hom07, HUR08]). Minor variations of the static force pressing the transducer against the sample or a thin coupling material optionally inserted between can lead to a significant scatter of the thermal response [DR03, ZDR$^+$03, CKLR07]. For these reasons, the overall reliability and even the reproducibility of vibrothermography might be low compared to other methods. Especially the inherent and hardly controllable non-linearity of the coupling between transducer and sample is a major issue that currently impedes the reliable implementation and a wider use of the method across industries [HUR08, MC08].

While being aware of the potential non-reproducibility of the ultrasound excitation, Bolu et al. [BGPH10] conducted a series inspection of 60 cracked aero-engine turbine blades with crack lengths between 0.3 and 6 mm. Prior to testing the whole set of blades a comprehensive repeatability study on four blades was carried out to assess the robustness of the defect detection process. All 300 preliminary test runs, including reclamping of the sample and renewal of the coupling material after different numbers of inspections, revealed the existing flaws. Finally, the cracks in 52 out of 60 turbine blades have been detected three times in a row. Consequently, the authors suggest vibrothermography as a rapid screening technique, which is applied prior to any (mandatory) standard method. Here, the volume of turbine blades subjected to the time-consuming and expensive cleaning process associated with fluorescent penetrant inspection (FPI) could be considerably decreased. In contrast, Bode [Bod11] reported an entirely satisfying reliability of vibrothermography since all known 21 cracks in a turbine disk with lengths ranging from 0.4 to 2.6 mm have been detected in an ad-hoc experiment. Moreover, some thermographic crack indications (true calls) gave no FPI indications. Hassan found cracks in aero-engine blades that were not detected before using standard techniques like EC inspection and FPI, which demonstrates the superiority of ultrasound excited thermography in detecting tight fatigue cracks in tight geometries [HHW+07].

It has been shown in the past that under certain conditions the detection threshold of ultrasound excited thermography can be far below flaw sizes of 0.5 to 1 mm, which is technically recognised as the transition between crack initiation and propagation stages. Homma [Hom07] clearly detected a 200 µm long high cycle fatigue crack in a steel specimen. The crack intentionally originated from the corner of an indentation left from a Vickers hardness test. Based on his experimental setup, Homma estimated the detection limit of vibrothermography below 100 µm. Favro et al. [FTH+01] presented thermograms indicating small cracks[6] in two notched titanium specimens. According to the authors, the crack lengths have been measured optically to be 54 and 20 µm, respectively.

Although several case studies attested ultrasound excited thermography a high sensitivity, there is no accepted probabilistic model of the defect detectability available yet. Based on the current status of vibrothermography research, it appears extremely difficult to develop a universally applicable probability-of-detection (POD) curve depending only on the crack size. According to Holland, at least the dependence of heating on the vibration level or the crack motion has to be considered. Due to the fact that the excited vibration is different at any point of the specimen, a proper POD study would have to evaluate the probability of detection for cracks of every possible position and orientation, which is virtually impossible [HUR08]. In order to qualify vibrothermography as a mainstream NDT method, a sophisticated procedure is required that allows to extrapolate the POD from a limited number of crack motion and heating measurements. The differential motion of crack faces, however, cannot yet be meaningfully represented as a scalar [HUR08]. In case of the titanium specimens described in [HUO+11], which were subjected to bending at a peak-to-peak stress of 100 MPa, Holland et al. found the POD for 1.5 mm long surface cracks to be about 98 % and for a length of 1 mm still greater than 60 %.

Clearly, the vibrothermographic flaw detection limit strongly depends on the chosen excitation method and the applied data processing. The post-processing of infrared records usually aims at the suppression of remaining camera noise and the enhancement of the defect contrast [Mal01]. Here, unspecific denoising and sharpening filters known from optical image processing are commonly

[6] Judging by the presented thermograms, these cracks seem to be corner flaws.

used with infrared images [HR10]. Regarding a tone-burst excitation, more advanced image processing techniques incorporate the physics of crack heating to develop a more specific and thus improved filtering. Li and colleagues constructed a matched filter based on an empirical time signal and a typical spatial pattern achieved from a crack's thermal response. Applying the 3D matched filter to IR records strongly increased the signal-to-noise ratio (SNR) and therefore the POD. Furthermore, an SNR-based detection criterion is proposed that allows for an automatic crack detection [LHM10a, LHM10b]. Holland and Renshaw recently addressed the issue of blurred infrared signatures caused by heat conduction. Based on the heat diffusion equation they derived an image processing technique, which compensates the effects of lateral heat diffusion. In the 2D case, e.g. for thin metal sheets, the procedure exactly yields the quantitative heat source intensity distribution. In contrast to the raw infrared sequence, temporal and spatial resolution of the heating process are dramatically increased. With three-dimensional heat flow the technique still enhances the qualitative localisation of heat sources in space and time [HR10].

A rather special excitation and data processing technique proposed by Busse and Rantala et al. [BWK92, RWB96, RWSB96], which is widely known as ultrasound lock-in thermography, greatly enhances the overall sensitivity of thermal signal detection (see Chapter 3 for a detailed description). Here, the ultrasound power is sinusoidally modulated at a low frequency. Consequently, the thermal response of a dissipating defect corresponds to this temporal modulation [DBW99]. Transferring the infrared record to the frequency domain allows to analyse the amplitude (and phase) of the defect signal at the known modulation frequency. Any noise, which is not related to the modulation but spread over all frequencies, is effectively filtered out so the signal-to-noise ratio greatly increases. The drawback of this method is clearly seen in the longer test duration required for several low-frequency modulation cycles.

3 Review of ultrasound excited thermography applied to metals

---

Owing to the interdisciplinarity of ultrasound excited thermography, this chapter contains an outline of the involved technical subdisciplines focusing on the inspection of metal structures. After giving an introduction to the excitation of plate waves and vibrations a key section of this chapter explicitly deals with the potential heat generation mechanisms. Here, the modelling of frictional heating is addressed in detail. The fundamentals of heat transfer are described next, with an emphasis on heat conduction including the propagation of thermal waves. Subsequently, the procedure of remote temperature measurements by means of a focal plane array infrared camera is illustrated. Finally, the most accepted approaches for vibrothermographic crack detection on metal components are reviewed.

3.1 Generation of structure-borne sound and vibration of plates

In this section basic concepts of mechanics are interpreted to describe the succession of insonification, wave propagation and resonant vibration. After giving a brief introduction to sound waves in an infinite solid the classical plate theory is introduced to estimate wavelength and phase velocity of guided waves including a discussion of its validity in comparison to the Lamb wave theory. Lastly, approximate formulas for the computation of free vibrations are derived and used further to predict the reaction of a damped plate to forced vibrations.

3.1.1 Orthogonal insonification and wave propagation in solids

First of all, the process of insonification by means of ultrasound transducers is considered, since this type of mechanical excitation is used in the present work. Throughout the literature of the last ten years, high-power weld generators with piezoelectrically driven converters are most commonly employed. The ultrasonic weld tool, comprising the converter with optional booster and sonotrode, is designed to perform only vibrations along its longitudinal axis. Any transversal oscillation, that means displacements orthogonal to the converter's axis but also torsional displacements, are unwanted and need to be avoided for the sake of a long service life of the weld tool without early fatigue damage.

With each single cycle of compression and expansion the driving piezo stack sends a pressure wave in the direction of the transducer's longitudinal axis. Ultrasonic converters, boosters and sonotrodes are tuned to be operated at specific eigenfrequencies such as 20, 40 or 60 kHz. When resonance is achieved, the longitudinal wave is reflected at the free end and interferes with the next incident waves so that the superposition of particle displacements results in a standing wave. Here, fixed positions along the weld tool axis can be found whose displacements are constantly zero. These are referred to as the nodal points of vibration. Other characteristic points, which are known as the vibration antinodes, exhibit the maximum displacement amplitude.

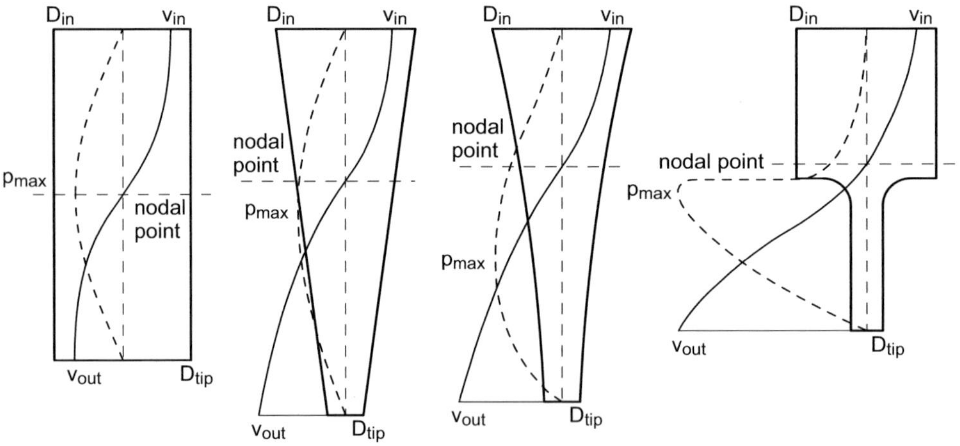

Figure 3.1: Cylindrical, conical, exponential and stepped sonotrode with characteristic functions of pressure and velocity amplitude [Bra05].

Figure 3.1 exemplarily illustrates four typical sonotrode shapes together with the functions of the displacement and sound pressure amplitude over the longitudinal axis. A tapered cross section with an input diameter greater than the tip diameter leads to a magnification of the displacement/velocity amplitude at the free end.

When the ultrasound converter or sonotrode is orthogonally brought into contact with the solid body to be investigated, ultrasonic waves can propagate through the interface and spread into the body. In the general case the incident longitudinal wave is splitted at the boundary into a reflected and a transmitted wave. The percentages of reflected and transmitted sound pressure p_r and p_t with respect to the incident pressure p_i are known as the reflection coefficient R and transmission coefficient T [KK86]:

$$R = \frac{p_\mathrm{r}}{p_\mathrm{i}} = \frac{Z_2 - Z_1}{Z_2 + Z_1} \quad , \qquad T = \frac{p_\mathrm{t}}{p_\mathrm{i}} = \frac{2Z_2}{Z_2 + Z_1} \quad . \tag{3.1}$$

Here, Z denotes the specific acoustic impedance of the material 1 (incident and reflected wave) and material 2 (transmitted wave), which depends on the density ρ and the actual longitudinal sound velocity c_l:

$$Z = \rho \cdot c_\mathrm{l} \tag{3.2}$$

In any case, p_i, p_r, p_t, R and T satisfy the sound pressure balance

$$p_\mathrm{i} + p_\mathrm{r} = p_\mathrm{t} \quad , \qquad 1 + R = T \quad . \tag{3.3}$$

For the material combination of titanium ($\rho = 4500\,\mathrm{kg/m^3}$, $c_\mathrm{l} = 5080\,\mathrm{m/s}$) and steel ($\rho = 7850\,\mathrm{kg/m^3}$, $c_\mathrm{l} = 5180\,\mathrm{m/s}$), equations (3.1) to (3.3) yield a transmission coefficient $T = 1.28$ (and $R = 0.28$). Thus, about 30 % of the perpendicularly insonified pressure is reflected at the interface. As proposed by Zweschper and colleagues [ZRDB04] a thin layer of a coupling material can help to harmonise the different impedances of the horn material Z_1 and the insonified structure Z_2 and therefore improve the sound transmission. An optimal impedance matching is achieved if the coupling material's impedance Z_c complies with the geometric mean of both values Z_1 and Z_2:

$$Z_\mathrm{c} = \sqrt{Z_1 \cdot Z_2} \tag{3.4}$$

Besides of the density or longitudinal waves, shear or transversal waves occur in solid materials. In an infinite elastic medium only these two types of waves exist. Once excited by harmonic or pulse-like forces, shear and pressure waves propagate through the medium. As indicated in Figure 3.2, the spatial distance between a node and the succeeding antinode is one fourth of the wavelength. Maximum stresses occur at the displacement nodes and zero stresses occur at displacement antinodes.

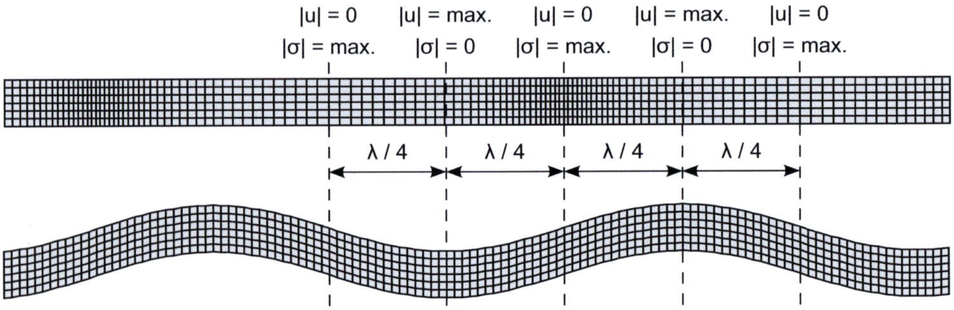

Figure 3.2: Snapshot of a longitudinal and a transversal wave in an infinite medium; u denotes the longitudinal or transversal displacement and σ is the axial or shear stress, respectively.

Wavelength λ, frequency f and circular frequency ω are generally related to the sound velocity c through

$$c = \frac{\omega}{2\pi}\lambda = f \cdot \lambda \quad , \tag{3.5}$$

whereas the speed of sound only depends on the type of wave and the elastic properties of the solid. In case of a macroscopically isotropic material such as steel, the longitudinal and transversal wave velocities c_l and c_t are given by Krautkrämer [KK86]:

$$c_\mathrm{l} = \sqrt{\frac{E}{\rho}\frac{1-\nu}{(1+\nu)(1-2\nu)}} \quad , \qquad c_\mathrm{t} = \sqrt{\frac{E}{\rho}\frac{1}{2(1+\nu)}} = \sqrt{\frac{G}{\rho}} \tag{3.6}$$

Here, E, G and ν denote the Young's modulus, the shear modulus and the Poisson's ratio, respectively. It should be noted that c_l and c_t in an infinite medium are independent of the wavelength or frequency. Thus, the basic wave types are non-dispersive, and c_l and c_t are phase velocities. For steel, the transversal wave velocity is about 55 % of the longitudinal wave velocity. Since the material properties are depending on temperature, c is also temperature dependent. Furthermore, a non-zero stress state, e.g. due to a static preload, also affects the sound velocity [KK86].

If the incident wave is not directed perpendicular to an interface but inclined by a certain angle α, a mode conversion can arise at the boundary. The directions of the incoming and all reflected or transmitted waves obey Snell's law of refraction.

$$\frac{\sin\alpha_{\text{I}}}{\sin\alpha_{\text{II}}} = \frac{c_{\text{I}}}{c_{\text{II}}} \tag{3.7}$$

Here, wave I denotes the incident wave inclined by α_{I} with respect to the surface normal and wave II represents an arbitrarily reflected or refracted wave. In both cases the direction of wave II is determined by the sound velocities c_{I} and c_{II}, which depend on the material and the type of wave. The partition of the sound pressure is very complex and cannot be discussed here. A comprehensive summary of the main characteristics and the dependencies on angle, sound velocity and acoustic impedance can be found in the work of Krautkrämer [KK86].

Even if the weld tool is attached perfectly orthogonal to the underlying structure, it cannot be avoided that an oblique insonification is initiated from the horn tip edges. The diffraction of sound can be explained by Huygens principle, which states that any position of the current wave front can be interpreted as the origin of a new semi-circular (2D) or semi-spherical (3D) elementary wave. The diffracted wavefront is then given by the envelope of all possible elementary wavefronts. This way, even a plane, uniaxially oscillating face such as the sonotrode tip spreads a convex wavefront, which is flat below but curved at its boundaries.

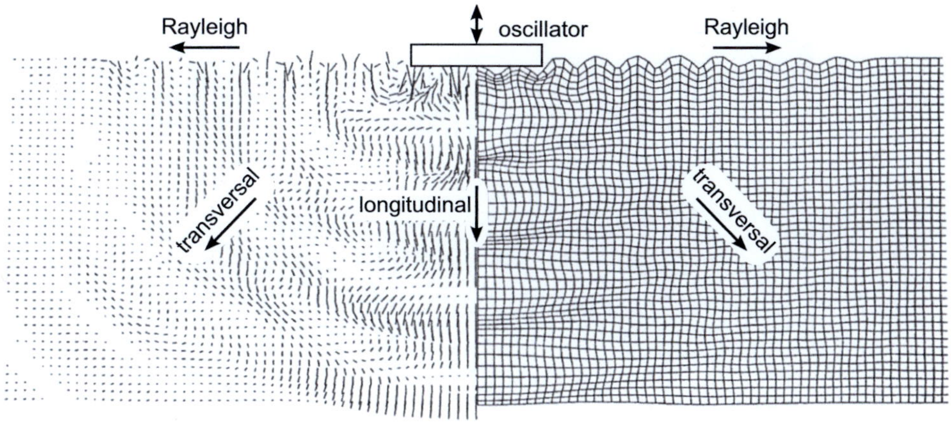

Figure 3.3: Wave propagation into the solid half-space caused by a small longitudinal wave transmitter; left: vector representation; right: grid displacements; re-composed image from Krautkrämer and Krautkrämer [KK86].

Figure 3.3 illustrates the complex situation of a uniaxially, point-like excitation of a solid half-space. Primarily, a longitudinal wave propagates in the normal direction as expected. But due to the edges of the oscillator, which represent geometrical discontinuities and therefore sources of diffraction, an additional wave develops that propagates radially along the free surface of the solid. This type of surface wave is referred to as Rayleigh wave. The resulting curved wave front causes shear displacements in the bulk material away from the oscillator's longitudinal axis so that a third wave type occurs, the transversal wave.

3.1.2 Wave propagation and standing waves in plates

Surface or Rayleigh waves have a rather small penetration depth in the order of one wavelength. If one considers a wave guide instead of the half-space and reduces the plate thickness further, the Rayleigh wave turns into a Lamb wave. These plate waves, which have been successfully used in structural health monitoring, occur in two basic shapes that have to be distinguished. Figure 3.4 depicts the fundamental types, the symmetric S0 mode and the antisymmetric A0 mode. The displacement vectors indicate the characteristic elliptic movements of the surface particles. In contrast to the zero modes, the existence of higher order modes denoted as S1, S2, S3, ... and A1, A2, A3, ... depends on the relationship between plate thickness and wavelength.

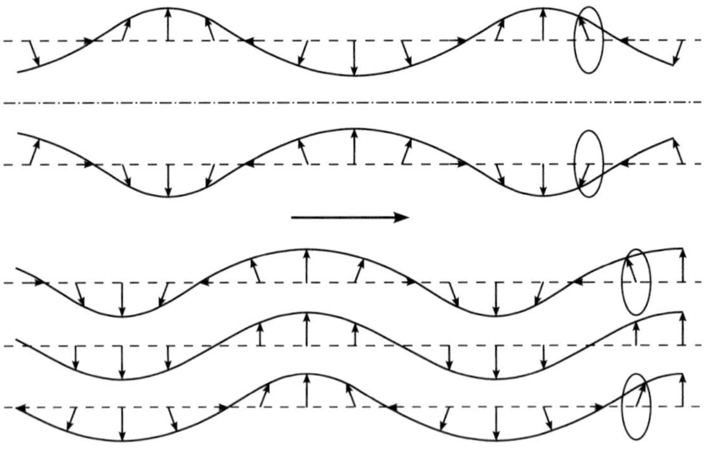

Figure 3.4: First symmetric (S0) and antisymmetric (A0) Lamb wave modes with surface displacement vectors based on [KK86].

Figure 3.5 shows the dispersion curves of the first four A and S modes computed with the software Vallen Dispersion R2010 [Val10] for a 10 mm thick steel plate. These functions represent the frequency dependent phase velocity and reveal that in steel specimens with a thickness of about 10 mm, which are mainly investigated in this work, only the fundamental modes A0 and S0 appear at relevant low ultrasound frequencies.

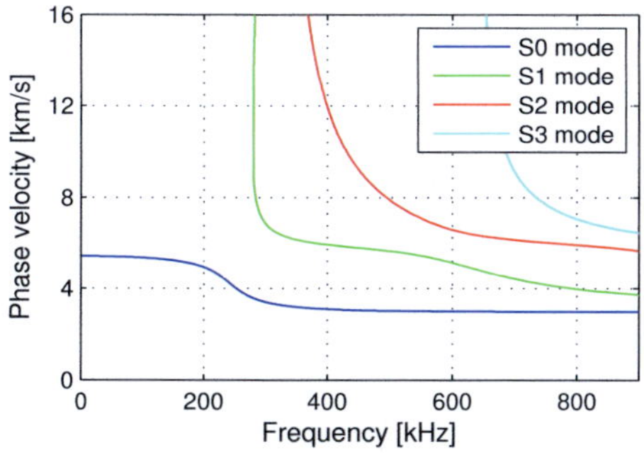

(a) Symmetric Lamb wave modes S0 to S3

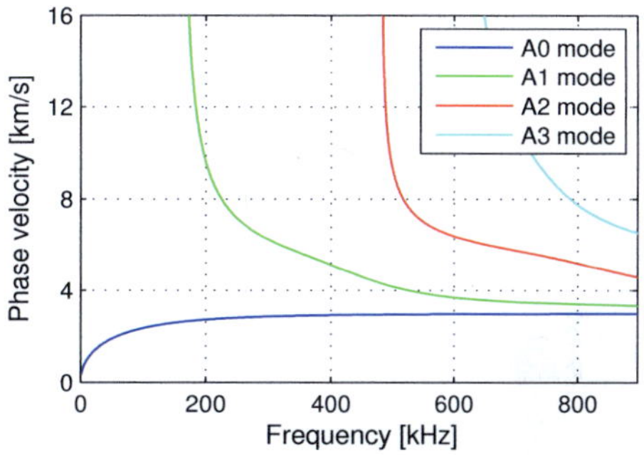

(b) Antisymmetric Lamb wave modes A0 to A3

Figure 3.5: Phase velocities for symmetric and antisymmetric Lamb wave modes in 10 mm thick steel plates computed with Vallen Dispersion R2010 [Val10].

The outlines of the deformed plate section in Figure 3.4 closely resemble the displacements of membrane and flexural vibration modes based on engineering mechanics as shown in Figure 3.6. Indeed, the wave propagation can be approximately described by the classical plate theory.

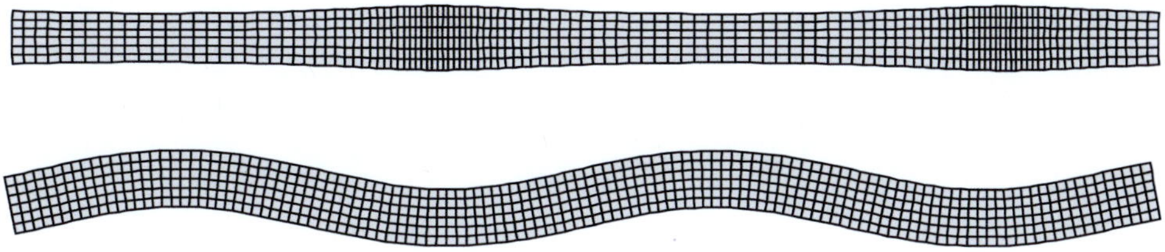

Figure 3.6: Extensional and flexural plate vibration mode shapes according to the classical plate theory.

The equations of motion for a thin plate/membrane defined in the coordinate system given in Figure 3.7 are derived by Wauer [Wau08]. In the following, constant material properties ρ, E and ν as well as a constant plate thickness h are assumed. Material damping is considered through the term d_a and the distributed loads p, q and r are directed in correspondence to the displacement components

u, v and w, respectively. Membrane or in-plane vibrations are governed by the following set of coupled equations:

$$\rho h\left(\frac{\partial^2 u}{\partial t^2}+d_a\frac{\partial u}{\partial t}\right)-\frac{E\,h}{1-\nu^2}\left[\frac{\partial^2 u}{\partial x^2}+\nu\frac{\partial^2 v}{\partial x\,\partial y}+\frac{1-\nu}{2}\left(\frac{\partial^2 u}{\partial y^2}+\frac{\partial^2 v}{\partial x\,\partial y}\right)\right]=p(x,y,t)$$

$$\rho h\left(\frac{\partial^2 v}{\partial t^2}+d_a\frac{\partial v}{\partial t}\right)-\frac{E\,h}{1-\nu^2}\left[\frac{\partial^2 v}{\partial y^2}+\nu\frac{\partial^2 u}{\partial x\,\partial y}+\frac{1-\nu}{2}\left(\frac{\partial^2 u}{\partial y\,\partial x}+\frac{\partial^2 v}{\partial x^2}\right)\right]=q(x,y,t) \qquad (3.8)$$

Bending or out-of-plane vibrations are described by

$$\rho h\left(\frac{\partial^2 w}{\partial t^2}+d_a\frac{\partial w}{\partial t}\right)+\frac{E\,h^3}{12\,(1-\nu^2)}\nabla^2\nabla^2 w=r\,(x,y,t) \quad . \qquad (3.9)$$

Here, the expression with Laplace operators $\Delta=\nabla^2$ can be further resolved to yield

$$\nabla^2\nabla^2 w=\left[\frac{\partial^2}{\partial x^2}+\frac{\partial^2}{\partial y^2}\right]\left[\frac{\partial^2}{\partial x^2}+\frac{\partial^2}{\partial y^2}\right]w=\frac{\partial^4 w}{\partial x^4}+2\frac{\partial^4 w}{\partial x^2\,\partial y^2}+\frac{\partial^4 w}{\partial y^4} \quad . \qquad (3.10)$$

In order to solve the above equations, boundary and initial conditions are required. The boundary conditions at the edges of the plate correspond to those known from statics: Clamped, pivoted or (force-)free. As initial conditions at time $t=0$ commonly the displacement and velocity field are being specified, which have to satisfy the boundary conditions. Here, the initial value problem also describes all transient phenomena, which in general quickly decay due to damping. The homogeneous equation of motion describes the free oscillations.

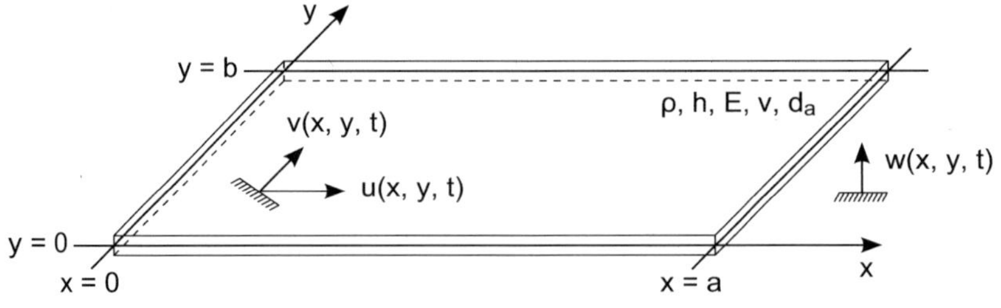

Figure 3.7: Rectangular plate/membrane in a Cartesian coordinate system [Wau08].

In the following, only the stationary vibrations are of interest, which are caused by a periodic, ideally harmonic excitation. These periodic solutions are strongly determined by the solution of the boundary value problem. Thus, the free oscillations will be analysed first. In the case of resonant excitation, stationary vibration nodes and antinodes are formed. This applies to both in-plane extension/shear vibrations as well as flexural vibrations, which coexist independently of each other. According to the linear theory, in-plane and out-of-plane displacements can be superimposed without any restrictions.

If the damping and excitation terms d_a, p and q are set to zero, equation (3.8) yields the homogeneous set of equations for the free membrane vibrations.

$$\rho h \frac{\partial^2 u}{\partial t^2} - \frac{E h}{1-\nu^2}\left[\frac{\partial^2 u}{\partial x^2} + \nu \frac{\partial^2 v}{\partial x \partial y} + \frac{1-\nu}{2}\left(\frac{\partial^2 u}{\partial y^2} + \frac{\partial^2 v}{\partial x \partial y}\right)\right] = 0$$

$$\rho h \frac{\partial^2 v}{\partial t^2} - \frac{E h}{1-\nu^2}\left[\frac{\partial^2 v}{\partial y^2} + \nu \frac{\partial^2 u}{\partial x \partial y} + \frac{1-\nu}{2}\left(\frac{\partial^2 u}{\partial y \partial x} + \frac{\partial^2 v}{\partial x^2}\right)\right] = 0 \tag{3.11}$$

Correspondingly, equation (3.9) is reduced to the homogeneous equation of motion describing free plate vibrations.

$$\rho h \frac{\partial^2 w}{\partial t^2} + \frac{E h^3}{12(1-\nu^2)} \nabla^2 \nabla^2 w = 0 \tag{3.12}$$

An estimation of the bending eigenfrequencies, which are deliberately excited by placing the weld-tool orthogonal to the structure, can be made in analytic form for a rectangular plate with dimensions a and b spanning in the x and y directions.[1] Assuming articular supports at the four edges, deflection and curvature remain zero at all time.

$$w(x,y,t) = \Delta w(x,y,t) = 0 \qquad \text{for} \quad x = \begin{cases} 0 \\ a \end{cases} \quad \text{and} \quad y = \begin{cases} 0 \\ b \end{cases} \tag{3.13}$$

A common approach is to separate time and spatial dependence in a multiplicative approach:

$$w(x,y,t) = \hat{w}(x,y) \cdot e^{i\omega t} \tag{3.14}$$

Now, $\hat{w}$ represents the distribution of the displacement amplitude.

[1]Formulas for eigenfrequencies of a large variety of plate and shell geometries are given by Leissa [Lei69, Lei73].

If this expression is inserted into the homogeneous differential equation (3.12), one finds

$$\Delta\Delta\hat{w}(x,y) - \frac{\rho\, h\, \omega^2}{B}\hat{w}(x,y) = 0 \quad , \tag{3.15}$$

where B denotes the (real) plate stiffness:

$$B = \frac{E\, h^3}{12\,(1-\nu^2)} \tag{3.16}$$

Equation (3.15) states an eigenvalue problem, which is commonly rearranged introducing the circular wavenumber

$$\kappa = \frac{\omega}{c} = \frac{2\,\pi}{\lambda} \qquad \text{and} \qquad \kappa^4 = \frac{\rho\, h\, \omega^2}{B} \tag{3.17}$$

to yield

$$\Delta\Delta\hat{w}(x,y) - \kappa^4\, \hat{w}(x,y) = 0 \quad . \tag{3.18}$$

Considering the boundary conditions (simply supported along all four edges), an infinite number of solutions in the form of

$$\hat{w}_{mn}(x,y) = C_{mn} \cdot \varphi_{mn}(x,y) \tag{3.19}$$

can be found. Here, C_{mn} are still unspecified constants and φ_{mn} denote the eigenfunctions:

$$\varphi_{mn}(x,y) = \sin\frac{m\,\pi\,x}{a} \cdot \sin\frac{n\,\pi\,y}{b} \quad , \qquad m = 1,2,\ldots,\infty \quad , \qquad n = 1,2,\ldots,\infty \tag{3.20}$$

The connection between eigenfunction, circular wavenumber and circular frequency arises from the conditional equations

$$\kappa_{mn} = \pi\sqrt{\left(\frac{m}{a}\right)^2 + \left(\frac{n}{b}\right)^2} \qquad \text{and} \qquad \kappa_{mn}^4 = \frac{\rho\, h\, \omega_{mn}^2}{B} \quad . \tag{3.21}$$

Finally, the circular eigenfrequencies of the vibrating rectangular plate with m and n antinodes along directions x and y are computed through [Wau08, KSA06]:

$$\omega_{mn} = \pi^2\left[\left(\frac{m}{a}\right)^2 + \left(\frac{n}{b}\right)^2\right] \cdot \sqrt{\frac{B}{\rho\, h}} \tag{3.22}$$

The above equation is particularly useful for finding suitable frequency-wavelength pairs. If only the frequency range of interest is known at first, the corresponding bending wavelength λ_b can be estimated according to Kollmann [KSA06].

$$\lambda_b = \frac{2\pi}{\sqrt{\omega}} \sqrt[4]{\frac{B}{\rho h}} \tag{3.23}$$

The number of sine half-waves m and n that will develop in a plate with given dimensions in case of resonance is then determined. Finally, equation 3.22 is solved for the actual eigenfrequency. Interestingly, the density of plate bending eigenfrequencies is independent of the frequency, which is not the case for beam bending modes. According to Möser [MÖ7], the number of resonances ΔN existing within the range of frequencies Δf is assessed through

$$\Delta N = \frac{1}{2} A \sqrt{\frac{\rho h}{B}} \Delta f \quad , \tag{3.24}$$

where the factor A is the plate's area. So, at constant thickness, a twice as large plate has therefore twice as many natural frequencies.

The classical Kirchhoff-Love plate theory becomes increasingly inaccurate with higher frequency and thus smaller bending wavelength. The propagation velocity (phase velocity) of a bending wave according to the classical theory is derived from equations (3.23) and (3.5):

$$c_b = \sqrt{\omega} \sqrt[4]{\frac{B}{\rho h}} \tag{3.25}$$

Möser [MK10] proposes a correction to consider the effects of the previously neglected rotational inertia and shear deformation in a simplified manner. The reduced spreading rate is given as

$$c_{b,corr} = c_b \left[1 - 4 \left(\frac{h}{\lambda_b} \right)^2 \right] \quad . \tag{3.26}$$

Once the flexural wavelength becomes smaller than $6h$, the corrective term amounts about 10 % of the propagation velocity according to the classical plate theory. A direct comparison of the phase velocities predicted by the Lamb wave theory (Fig. 3.8a) and estimated by the Kirchoff-Love theory (Fig. 3.8b) indeed manifests significant differences in the relevant frequency range for steel plates with thicknesses about 10 mm. An evaluation at 20 kHz reveals that for a 6 mm thick plate the deviation from the "true" propagation speed is only 5.4 %, while for a 14 mm thick plate the discrepancy increases to 10.2 %. It becomes evident that for the considered plate thicknesses the classical plate theory is only approximately valid at the lower bound of ultrasonic frequencies but sufficient for a quick assessment of the bending wavelengths and thus the resonant frequencies.

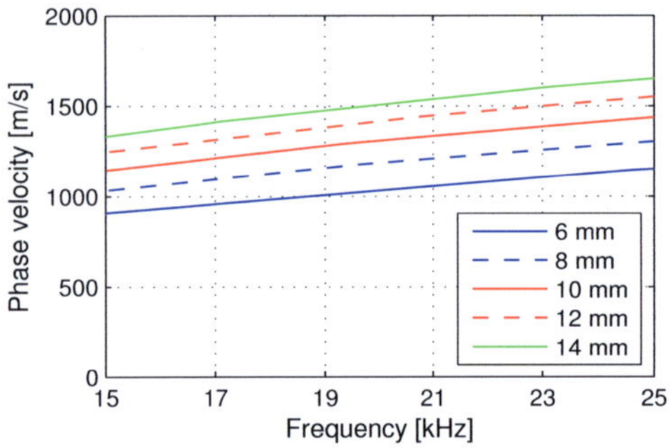

(a) Lamb wave theory (A0 mode)

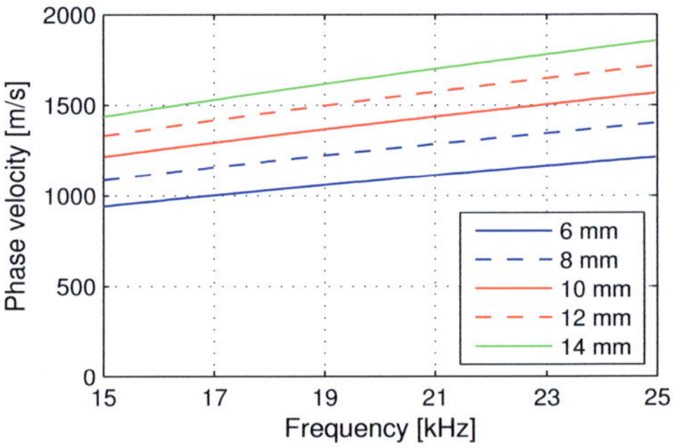

(b) Classical plate wave theory (Kirchhoff-Love)

Figure 3.8: Comparison of phase velocities of bending waves in the 20 kHz regime predicted by the classical plate theory and the Lamb wave theory.

3.1.3 Forced vibrations of a damped plate

The point-like ultrasound excitation of metal components, as it is used for vibrothermographic inspection, causes forced vibrations of the structure. These will arise only when the part is sufficiently "filled" with sound energy and the transient response turns into a steady state of vibration. The initial wave propagation from a point source is governed by the same flexural wave characteristics as those valid for plane waves. However, it should be noted that for energetic reasons the plate velocity and displacement amplitude will decay inversely proportional to the square root of the distance r from the point source [MÖ7]. If the maximum velocity amplitude is denoted as $\hat{v}_0$, the radially symmetric velocity field is described by

$$\hat{v} = \frac{\hat{v}_0}{\sqrt{r}} \mathrm{e}^{-\mathrm{i}\kappa r} \quad . \tag{3.27}$$

The final vibration field of a plate-type component subjected to periodic forces can be constructed from the solutions of the boundary value problem. Here, Kollmann's approach [KSA06] is presented, taking the material damping into account in form of the loss factor. Instead of the bending stiffness of the undamped plate, the complex plate stiffness

$$\underline{B} = (1 + \mathrm{i}\,\eta)\,B = \frac{(1 + \mathrm{i}\,\eta)\,E\,h^3}{12\,(1 - \nu^2)} \tag{3.28}$$

is now introduced. Hence, the equation of motion for the damped plate reads

$$\underline{B}\,\Delta\Delta\underline{w} + \rho\,h\frac{\partial^2 \underline{w}}{\partial t^2} = \underline{r}(x,y,t) \quad . \tag{3.29}$$

In practice, the yet unspecified excitation term $\underline{r}$ represents a harmonic excitation with angular frequency Ω and amplitude $\hat{r}$ acting normal to the plate surface:

$$\underline{r}(x,y,t) = \hat{r}(x,y)\,\mathrm{e}^{\mathrm{i}\,\Omega t} \tag{3.30}$$

It can be shown that the solution to the equation of motion (3.29) is found as the modal superposition of the known eigenfunctions $\varphi_{mn}(x,y)$ for the undamped plate

$$\underline{w}(x,y,t) = \sum_{m=1}^{\infty}\sum_{n=1}^{\infty} \underline{W}_{mn}\,\varphi_{mn}(x,y)\,\mathrm{e}^{\mathrm{i}\,\underline{\omega}_{mn} t} \quad , \tag{3.31}$$

where $\underline{W}_{mn}$ are now complex coefficients and $\underline{\omega}$ the complex eigen angular frequencies with

$$\underline{\omega}^2 = (1 + \mathrm{i})\,\omega^2 \quad . \tag{3.32}$$

Based on this framework, Kollmann gives the velocity distribution of the plate mid face for a harmonically varying single force with amplitude $\hat{F}$ acting at the point $(x_{\mathrm{exc}}, y_{\mathrm{exc}})$.

$$\hat{\underline{v}}(x,y) = \frac{4\hat{F}\,\Omega}{\rho\,a\,b\,h}\sum_{m=1}^{\infty}\sum_{n=1}^{\infty} \sin\left(\frac{m\,\pi\,x_{\mathrm{exc}}}{a}\right)\sin\left(\frac{n\,\pi\,y_{\mathrm{exc}}}{b}\right)$$
$$\sin\left(\frac{m\,\pi\,x}{a}\right)\sin\left(\frac{n\,\pi\,y}{b}\right)\cdot\frac{\omega_{mn}^2 - \Omega^2 - \mathrm{i}\,\eta\,\omega_{mn}^2}{(\omega_{mn}^2 - \Omega^2)^2 + \eta^2\,\omega_{mn}^4} \tag{3.33}$$

Especially in case of the point force excitation the mechanical admittance of the plate as a measure of the willingness to comply with the alternating load and perform vibrations is of great interest. Therefore, as a last point, the concept of admittance $\underline{\mathrm{h}}$ should be introduced, which can also be understood as a transfer function presenting a cause-effect relationship. In general, the admittance can be defined as the ratio of velocity at any point (x,y) and the excitation amplitude perpendicular to the plate at a fixed point $(x_{\text{exc}}, y_{\text{exc}})$.

$$\underline{h}\,(x_{\text{exc}}, y_{\text{exc}}, x, y) = \frac{\hat{\underline{v}}(x,y)}{\hat{F}\,(x_{\text{exc}}, y_{\text{exc}})} \tag{3.34}$$

Kollmann gives the absolute value of the input admittance $\underline{h}_{\text{i}}$ $(x = x_{\text{exc}},\ y = y_{\text{exc}})$ of the rectangular plate with four pivoted edges being subjected to a harmonic force at $(x_{\text{exc}}, y_{\text{exc}})$:

$$|\underline{h}_{\text{i}}\,(x_{\text{exc}}, y_{\text{exc}})| = \frac{4\,\Omega}{\rho\,a\,b\,h} \sum_{m=1}^{\infty} \sum_{n=1}^{\infty} \frac{\sin^2\left(\frac{m\,\pi\,x_{\text{exc}}}{a}\right)\sin^2\left(\frac{n\,\pi\,y_{\text{exc}}}{b}\right)}{\sqrt{\left(\omega_{mn}^2 - \Omega^2\right)^2 + \eta^2\,\omega_{mn}^4}} \tag{3.35}$$

In case of resonant excitation ($\Omega = \omega$), infinitely large velocity and therefore displacement amplitudes would occur if the vibrating system did not exhibit any damping effects. According to Möser [MÖ7], the loss factor η for steel is about 10^{-4}, whereas this value represents only the inner material damping. Any structural damping caused by bolted or riveted joints involving frictional slip is not covered.

3.2 Heat generation due to dissipative effects

In this section fundamentals and discussions of the essential heat generation mechanisms are presented, which – according to the present knowledge – occur regularly during ultrasound excited thermography when applied to metal components. Since dry friction seems to be the prevailing effect in most cases, refined approaches including the normal contact behaviour and a micro-slip model are considered within this key section.

3.2.1 Dry friction at crack interfaces

When two solid bodies come into normal contact frictional forces will arise that counteract any tangential relative displacement. Thus, to keep the two bodies sliding on each other mechanical work is to be done. The work rate or the frictional power is determined by the current friction force and the sliding velocity. Some small part of the expended frictional work will alter the bodies' surface structure and might lead to abrasive wear, but the majority of the frictional work will heat up the contact area. In most applications, excessive frictional heating is unwanted and needs to be compensated by cooling or prevented by lubrication of the contacting parts. Examples where frictional heat is purposely generated are friction stir welding and ultrasound welding and of course vibrothermography.

A general distinction has to be made concerning the phenomenological observation of heating due to dry friction. In fact, temperature increases can be identified on a micro and a macroscale [CH10]:

- The flash temperature of temporally and spatially distributed micro-contacts
- The mean volumetric temperature increase of solid bodies

From the microscopic point of view, even a polished steel surface has a characteristic distribution of asperities. If one body approaches another one, very few micro-contacts establish so that the true contact area A is only a small fraction of the apparent contact area A_0. Popov [Pop09] considers a roughly ground steel cube with an edge length of 10 cm being pressed with a force of 100 N against a steel plate and estimates the ratio A/A_0 to be 10^{-5}. In case of tangential sliding, the dissipated work is initially concentrated to these asperity peaks. Furthermore, heat conduction to the bulk material is hampered due to the geometry of asperities. Typically, only a small area connects the protruding asperity peaks to the body's volume. Seen in this way, measured micro-contact flash temperatures that exceed the melting point of the substrate material can be explained.

According to Czichos [CH10], the models of Blok (1937, line contact) and Archard (1958, point contact) are the most popular flash temperature theories in the field of tribology. The review paper of Bogdanovich and Tkachuk [BT09] includes a comparison of the main models for frictional heating and a discussion of further thermomechanical phenomena such as oxidation and thermal cracking. Ashby et al. [AAK91] developed approximate models for the calculation of bulk and flash heating during dry sliding and presented numerous surface temperature maps to illustrate the temperature variation depending on velocity and contact pressure. Sutter and Ranc [SR10] recently conducted thermographic measurements of flash temperatures greater than 1100 °C on steel specimens during high-speed sliding. Although the flash temperature theories are beyond the scope of this work, it should be noted that frictional heating on the microscale exceeds by far the macroscopic thermal response observed in non-destructive testing.

Regarding practical thermosonic flaw detection, where steady state oscillations and a time scale of seconds are realistic, two solid bodies or crack halves sliding on each other will generate a surface heat load somehow distributed over the apparent interface area A_0. In the general case, the total expended frictional work accumulated over a time period t_{exc} depends on the time histories of the friction force F and the sliding velocity v. The instantaneous frictional heating power is expressed as the frictional force times the relative velocity of the sliding parts. Since the direction of sliding is not relevant, only absolute values need to be considered. Integration of the instantaneous power

$$P(t) = |F(t)| \cdot |v(t)| \tag{3.36}$$

gives the total work W spent during the excitation period t_{exc}.

$$W(t_{exc}) = \int_0^{t_{exc}} P(t)\, \mathrm{d}t \tag{3.37}$$

In case of constant friction force and velocity, equation (3.37) yields the definition of mechanical work and simplifies to

$$W(t_{exc}) = |F| \cdot |v| \cdot t_{exc} = |F \cdot s_{cum}(t_{exc})| \quad , \tag{3.38}$$

where s_{cum} denotes the cumulated sliding displacement. Assuming a harmonic relative displacement with a velocity amplitude $\hat{v}$ and angular frequency ω, caused by an ultrasound excitation of duration t_{exc}, the cumulated distance

$$s_{cum}(t_{exc}) = \int_0^{t_{exc}} |v(t)|\ \mathrm{d}t = \int_0^{t_{exc}} |\hat{v} \cdot \sin(\omega \cdot t)|\ \mathrm{d}t \tag{3.39}$$

can be simplified if only values of t_{exc} are considered, which are multiples of a quarter cycle period. Then, the frictional work is approximated through

$$W(t_{exc}) = \int_0^{t_{exc}} P(t)\, \mathrm{d}t = |F| \cdot s_{cum}(t_{exc}) \approx |F| \cdot \frac{2\hat{v}}{\pi} \cdot t_{exc} = \overline{P} \cdot t_{exc} \tag{3.40}$$

and the average heat load per unit area is

$$\overline{q} = \frac{\overline{P}}{A_0} = \frac{|F|}{A_0} \cdot \frac{2\hat{v}}{\pi} = |\tau| \cdot \frac{2\hat{v}}{\pi} \quad . \tag{3.41}$$

Here, $\overline{P}$ denotes the temporal mean of the dissipation power and τ is the shear stress. The term $2\hat{v}/\pi$ represents the averaged absolute velocity (compare [HRBS06]). Note that the expended work is linearly depending on the velocity amplitude.

Up to this point, no friction model has been introduced, which prescribes the friction force. If one presumes validity of the classical Coulomb-Amonton friction law including an increased break-away force at stick conditions, two particular friction forces are defined for stick and slip using a static and a dynamic coefficient of friction μ_s and μ_d. The normal contact force F_n, which is defined positive although being a compressive force, acts as linear scale factor:

$$F_\mathrm{s} = \mu_\mathrm{s} \cdot F_\mathrm{n} \quad , \qquad F_\mathrm{d} = \mu_\mathrm{d} \cdot F_\mathrm{n} \qquad \text{for} \quad \mu_\mathrm{s} \geq \mu_\mathrm{d} > 0 \tag{3.42}$$

Static and kinetic friction forces F_s and F_d are always directed opposite to the applied tangential force F_t (or to the enforced sliding displacement) per definition. The friction law is then summarised by the following states and illustrated in Figure 3.9a.

$$F = \begin{cases} F_\mathrm{t} & \text{if } v = 0 \text{ and } |F_\mathrm{t}| \leq F_\mathrm{s} \\ F_\mathrm{s} & \text{if } v = 0 \text{ and } |F_\mathrm{t}| > F_\mathrm{s} \\ F_\mathrm{d} & \text{if } v \neq 0 \end{cases} \tag{3.43}$$

In order to avoid computational difficulties, Coulomb's non-smooth transition from stick to slip conditions at $v = 0$ is commonly approximated by differentiable functions such as the inverse tangent, hyperbolic tangent or logistic functions [RPnt, Wri06]:

$$\frac{F}{\mu_\mathrm{d} \cdot F_\mathrm{n}} = \begin{cases} \frac{2}{\pi} \cdot \arctan\left(\frac{v}{v_c}\right) \\ \left(1 - \mathrm{e}^{v/v_c}\right) / \left(1 + \mathrm{e}^{v/v_c}\right) \\ \tanh\left(\frac{v}{v_c}\right) \end{cases} \tag{3.44}$$

In any case, a tuning constant v_c is required to adjust the steepness of the curve at zero velocity. Figure 3.9a exemplarily depicts the three curves using the same value for v_c. A smooth expression for the drop of the friction coefficient is achieved by an exponential function given by Wriggers [Wri06]:

$$\frac{F}{\mu_\mathrm{d} \cdot F_\mathrm{n}} = 1 + \frac{(\mu_\mathrm{s} - \mu_\mathrm{d})\,\mathrm{e}^{-cv}}{\mu_\mathrm{d}} \tag{3.45}$$

Here, the constant c determines the decay rate of the friction coefficient.

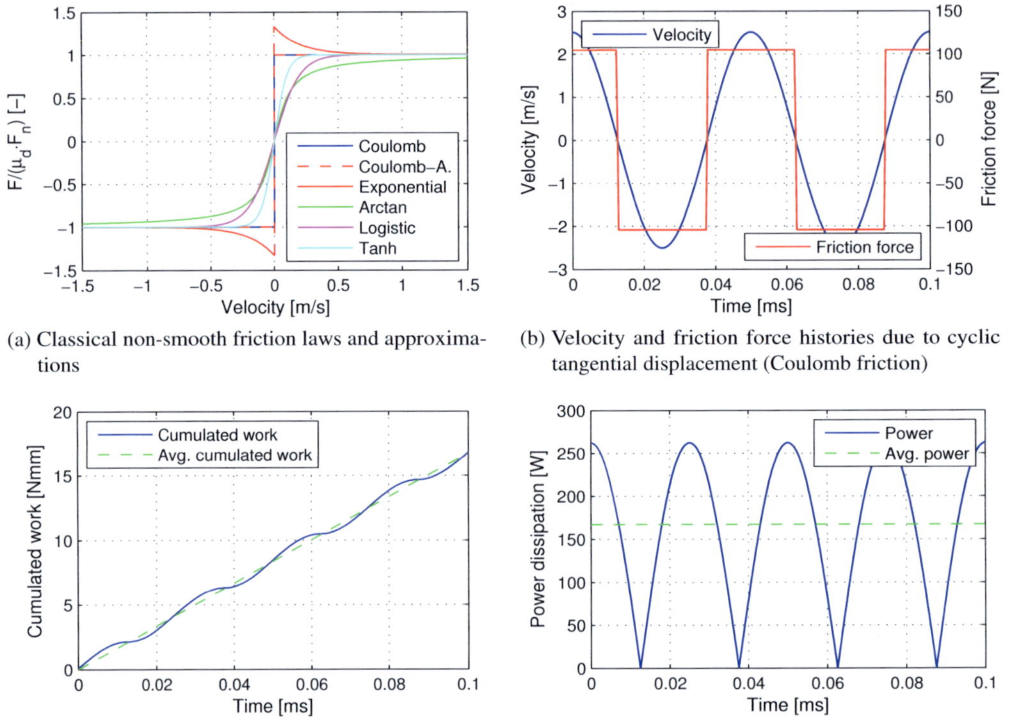

(a) Classical non-smooth friction laws and approximations

(b) Velocity and friction force histories due to cyclic tangential displacement (Coulomb friction)

(c) Cumulated work due to cyclic tangential displacement (Coulomb friction)

(d) Dissipated power due to cyclic tangential displacement (Coulomb friction)

Figure 3.9: Classical friction laws (a) and exemplary evolution of interface quantities friction force (b), cumulated work (c) and power dissipation (d) due to cyclic tangential displacement.

In order to analyse the power dissipation, a realistic contact situation is considered where two steel blocks with an apparent interface area A_0 of $10 \times 10\,\mathrm{mm}^2$ are statically pressed against each other so that the mean planes of the rough surfaces reduce their distance from an initial value d_0 (just touching) to d. As illustrated in Figure 3.10 the approach of the mean planes is then defined as:

$$z = d_0 - d \tag{3.46}$$

A constitutive equation is required that relates the approach z to the macroscopic normal pressure. For this purpose the model originally proposed by Hagman [Hag93] is introduced, which represents the contat pressure law in a common way according to Wriggers [Wri06]. Here, the shape of the asperities is assumed to be spherical with radius R and the height distribution of peaks is uniform.

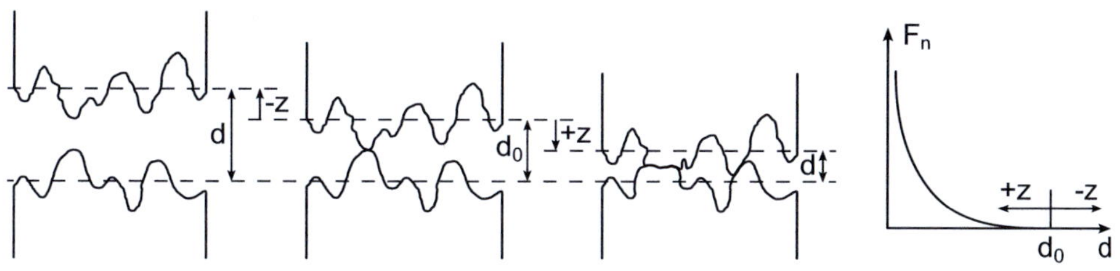

Figure 3.10: Normal contact of rough surfaces.

For a given approach z the normal load F_n can then be expressed as the sum of all single asperities in contact. Finally F_n is related to z through the power function (see [Olo95] for the derivation)

$$F_n = \frac{8}{15} E^* C A_0 R^{1/2} z^{5/2} \quad , \tag{3.47}$$

where E^* denotes the composite modulus of elasticity that allows for the treatment of two solid bodies made from different materials 1 and 2 with varying values of the Young's modulus E and the Poisson's ratio ν:

$$E^* = \left(\frac{1-\nu_1^2}{E_1} + \frac{1-\nu_2^2}{E_2} \right)^{-1} \tag{3.48}$$

The constant C in equation (3.47) is a surface characteristic and determines the number of contacts per unit area and approach z. According to the experiments on stainless steel reported in [Olo95], the values $C = 10^{10}$ contacts/(m^2 m) and $R = 10\,\mu$m are chosen for the analysis. Using these values, an approach of 20 µm corresponds to a macroscopic normal pressure of 3.48 N/mm^2.

First of all, the normal force is kept constant while one block is subjected to a sinusoidal tangential displacement with an amplitude of 20 µm and an ultrasound frequency $f = 20\,$kHz. Based on the classical Coulomb law (3.42, 3.43) with chosen values $\mu_s = \mu_d = 0.3$ the time dependent friction force, work and power dissipation are numerically calculated and illustrated in Figures 3.9b to 3.9d. Referring to equations (3.40) and (3.41), the average power $\overline{P}$ and heat load $\overline{q}$ are directly proportional to the velocity amplitude and the friction force that was presumed constant.

The situation changes if the normal displacement or the approach of the opposing faces is no longer constant but fluctuating over time. Assuming that the normal approach z also varies harmonically with the same frequency f allows to estimate the influence of a superimposed "clapping" displacement on the power dissipation. A more realistic simulation is achieved by considering a constant offset z_0, which corresponds to a compressive preload or to an initial gap, respectively (compare Fig. 3.10). Furthermore, a phase angle φ is taken into account that allows for a phase difference between tangential displacement u and normal displacement z:

$$u = \hat{u} \cdot \sin(2\pi \cdot f \cdot t) \quad , \qquad z = \hat{z} \cdot \sin(2\pi \cdot f \cdot t + \varphi) + z_0 \tag{3.49}$$

The frictional work is calculated and averaged as before, whereas equation (3.47) is evaluated for each time step of size $1/(f \cdot 160)$. Steps with a negative z value represent the gap status and do not contribute to the power dissipation. Until the normal displacement amplitude $\hat{z}$ is hold constant, the mean power dissipation is still increasing linearly with the tangential amplitude $\hat{u}$. Clearly, the variation of the offset level z_0 alters the "efficiency" of the heat generation as can be seen from Figure 3.11a.

If $\hat{z}$ is linearly related to $\hat{u}$ through $\hat{z} = g \cdot \hat{u}$, the mean power dissipation increases with $\hat{u}$ according to a power law of the form $\overline{P} = a \cdot \hat{u}^b$ with a and b being free coefficients. The characteristic exponent $b = 3.5$ has been retrieved by a curve fit using the non-linear least squares method, which is implemented in MATLAB 7.6 [The08]. Figure 3.11b depicts the results for g ranging from 0.4 to 1.6. In this load configuration any variation of the coefficient of friction μ and the phasing φ does not affect the exponent of the power function.

A change of the offset z_0, while keeping $\hat{z} = \hat{u}$, results in a threshold behaviour concerning the onset of power dissipation since there is no contact as long as the initial gap is not compensated by the normal displacement amplitude (Fig. 3.11c). If one shifts the curves achieved for negative values of z_0 back to the origin, so that the threshold is eliminated, exactly the same type of power law holds true. However, the exponent approaches unity if the positive offset z_0, which corresponds to a compressive preload, is increased. Thus, regarding the curve shape, the latter case of a high contact pressure cannot be definitely distinguished from the pure Coulomb friction without any normal force modulation. In this load configuration μ again has no influence and φ has little effect on the power law exponent as illustrated in Figure 3.11d.

Finally, the influence of the phase difference φ on the absolute dissipation is analysed. No matter in which way the normal displacement amplitude and the offset are varied, a phase difference of $\pm\pi/2$ results in the greatest power dissipation and values of 0 or $\pm\pi$ lead to minimal heat generation. This outcome was expected since the frictional work is maximised when the peak tangential velocity temporally coincides with the maximum normal force. The achieved curves for $\overline{P}$ in Figure 3.11e and 3.11f are symmetrical about $\varphi = 0$.

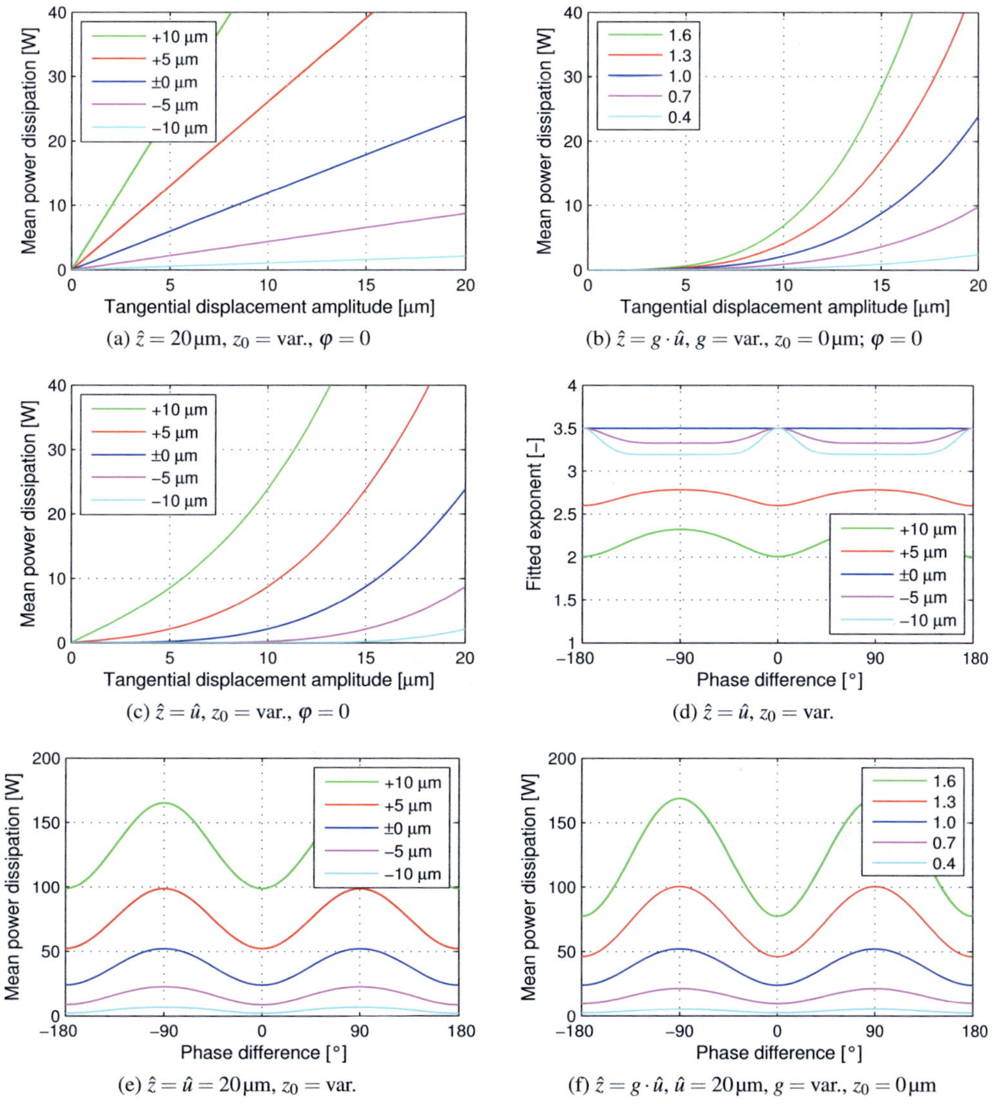

Figure 3.11: Mean power dissipation of two sliding steel blocks predicted by the Coulomb friction law with normal force variation according to Hagman's constitutive equations; influence of tangential and normal displacement amplitude and static preload on the power dissipation (a, b, c); influence of the phase difference on the power law exponent (d) and on the mean dissipation power (e, f).

In the previous model calculation a realistic normal contact behaviour was explicitly taken into account. If one considers tangential relative displacements in the micrometre range, which is a typical order of magnitude in thermosonic testing of metal components, the assumption of the conventional Coulomb friction law might be inappropriate. Especially in the controls community [ADC94] a phenomenon called micro-slip, pre-sliding displacement or Dahl effect is well known, which can be observed in quasi static or cyclic friction tests when the relative displacement is monitored with a high resolution. The experiments of Courtney-Pratt and Eisner [CPE57] on polished platinum-to-platinum specimens clearly revealed that a certain sliding distance is required in order to asymptotically reach the friction force predicted by Coulomb's law. A micro-slip region can be identified, in which the frictional load-displacement relationship is similar in shape to an elasto-plastic stress-strain curve. Anand [Ana93], who achieved very similar results with a steel-copper pairing, points out that the initial relationship between tangential traction and relative displacement is nearly linear with a certain shear modulus independent of the previous displacement history.

Physically, the micro-slip effect can be interpreted as the sticking of individual micro-contacts within the real contact area. Those asperities behave elastically and allow that the deformations vanish once the load is removed from the system [Wri06]. The classical Coulomb law, which prescribes that the maximum friction force F is immediately reached at the onset of a relative displacement is therefore not valid if small sliding distances in the micrometre range are considered.

The idea of splitting the sliding displacement into an elastic and a "plastic" part was picked up by Hagman, who developed his model for unidirectional sliding displacements. Later, the friction law has been extended to describe cyclic displacements and verified experimentally by Olofsson and Hagman [Olo95, OH97]. Sellgren and Olofsson succeeded in implementing the law into the commercial FE software ANSYS via user routines [SO96].

According to Olofsson, the friction force F within the micro-slip regime can be understood as the additive composition of an elastic force F_{spring} and a plastic force F_{slip}. Here, F_{spring} represents the total frictional force that results from contacting asperities, which are elastically deformed and do not undergo gross-slip. F_{slip} is the contribution from sliding asperities that have reached their limit tangential deflection. In [Olo95] both fractions of F and the sum are given as:

$$F_{\text{spring}} = \frac{8}{15}\mu E^* C A_0 R^{1/2}\left[z^{5/2} - \left(z - \frac{4G^* u}{\mu E^*}\right)^{5/2} - \left(\frac{4G^* u}{\mu E^*}\right)^{5/2}\right] \tag{3.50}$$

$$F_{\text{slip}} = \frac{8}{15}\mu E^* C A_0 R^{1/2} \left(\frac{4G^* u}{\mu E^*}\right)^{5/2} \tag{3.51}$$

$$F = F_{\text{spring}} + F_{\text{slip}} = \frac{8}{15}\mu E^* C A_0 R^{1/2} \left[z^{5/2} - \left(z - \frac{4G^* u}{\mu E^*}\right)^{5/2}\right] \tag{3.52}$$

In the above equations the constant G^* denotes the composite shear modulus, which is defined as

$$G^* = \left(\frac{2-\nu_1}{G_1} + \frac{2-\nu_2}{G_2}\right)^{-1} \quad . \tag{3.53}$$

Since there is no further differentiation of static and kinetic friction in the Hagman model, the general coefficient of friction $\mu = 0.3$ is chosen again. The evolution of the friction force F is independent of the sliding velocity but depends on the relative displacement u. With the given relation between F and u the shear stiffness is also determined and can be found by the derivation of F with respect to u.

$$K = \frac{\mathrm{d}F}{\mathrm{d}u} = \frac{16}{3} C A_0 G^* R^{1/2} \left(z - \frac{4G^* u}{\mu E^*}\right)^{3/2} \tag{3.54}$$

Macroscopic sliding sets in when all asperities start to slide. This is the case if the relative displacement exceeds the value $u_{\text{micro,max}}$, which represents the length of the micro-slip zone and depends on the friction coefficient, the material properties and the surface approach or the normal load. Thus, the validity of equations (3.50) to (3.52) is limited to values $u \leq u_{\text{micro,max}}$ with

$$u_{\text{micro,max}} = \frac{z\mu E^*}{4G^*} = \frac{\mu E^*}{4G^*} \cdot \left(\frac{15 F_{\text{n}}}{8E^* C A_0 R^{1/2}}\right)^{2/5} \quad . \tag{3.55}$$

Olofsson further developed the model to describe cyclic displacement under constant normal load. Being F^* and u^* the initially reached friction force and relative displacement, respectively, unloading and loading in reverse direction up to $-u_{\text{micro,max}}$ is described by

$$F_u = F^* - \frac{16}{15}\mu E^* C A_0 R^{1/2} \left[z^{5/2} - \left(z - \frac{2G^*(u^* - u)}{\mu E^*}\right)^{5/2}\right] \tag{3.56}$$

Reloading to u^* completes a hysteresis loop, where the friction force is

$$F_r = -F_u(-u) \tag{3.57}$$

The mechanical work spent during one micro-slip cycle is given by Olofsson [Olo95] as:

$$W_{\text{micro}} = \int_{-u^*}^{u^*} (F_u - F_r)\,\mathrm{d}u$$
$$= \frac{4}{7}\left[14\,\mu\,F_{\text{n}}\,u^* + \frac{F_{\text{n}}\,(\mu\,E^*\,z - 4\,G^*\,u^*)^{7/2}}{\mu^{3/2}\,E^{*5/2}\,z^{5/2}\,G} - \frac{\mu^2\,E^*\,z\,F_{\text{n}}}{G^*} - 7\,F^*\,u^*\right] \quad (3.58)$$

Recalling the model analysis of the two blocks, Figure 3.12a illustrates the evolution of F, F_{spring}, F_{slip} and K with increasing sliding displacement. Note that at $u_{\text{micro,max}} \approx 7.3\,\mu\text{m}$ the same total friction force is reached as with the pure Coulomb friction (compare Fig. 3.9b). Passing the micro-slip zone can be modelled by "switching" to the conventional Coulomb law. Due to the smooth transition to zero stiffness and full friction force, there is no discontinuity in the evolution of F. Figure 3.12b reveals that the jumps caused by the strong non-linearity of the Coulomb law are softened by the micro-slip approach. Plotting the friction force histories against time for the case of pure micro-slip with $\hat{u} = u_{\text{micro,max}}$ and a combined Coulomb/micro-slip situation with $\hat{u} = 20\,\mu\text{m}$ clarifies the temporal delay of the resisting force compared to the pure Coulomb friction response. The corresponding force-displacement hysteresis loops are illustrated in Figure 3.12c and have been evaluated in terms of the expended frictional work per cycle. As long as the total sliding range does not exceed the cyclic micro-slip length $2 \cdot u_{\text{micro,max}}$, equation (3.58) is valid. Larger relative displacements can be split up so that the effective Coulomb displacement amplitude is reduced by $u_{\text{micro,max}}$. Then, the total friction work per cycle is

$$W_{\text{cylce}} = W_{\text{micro}} + W_{\text{coulomb}} = W_{\text{micro}} + 4 \cdot \left(\hat{u} - u_{\text{micro,max}}\right) \cdot \mu\,F_{\text{n}} \quad (3.59)$$

and the mean power dissipation is

$$\overline{P} = f \cdot W_{\text{cycle}} \quad . \quad (3.60)$$

Figure 3.12d shows a comparison of the mean power dissipation depending on the displacement amplitude, once computed with and without consideration of the micro-slip phenomenon. The non-linear branch indicating the micro-slip zone can be approximated by the same type of power function as used earlier $\overline{P} = a \cdot \hat{u}^b$. Here, an exponent of 2.65 generally leads to the best fit.

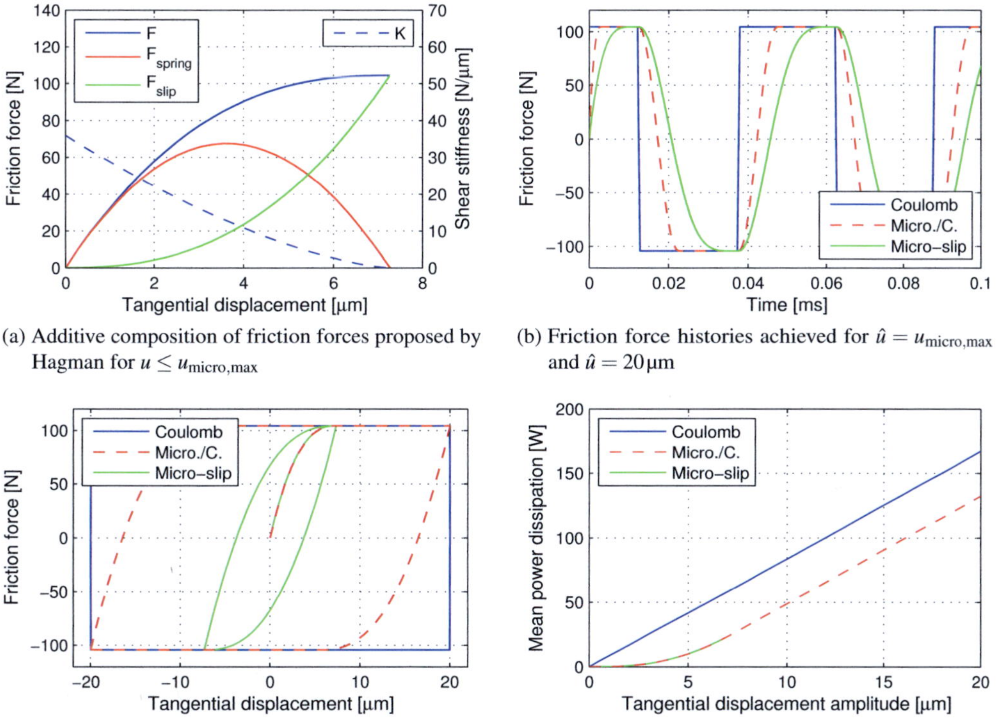

(a) Additive composition of friction forces proposed by Hagman for $u \leq u_{\text{micro,max}}$

(b) Friction force histories achieved for $\hat{u} = u_{\text{micro,max}}$ and $\hat{u} = 20\,\mu\text{m}$

(c) Hysteresis loops

(d) Evolution of the mean power dissipation

Figure 3.12: Composition of the friction force according to Hagman's micro-slip model (a); comparison of the classical Coulomb and the Hagman model regarding friction force (b), hysteresis loops (c) and mean power dissipation (d) due to cyclic displacement in tangential direction.

3.2.2 Bulk heating due to material damping

A real oscillating system is constantly exposed to energy absorbing and therefore damping effects. The damping forces are directed opposite to the displacement and ensure that after a certain period of vibration an activated oscillator comes back to rest in a static equilibrium or that a constant energy input is needed to maintain a desired oscillation amplitude. In general, the material damping of metals is very small compared to plastics (see for example [Pet96]). While damping phenomena are negligible for quasi-static and very slow loading and unloading procedures, material damping becomes relevant when a structure is subjected to dynamic force or displacement excitation. Damping in metals manifests in the form of a hysteresis loop in the force-displacement or stress-strain diagram, even if the yield strength is not exceeded. A variety of rheological models have been proposed in order to phenomenologically describe material damping and time-dependent material behaviour in general, such as creep or relaxation processes but also plasticity. A comprehensive discussion of the basic models as for instance the Maxwell, Voigt-Kelvin, Prandtl-Reuss and Coulomb models can be found in [Pet96].

The Hysteresis curves occurring in steel can be approximately quantified by the Voigt-Kelvin model depicted in Figure 3.13a. The mechanical model includes a discrete Hookean spring element with stiffness k and a single Newtonian damper element with constant c connected in parallel. So, the Voigt-Kelvin model is capable of representing linear elasticity, viscosity and viscoelasticity.

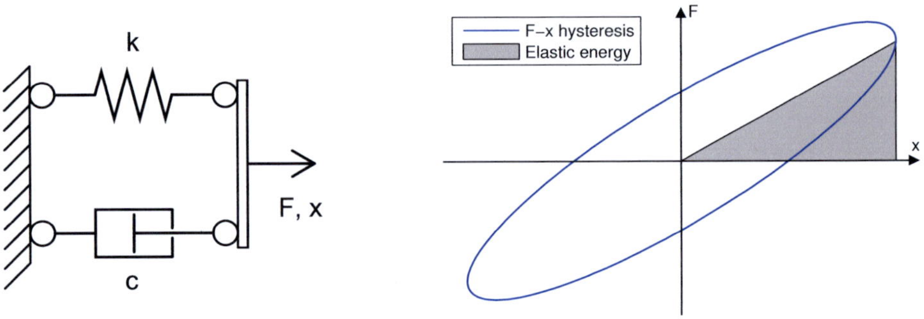

(a) Voigt-Kelvin model (b) Hysteresis loop and elastic strain energy

Figure 3.13: Voigt-Kelvin model and elliptical force-discplacement hysteresis loop.

Due to the parallel connection, the displacements of the elastic spring x_e and viscous damper x_d are equal at all times ($x(t) = x_\mathrm{e} = x_\mathrm{d}$). In contrast, the applied force is split into the spring force F_e and the damper force F_d as follows: [KSA06]

$$F(t) = F_\mathrm{e} + F_\mathrm{d} = kx + c\frac{\mathrm{d}x}{\mathrm{d}t} \tag{3.61}$$

The dissipated energy of the Voigt-Kelvin model can be derived by assuming a harmonic displacement response due to an excitation with the angular frequency Ω and amplitude $\hat{x}$:

$$x(t) = \hat{x}\cos(\Omega t) \quad , \qquad \frac{\mathrm{d}x}{\mathrm{d}t} = -\hat{x}\Omega\sin(\Omega t) \tag{3.62}$$

Next, the differential work of the damper $\mathrm{d}W_\mathrm{d}$ done along the differential displacement $\mathrm{d}x$

$$\mathrm{d}W_\mathrm{d} = F_\mathrm{d}\,\mathrm{d}x = F_\mathrm{d}\,\frac{\mathrm{d}x}{\mathrm{d}t}\,\mathrm{d}t \tag{3.63}$$

is integrated over one period with cycle duration $t_\mathrm{cycle} = 2\pi/\Omega$. Plugging in the velocity dependent damper force F_d leads to the dissipated work, which is fully converted to heat [Gei10, BSMM97]:

$$W_\mathrm{d,v} = \int_0^{t_\mathrm{cycle}} F_\mathrm{d}\frac{\mathrm{d}x}{\mathrm{d}t}\,\mathrm{d}t = \int_0^{2\pi/\Omega} c\left(\frac{\mathrm{d}x}{\mathrm{d}t}\right)^2\,\mathrm{d}t = \int_0^{2\pi/\Omega} c\hat{x}^2\,\Omega^2\,\sin^2(\Omega t)\,\mathrm{d}t = \pi\,c\,\Omega\,\hat{x}^2 \tag{3.64}$$

The converted energy is graphically represented by the area of the hysteresis loop shown in Figure 3.13b. In general, the area of an ellipse A is given as the product of the semi axes a and b multiplied by a factor π. Obviously, the resulting term for $W_\mathrm{d,v}$ directly corresponds to the formula $A = \pi\,a\,b$ with $a = c\,\Omega\,\hat{x}$ and $b = \hat{x}$. Thus, the energy dissipated per cycle grows linearly with the frequency and the square of the displacement amplitude. Since the elastic strain energy of the spring

$$U = \frac{1}{2}F_\mathrm{e}\,\hat{x} = \frac{1}{2}k\hat{x}^2 \tag{3.65}$$

is also increasing with $\hat{x}^2$, the viscous loss factor

$$\eta = \frac{W_\mathrm{d,v}}{2\pi U} = \frac{c\Omega}{k} \tag{3.66}$$

is linearly depending on the excitation frequency.

With increasing frequency the force temporally advances the displacement. Force and displacement are no longer in-phase but shifted against each other by a phase angle φ between 0 and $\pi/2$. The maximum phase difference is obtained when the damper is very stiff and the spring effectively bears no load [Pet96]. It can be shown that the phasing is directly related with the loss factor through ([KSA06])

$$\tan\varphi = \frac{c\Omega}{k} = \eta \quad . \tag{3.67}$$

Hence, the dissipated work per cycle due to viscous damping can be rewritten as

$$W_\mathrm{d,v} = \pi\,k\,\eta\,\hat{x}^2 \quad . \tag{3.68}$$

With many real engineering materials the damping is only marginally affected by the loading frequency. The frequency dependence can be formally compensated by declaration of a new damping quantity, which is inversely proportional to the frequency. When redefining the damping constant in the form of $c = \gamma/\Omega$, the hysteretic damping is found with the dissipated energy being

$$W_{\mathrm{d,h}} = \pi \gamma \hat{x}^2 \quad . \tag{3.69}$$

For the analysis of continuous bodies it is advantageous to incorporate viscous damping through the dynamic modulus of elasticity. The transfer from the discrete Voigt-Kelvin model to the (one-dimensional) continuum can be achieved if the force F is substituted by the stress σ, displacement x by strain ϵ, the spring stiffness k by the real modulus of elasticity E and the damper constant c by the expression $E \cdot \vartheta$. In the following, only harmonic variations of stress and strain are regarded. First of all, the viscoelastic material law applies with the auxiliary variable ϑ:

$$\sigma = E\,\varepsilon + E\,\vartheta \frac{\mathrm{d}\varepsilon}{\mathrm{d}t} = E\left(\varepsilon + \vartheta \frac{\mathrm{d}\varepsilon}{\mathrm{d}t}\right) \tag{3.70}$$

Through the definition of the complex strain

$$\underline{\varepsilon} = \hat{\varepsilon}\,\mathrm{e}^{\mathrm{i}\Omega t} \tag{3.71}$$

the material law follows from equation (3.70) introducing the complex modulus $\underline{E}$:

$$\underline{\sigma} = E\,\underline{\varepsilon} + E\,\vartheta\,\mathrm{i}\Omega\underline{\varepsilon} = E\,(1 + \mathrm{i}\,\vartheta\,\Omega)\,\underline{\varepsilon} = \underline{E} \cdot \underline{\varepsilon} \tag{3.72}$$

According to the substitutions made above, the dissipated energy per unit volume is

$$W_{\mathrm{d,v}} = \pi E\,\vartheta\,\Omega\,\hat{\varepsilon}^2 \quad . \tag{3.73}$$

A comparison with equation (3.68) reveals the relation between ϑ and the loss factor η.

$$\eta = \vartheta \cdot \Omega \tag{3.74}$$

This leads to the common definition of the complex modulus of elasticity that was already used in Section 3.1.3.

$$\underline{E} = (1 + \mathrm{i}\,\eta)\,E \tag{3.75}$$

The hypothetical adiabatic temperature change of a material portion with density ρ and specific heat capacity C_p, resulting from energy dissipation due to material damping is therefore

$$\Delta T_{\mathrm{d,v}} = \frac{W_{\mathrm{d,v}}}{\rho\, C_p} \quad . \tag{3.76}$$

For the sake of completeness, it should be mentioned that the thermoelastic effect does not contribute to the long-term heating of a dynamically loaded structure, directly. Instead, a reversible temperature change occurs during elastic loading and unloading. The temperature variation under adiabatic conditions, which provides the basis for thermoelastic stress analysis [HC91], is given as follows.

$$\Delta T_{\mathrm{e}} = -K\, T_0\, \Delta\sigma_{\mathrm{I}} = -\frac{\alpha_{\mathrm{T}}}{\rho\, C_p} T_0\, \Delta\sigma_{\mathrm{I}} \tag{3.77}$$

The temperature variation is proportional to the sum of the principle stresses (the first stress invariant σ_{I}), the initial temperature T_0 and the thermoelastic constant K, which comprises the coefficient of linear thermal expansion α_T, the density and the specific heat.[2] A derivation of the classical theory can be found in [Sne72]. Compressive stresses result in heating and tensile stresses provoke cooling of the material portion. Thus, there is no net heating of a harmonically loaded structure. However, due to heat conduction in real structures, a heat flow from compressed to elongated regions is initiated within each half loading cycle. Since the exchange of heat is an irreversible process, the thermoelastic effect represents a source of material damping – the thermoelastic damping.

It is beyond the scope of this work to further distinguish any (micro)physical effects like relocation of crystallographic defects, movement of dislocations and grain boundary viscosity, which contribute to the material damping but cannot be identified by macroscopic measurements.

3.2.3 Plasticity at crack tip regions and crack growth

As long as the structure under investigation is harmonically loaded within the elastic regime, stress and strain are linearly related if the phase shift due to material damping is taken into account. The situation changes once the yield stress is exceeded. Then, (cyclic) plastic deformations occur and the stress and strain histories are no longer proportional. Plotting the data pairs in a stress-strain diagram reveals a hysteresis loop that significantly differs from that of an elastically damped structure.

[2] The cited equation describes only the linear thermoelastic coupling. The temperature dependence of the material properties and also applied mean stresses (or residual stresses) lead to higher order coupling terms, which are considered in a revised theory of thermoelasticity [WJS87, WSD88].

The cyclic material behaviour of steel is mainly determined by its cyclic stress-strain curve, which is derived from fatigue tests at different loading levels. The reversal points of all possible hysteresis loops with zero mean stress are lying on this characteristic curve. Figure 3.14a illustrates a typical hysteresis curve based on a Ramberg-Osgood power law approximation of a real cyclic stress-strain curve. The corresponding temporal variations of the strain and the stress, which is understood as the dependent variable, is shown in Figure 3.14b.

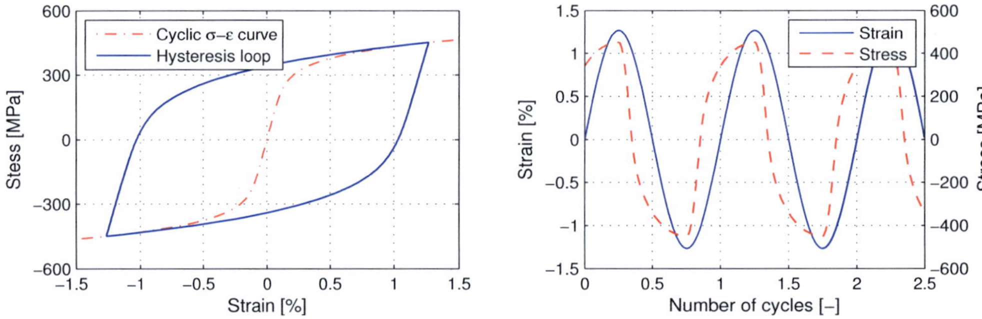

(a) Elastic-plastic hysteresis and cyclic σ-ε curve (b) Harmonic strain variation and dependent stress

Figure 3.14: Elastic-plastic hysteresis loop and corresponding histories of stress and strain.

Similar to the aforementioned rheological models that phenomenologically describe linear material damping, ductility models with an arbitrary number of parameters can be set up and adjusted in order to simulate measured elasto-plastic material behaviour.

The Coulomb model depicted in Figure 3.15a, comprising a friction element with the maximum friction force R and a Hookean spring with stiffness k connected in parallel, leads to a rectangular hysteresis loop for $k = 0$ (compare Fig. 3.12c in Section 3.2.1). With a spring stiffness $k > 0$ the simple model can only roughly approximate the curve shown in Figure 3.14a, whereas the loop turns into a parallelogram. Hence, the model describes a rigid/plastically hardening material. Concerning the dissipated energy per cycle, which corresponds to the area of the hysteresis, the Coulomb model can represent an upper limit if the realistic loop is enveloped by the parallelogram matching the points of reversal. Being independent of k, the dissipated work per cycle is

$$W_{\mathrm{d,C}} = 4R\hat{x} \quad , \tag{3.78}$$

where $\hat{x}$ is the imposed displacement amplitude.

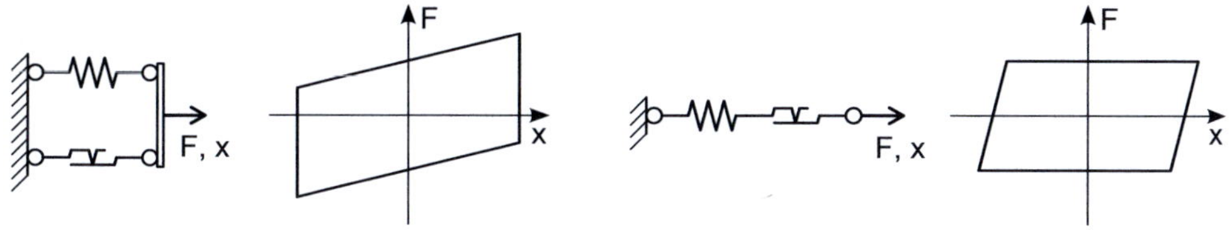

(a) Coulomb model (b) Prandtl-Reuss model

Figure 3.15: Basic rheological models and their load-displacement hysteresis loops.

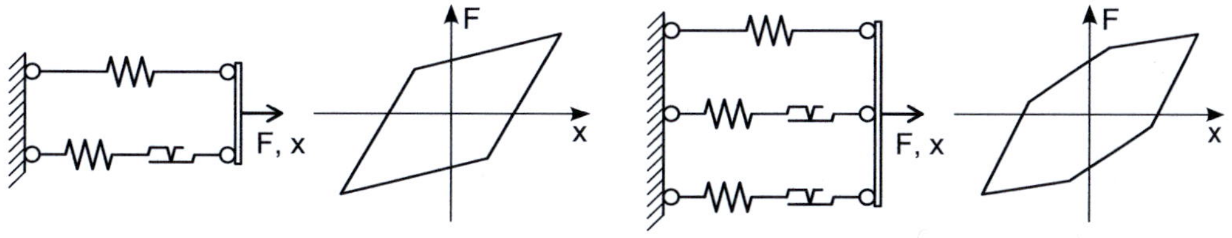

(a) Combined Coulomb and Prandtl-Reuss model (b) Masing model

Figure 3.16: Combined rheological models and their load-displacement hysteresis loops.

With a friction element and a Hookean spring connected in series the Prandtl-Reuss model shown in Figure 3.15b is capable of describing the cyclic behaviour of an ideal elastic/ideal plastic material. Here, the dissipated energy per cycle is not only depending on the maximum friction force R but also on the spring stiffness k.

$$W_{\mathrm{d,P-R}} = 4R\left(\hat{x} - \frac{R}{k}\right) \tag{3.79}$$

More sophisticated, multi parameter models can be designed by connecting several Coulomb elements and another spring in series or by connecting several Prandtl-Reuss elements and another spring in parallel (compare Fig. 3.16). The resulting Iwan or Masing models are able to reproduce realistic hysteresis loops. However, due to the underlying degressive force-displacement characteristic, the cyclically dissipated energy W_{d} of all mentioned ductility models grows linearly with the displacement amplitude $\hat{x}$ or the strain amplitude $\hat{\varepsilon}$, respectively, if transferred to a continuum. In contrast to the viscous material damping, the plastic dissipation is independent of the load frequency.

The crack tip loading can be quantified by means of the classical concepts of linear elastic fracture mechanics (LEFM). Based on the elasticity theory, characteristic stress and displacement fields near the crack tip are assigned to each of the fundamental crack opening modes I, II and III, which are illustrated in Figure 3.17a. Mode I describes the in-plane cleavage of the crack (tensile opening). The modes II and III represent the in-plane and out-of-plane shear (sliding and tearing), respectively. It should be noted that for thin plates two additional modes i and ii, which result from bending and torsional loads, can be defined and used within the framework of fracture mechanics [Kun08, DA02]. However, these modes are not elementary since they can be attributed to modes I and II, whereas the magnitude and direction of the respective near-tip field varies along the crack front.

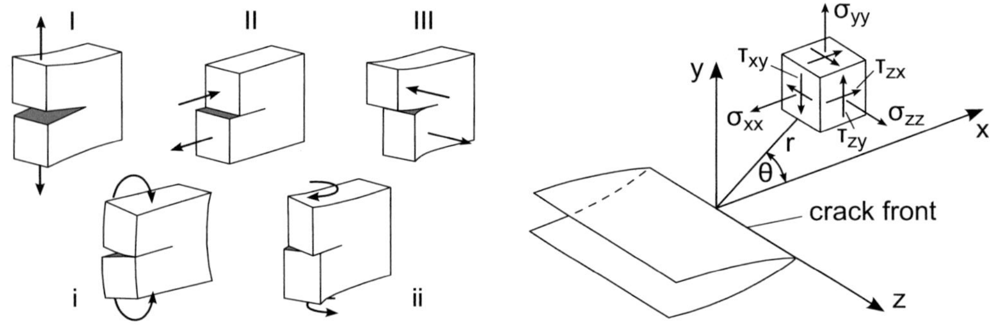

(a) Fundamental crack modes I, II, III, and plate specific modes i and ii according to [DA02]

(b) Crack tip/front coordinate system and denotation of stresses

Figure 3.17: Basic and plate specific crack modes and coordinate system at the crack front.

The assumption of an ideally sharp crack, which is the separation of crack faces without any gap between, leads to a stress singularity at the tip of type $1/\sqrt{r}$ with r being the distance from the tip. All displacement components u, v, w in directions x, y and z tend to zero when approaching $r = 0$ [GS07, Kun08]. A commonly used approximation of the crack tip displacement fields for mode I, II and III is given by [RS12] based on the polar coordinate system depicted in Figure 3.17b.

$$\text{I:} \quad \begin{Bmatrix} u \\ v \end{Bmatrix} = \frac{K_{\mathrm{I}}}{2G}\sqrt{\frac{r}{2\pi}} \begin{Bmatrix} \cos\frac{\theta}{2}\left(\kappa - 1 + 2\sin^2\frac{\theta}{2}\right) \\ \sin\frac{\theta}{2}\left(\kappa + 1 - 2\cos^2\frac{\theta}{2}\right) \end{Bmatrix} , \qquad w = 0 \tag{3.80}$$

$$\text{II:} \quad \begin{Bmatrix} u \\ v \end{Bmatrix} = \frac{K_{\mathrm{II}}}{2G}\sqrt{\frac{r}{2\pi}} \begin{Bmatrix} \sin\frac{\theta}{2}\left(\kappa + 1 + 2\cos^2\frac{\theta}{2}\right) \\ \cos\frac{\theta}{2}\left(\kappa - 1 - 2\sin^2\frac{\theta}{2}\right) \end{Bmatrix} , \qquad w = 0 \tag{3.81}$$

$$\text{III:} \quad \begin{Bmatrix} u \\ v \end{Bmatrix} = 0 , \qquad w = \frac{2K_{\mathrm{III}}}{G}\sqrt{\frac{r}{2\pi}}\sin\frac{\theta}{2} \tag{3.82}$$

In the above equations the conditions of plane stress (at the edge of a body) and plane strain (in the interior of a body) are to distinguished through the parameter κ:

$$\kappa = \begin{cases} (3-\nu)\,/\,(1+\nu) & \text{plane stress} \\ 3-4\nu & \text{plane strain} \end{cases} \tag{3.83}$$

The magnitude of each near-tip field is solely determined by the corresponding stress intensity factor (SIF) K_{I}, K_{II} or K_{III}. The SIF depends on the object's geometry including the crack as well as the applied load and displacement boundary conditions. Under mixed mode loading the three fundamental components of K can be separated and evaluated individually according to the superposition principle. The local stress and displacement fields as cited above cease to have validity outside a small region centred at the crack tip. According to Kuna [Kun08], the so called K-dominated zone is confined to the radius r_K, which is about 2 to 10 % of the crack depth a. The value of K provides a unique measure of the local crack tip loading and fully "controls" all fracture processes.

The stress intensity factors are mathematically defined as the limit of their non-zero stress component along the crack direction ($\theta = 0$) as r approaches zero.

$$\begin{Bmatrix} K_{\text{I}} \\ K_{\text{II}} \\ K_{\text{III}} \end{Bmatrix} = \lim_{r\to 0} \sqrt{2\pi r} \begin{Bmatrix} \sigma_{yy}(\theta = 0) \\ \tau_{xy}(\theta = 0) \\ \tau_{yz}(\theta = 0) \end{Bmatrix} \tag{3.84}$$

Analytical solutions exist only for a few standard problem geometries. In the majority of practical cases, numerical methods are employed in order to estimate the actual SIF or its range $\Delta K = K_{\text{max}} - K_{\text{min}}$, respectively. Especially with complex shaped three-dimensional crack configurations, where all SIF components typically vary along the crack front, computational methods are increasingly in demand to predict the crack propagation and the remaining safety margin against instable crack growth.

The assumption of a perfectly sharp crack tip and the stress singularity does not hold in reality since in any material, either brittle or ductile, a process zone will originate, in which plastic yield or rupture occurs. At this point, the elasticity theory is no longer valid. In metals the yield stress will be exceeded locally and a plastic region develops resulting in a stress redistribution and blunting of the crack tip.

As long as the external loading is moderate so that the plastic zone is small compared to the crack depth, the component's dimensions and the remaining ligament, the concepts of linear elastic fracture mechanics are still applicable. The described conditions are referred to as small scale yielding and represent the transition from linear elastic to elastic-plastic fracture mechanics (EPFM).

The extent of the plastic zone can be estimated through evaluation of the near-tip stress solution using the von Mises yield criterion. As a result, the contour depending on θ is found, where the equivalent stress matches the yield strength.

For the three basic crack modes the plastic zone radius r_p is found in [Kun08], [Ber06] and [GS07], respectively:

$$\text{I:}\quad r_\mathrm{p}(\theta) = \frac{1}{2\pi}\left(\frac{K_\mathrm{I}}{\sigma_\mathrm{y}}\right)^2 \cos^2\frac{\theta}{2} \begin{cases} \left(3\sin^2\frac{\theta}{2}+1\right) & \text{plane stress} \\ \left(3\sin^2\frac{\theta}{2}+(1-2\nu)^2\right) & \text{plane strain} \end{cases} \tag{3.85}$$

$$\text{II:}\quad r_\mathrm{p}(\theta) = \frac{1}{2\pi}\left(\frac{K_\mathrm{II}}{\sigma_\mathrm{y}}\right)^2 \begin{cases} \left(3-\frac{9}{4}\sin^2\theta+\sin^2\frac{\theta}{2}\right) & \text{plane stress} \\ \left(3-\frac{9}{4}\sin^2\theta+\sin^2\frac{\theta}{2}(1-2\nu)^2\right) & \text{plane strain} \end{cases} \tag{3.86}$$

$$\text{III:}\quad r_\mathrm{p}(\theta) = \frac{1}{\pi}\left(\frac{K_\mathrm{III}}{\tau_\mathrm{y}}\right)^2 \cos\theta \tag{3.87}$$

Assuming a static load according to mode I, which results in a stress distribution σ_{yy} along the crack direction ($\theta = 0$) as shown in Figure 3.18a, clarifies that the true plastic zone d_p must be greater than r_p. Simply limiting the stress to the yield strength σ_y would result in a violation of the global equilibrium with respect to the external load. Therefore, a stress redistribution occurs that extends the plastic zone by a factor of two.

A totally different estimation of the plastic zone size is based on Dugdale's strip yield model, which was developed for thin tensile specimens under plane stress mode I loading [Kun08, GS07]. Without giving any explanation of the model itself, it should be noted that it results in a plastic zone of the same order of magnitude as predicted by equation (3.85):

$$d_\mathrm{p} \approx \frac{\pi}{8}\left(\frac{K_\mathrm{I}}{\sigma_\mathrm{y}}\right)^2 \tag{3.88}$$

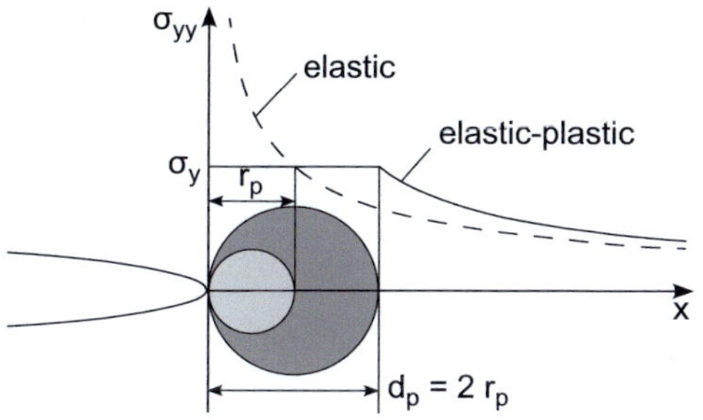

(a) Elastic and elastic-plastic stress distribution

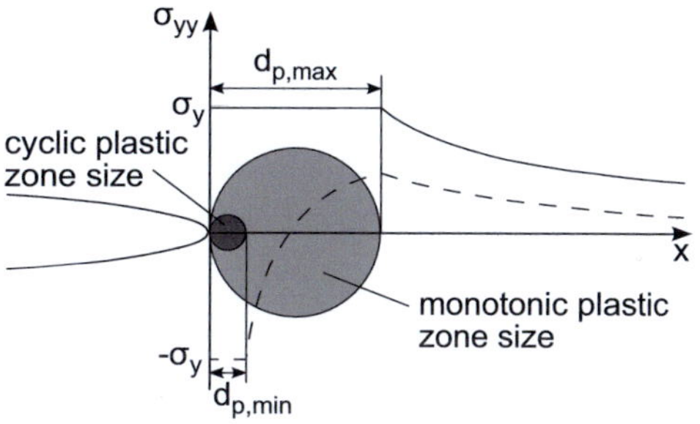

(b) Monotonic and cyclic plastic zone

Figure 3.18: Irwin's plastic zone correction due to elastic-plastic material behaviour and comparison of the monotonic and cyclic plastic zone size under mode I loading.

Considering Irwin's correction of the plastic zone size for $\theta = 0$, the different leading terms $1/\pi = 0.32$ and $\pi/8 = 0.39$ are in a good agreement, whereas the Dugdale model proposes a 23 % larger plastic zone in front of the crack. The above cited equations according to Irwin and Dugdale hold for monotonic loading conditions.

Both estimations can be further developed towards cyclic loading. Suresh and Kuna [Sur98, Kun08] illustrate the formation of a primary plastic zone at the first occurrence of the maximum load $K_{\mathrm{I,max}}$. The subsequent load reduction by an amount ΔK_{I} to a tensile load $K_{\mathrm{I,min}}$ results in a reversed plastic zone ahead of the fatigue crack. Assuming elastic/perfectly plastic material, the secondary plastic zone alters the original stress distribution and leaves compressive residual stresses in front of the crack tip even during the succeeding load cycles. The extent of the primary and secondary plastic zones $d_{\mathrm{p,max}}$ and $d_{\mathrm{p,min}}$ as well as their ratio depending on R are given by Kuna, based on the Dugdale model:

$$d_{\mathrm{p,max}} = \frac{\pi}{8}\left(\frac{K_{\mathrm{I,max}}}{\sigma_{\mathrm{y}}}\right)^2 \quad , \quad d_{\mathrm{p,min}} = \frac{\pi}{8}\left(\frac{-\Delta K_{\mathrm{I,max}}}{2\,\sigma_{\mathrm{y}}}\right)^2 \quad , \quad \frac{d_{\mathrm{p,min}}}{d_{\mathrm{p,max}}} = \frac{1}{4}(1-R)^2 \tag{3.89}$$

An experimental estimation of the plastic zone size under ultrasonic loading might be achieved by an inverse approach. If the three-dimensional displacement field of the crack region is scanned along a sufficiently dense grid, for example by means of laser vibrometry, equations (3.80) to (3.82) can be solved for the stress intensity factors. These in turn might be used to predict the extent of the cyclic plastic zone if the conditions of small scale yielding are met (compare [Hom07]). Clearly, if modern full-field optical strain measurement techniques like laser speckle and interferometry methods overcome the challenge of sampling vibrations at ultrasound frequencies, providing both a high spatial and temporal resolution, point measurements become obsolete.

In the framework of continuum mechanics the volumetric plastic work and the instantaneous plastic dissipation or the plastic work rate of a uniaxially loaded body are

$$W_{\mathrm{v,p}} = \sigma\,\varepsilon_{\mathrm{p}} \quad , \qquad P_{\mathrm{v,p}} = \frac{\mathrm{d}W_{\mathrm{v,p}}}{\mathrm{d}t} = \sigma\frac{\mathrm{d}\varepsilon_{\mathrm{p}}}{\mathrm{d}t} \quad . \tag{3.90}$$

The plastic dissipation formally resembles the dissipated power due to dry friction (3.36), [Dru54]. It is well known that the plastic deformation of metals is accompanied by a bulk heat generation. Indeed, the major part of the volumetric deformation work is converted to heat. The effective thermoplastic heating power $P_{\mathrm{v,p,eff}}$ is given by Bhalla et al. [BZH03]:

$$P_{\mathrm{v,p,eff}} = \beta\,P_{\mathrm{v,p}} = \beta\,\sigma\frac{\mathrm{d}\varepsilon_{\mathrm{p}}}{\mathrm{d}t} \tag{3.91}$$

Thus, the hypothetical adiabatic temperature change of a material portion undergoing plastic deformation is

$$\Delta T_{\mathrm{p}} = \beta\,\frac{W_{\mathrm{v,p}}}{\rho\,C_p} \quad . \tag{3.92}$$

The factor $\beta = P_{\mathrm{v,p,eff}}/P_{\mathrm{v,p}} \leq 1$ specifies the relative fraction of the total dissipated work, which is converted to heat, and $1-\beta$ gives the fraction, which is used for microstructural changes of the material, also known as the stored energy of cold work [BHT73]. For an estimation of the plastic heating β is commonly assumed to be constant with a value close to unity, about 0.9, for most metals. In fact, the plastic work heat fraction β is not a constant but depends on the strain amplitude, the strain rate, the temperature and the hardening state [HRRR00, RRHR01, BZH03]. Clearly, the stored energy of cold work represents a complex material characteristic that will have significant influence on the plastic heating of crack tips during mechanical loading in general and also during high-power ultrasound excitation.

In case of a material for which the cyclic stress-strain curve obeys a Ramberg-Osgood law (compare Section 4.1.2), the shape of the hysteresis loops is approximately known by assuming the Masing behaviour. Then, an expression for the plastic energy dissipation rate can be found, which depends only on the cyclic stress and strain ranges and the hardening modulus and coefficient [LM87, KDDW07, MRKS02].

An alternating plastic deformation inevitably leads to the accumulation of microstructural material damage. Therefore, if the relative displacements of the crack flanks are large enough so that a cyclic plastic zone develops on the macro-scale, the ultrasound excitation will result in the progression of the initial crack. Within the concept of linear elastic fracture mechanics, including small scale yielding, the rate of stable crack growth under mode I loading is predicted by Paris' law, also known as the Paris-Erdogan law:

$$\frac{\mathrm{d}a}{\mathrm{d}N} = C_{\mathrm{P}}\,(\Delta K_{\mathrm{I}})^{m_{\mathrm{P}}} \qquad \text{for} \quad \Delta K_{\mathrm{I,th}} \ll \Delta K_{\mathrm{I}} \ll \Delta K_{\mathrm{I,c}} = (1-R)\,K_{\mathrm{I,c}} \tag{3.93}$$

Here, N denotes the number of cycles and C_{P} and m_{P} are material properties that characterise the resistance against cracking. Equation 3.93 describes a straight line in a log-log plot of the crack propagation velocity over the applied stress intensity factor range. The lower bound of the so called Paris regime is marked by the threshold level $\Delta K_{\mathrm{I,th}}$, below which no measurable damage occurs ($\mathrm{d}a/\mathrm{d}N$ tends to zero). A critical cyclic stress intensity factor $\Delta K_{\mathrm{I,c}}$ depending on the fracture toughness $K_{\mathrm{I,c}}$ represents the upper bound. Once $\Delta K_{\mathrm{I,c}}$ is exceeded, a sudden fracture due to the rapid increase of the crack growth rate can be expected ($\mathrm{d}a/\mathrm{d}N$ approaches infinity).

The transitions near $\Delta K_{\mathrm{I,th}}$ and $\Delta K_{\mathrm{I,c}}$ have been incorporated in more sophisticated laws, for example the Erdogan-Ratwani equation, which covers the complete range of stress intensities [Hai02]:

$$\frac{\mathrm{d}a}{\mathrm{d}N} = C_{\mathrm{E}}\frac{\left(\Delta K_{\mathrm{I}} - \Delta K_{\mathrm{I,th}}\right)^{m_{\mathrm{E}}}}{(1-R)\,K_{\mathrm{I,c}} - \Delta K_{\mathrm{I}}} \qquad \text{for} \quad \Delta K_{\mathrm{I,th}} \leq \Delta K_{\mathrm{I}} \leq \Delta K_{\mathrm{I,c}} \tag{3.94}$$

Note that the used material constants C_{E} and m_{E} may differ from those of the classical Paris-Erdogan law, C_{P} and m_{P}.

Clearly, the concept of stress intensities together with crack propagation laws and the associated material properties provide the basis for estimating the damage progression of an initially present fatigue crack. However, there are some effects that are not considered by Paris' law in its original form (3.93) and its modification (3.94).

A phenomenon of real fatigue cracks, which was first investigated by Elber [Elb71], is the plasticity induced crack closure effect. During fatigue experiments with a fluctuating tensile load ($R = 0$) Elber found that the crack faces are in close contact even before the minimal load is applied. When the next loading phase begins the crack remains closed until a certain opening load level σ_{op} or K_{op} is exceeded. The crack closure effect can be explained if one considers the plastically deformed and finally disrupted crack faces just behind the crack tip. Since their morphology is irreversibly altered, the crack flanks are no longer compatible to each other and prevent the cracked body from a full recovery of its displacements during unloading [Kun08]. As an outcome, the stress intensity range, which is mainly responsible for crack growth, is being reduced to

$$\Delta K_{\text{eff}} = K_{\text{max}} - K_{\text{op}} \quad . \tag{3.95}$$

Consequently, Paris' law can be adjusted to approximately account for plasticity induced crack closure [Sur98]:

$$\frac{\mathrm{d}a}{\mathrm{d}N} = C_P \left(\Delta K_{\text{eff}}\right)^{m_P} \tag{3.96}$$

Crack closure in turn affects the size of the cyclic plastic zone since during unloading to the minimum load level the contact stresses arising along the crack faces relieve the crack tip. Thus, according to Haibach [Hai02], the compressive plastic zone is even smaller than predicted by equation (3.89).

Additional mechanisms of crack closure are explained by Suresh [Sur98]. These are for example oxide-induced, roughness-induced, fluid-induced and transformation-induced crack closure. Furthermore, significant crack deflection and crack-bridging by fibres or trapped particles promote the retardation of fatigue crack growth during constant amplitude loading. The aforementioned crack retardation mechanisms especially affect the crack propagation rate at near-threshold stress intensities.

Another important aspect to consider is the load history that led to crack initiation and propagation up to the current crack depth, which is to be detected. It is well known, that low amplitude loading (HCF or VHCF regime) will result in rather smooth crack faces and a sharp crack front. In contrast, higher load amplitudes (LCF regime) lead to a rugged relief of the crack flanks and significant blunting of the crack tip due to the enlarged plastic zone. Hence, the crack tip behaves more like a notch with a less severe stress concentration than the originally sharp crack tip [Sur98]. A summary of advanced models, which were developed to consider the influence of the load sequence can by found in [Hai02].

With regard to the above mentioned crack retarding effects, the prognosis of ultrasound vibration-induced damage seems to be hardly possible. The actual load situation at the crack front will be rather complex in practice when considering mixed mode loading, residual stress fields, crack closure effects and crack tip blunting.

3.3 Heat transfer

This section briefly reviews the basic concepts of the three heat transfer mechanisms conduction, convection and radiation. The emphasis is on conductive and radiative heat exchange. Conduction has by far the greatest impact on the maximum temperature increase and the temperature distribution of a solid body when heated locally. Although both radiation and convection do not significantly affect the thermal response in practice, electromagnetic radiation should be outlined here, since it is the physical effect that enables infrared thermography.

3.3.1 Conduction

The term heat conduction comprises the macroscopic observation and description of physical effects such as the transfer of kinetic energy from one electron to another. Similar to the mechanical waves described earlier, vibration energy spreads into the body at the molecular or atomic scale [GM01].

Heat conduction in a solid body immediately sets in as soon as it is heated up or cooled locally. This way, the inner material is permanently equalising any spatial temperature differences. Fourier's law states that the directed heat flux $\vec{q}$ is proportional to the magnitude of the current temperature gradient. Since heat is transferred from warmer regions to cooler regions, the flux is directed opposite in sign with respect to the mathematically defined temperature gradient. In the three-dimensional case, Fourier's law reads [LILV03]

$$\vec{q} = -k \nabla T \quad . \tag{3.97}$$

The constant of proportionality k is the thermal conductivity that might be depending on the spatial direction (anisotropic material) as well as on temperature. Based on the first law of thermodynamics, Baehr and Stephan [BS06] derive the heat equation of a solid body with constant density. Its solution yields the time dependent temperature and heat flux at every position $\vec{x}$. Taking volumetric heat sources P_v explicitly into account leads to the following differential equation:

$$\rho C_p(T) \frac{\partial T}{\partial t} = \operatorname{div}\left[k(T)\, \operatorname{grad} T\right] + P_\mathrm{v}(\vec{x}, t) \tag{3.98}$$

Here, C_p denotes the specific heat capacity at constant pressure, which is also a temperature dependent material property in the general case.

If both heat capacity and thermal conductivity are assumed to be constant, the differential operator $\operatorname{div}[k(T)\,\operatorname{grad} T]$ turns into the Laplace operator

$$k \operatorname{div} \operatorname{grad} T = k \nabla^2 T \tag{3.99}$$

and the above heat equation (3.98) in its general form simplifies to

$$\frac{\partial T}{\partial t} = \alpha \nabla^2 T + \frac{P_\mathrm{v}}{\rho C_p} \quad . \tag{3.100}$$

At this point, the material's thermal diffusivity α is introduced, which can be understood as a measure of how quickly heat flows through it:

$$\alpha = \frac{k}{\rho C_p} \tag{3.101}$$

Another heat transfer related material characteristic is known as the thermal effusivity e. It is a measure of the material's ability to exchange thermal energy with its surroundings, e.g. when contacting a different material.

$$e = \sqrt{k \rho C_p} \tag{3.102}$$

Table 3.1 summarises the thermal properties of some metals and clarifies the broad spectrum of the thermal conductivity and diffusivity.

The heat equation (3.100) describes only the heat flow within the body. In case of a non-stationary heating or cooling process, a temporal initial condition (e.g. at $t = 0$) and spatial boundary con-

Table 3.1: Thermomechanical properties of different metals and alloys.

Metal/alloy	ρ [kg/m^3]	C_p [J/(kg K)]	k [W/(m K)]	α [10^{-6} m^2/s]
Cr-Ni steel [BS06]	7800	500	15	3.85
Titanium [BS06]	4500	522	22	9.37
Steel S355J2+N [PW05]	7837	461	40.4	11.18
Low-carbon steel [BS06]	7850	465.5	50[1]	13.68[1]
Aluminium [BS06]	2700	888	237	98.85
Copper [BS06]	8930	382	399	116.97

[1]Mean value of the range originally provided by Baehr and Stephan [BS06]

ditions are required to complete the thermal problem statement and allow for computation of an unambiguous temperature field. Baehr and Stephan [BS06] categorise the spatial conditions as follows:

- First of all, the body's surface temperature itself can be given as a function of time and space.
- Secondly, the heat flux normal to the body's surface can be prescribed. This is most appropriate if a known external heat load is applied to it. With n being the surface normal

$$q = -k\frac{\partial T}{\partial n} \tag{3.103}$$

has to be satisfied. In case of a free adiabatic surface ($q = 0$), the spatial derivative $\frac{\partial T}{\partial n}$ equals zero. This condition is also met at any symmetry plane inside the solid body if all other boundary conditions and the heat loads are also symmetrical.

- The third and most complex conditions occur if several bodies are in mechanical contact with each other. Only in case of a perfect thermal contact, the heat flux and the temperature do not show a jump at the interface. Otherwise, an additional heat transfer coefficient that characterises the thermal contact needs to be regarded, which leads to a discontinuity of the temperature field. However, this third kind of boundary condition is not considered within the present work.

In vibrothermography potential heat sources based on the different mechanical effects described in Section 3.2 are typically activated pulse-like with the pulse width d or periodically at the modulation frequency f_{mod}. Due to pulsed or modulated activation, thermal waves originate from hot spots, propagate into the bulk material and finally reach the surface where they are detected. Any plate structure with a vertical or tilted crack of finite length actually states a complex 3D thermal diffusion problem, especially if heat sources are buried in a certain depth. However, the most important characteristics of thermal waves have been studied by considering a 1D heat conductor: Thermal waves are found to be highly attenuated as well as dispersive.

The initial temperature amplitude at the pulsating source decreases to a value of $1/\mathrm{e} = 37\,\%$ within a travel distance μ_{th} known as the thermal diffusion length [Mal01, MGB11]:

$$\mu_{\mathrm{th}} = \sqrt{\frac{k}{\pi f_{\mathrm{mod}} \rho C_p}} = \sqrt{\frac{\alpha}{\pi f_{\mathrm{mod}}}} \tag{3.104}$$

Equation 3.104 reveals that the penetration range of the thermal wave is depending on the given material properties and the chosen modulation frequency. In case of a pulse activation, f_{mod} can be estimated as $1/(2d)$ assuming the pulse period to be $2d$.

Table 3.2 presents theoretical values of the 1D thermal diffusion length in steel S355J2+N for a wide range of modulation frequencies. The dramatic decrease of μ_{th} with growing values of f_{mod} has a major impact on the detection capabilities of buried flaws using the common excitation and processing methods described in Section 3.5.

Table 3.2: Thermal diffusion length in low-carbon steel S355J2+N[1].

f_{mod}	[Hz]	0.01	0.02	0.05	0.1	0.2	0.5	1.0	2.0	5.0
μ_{th}	[mm]	18.86	13.34	8.44	5.97	4.22	2.67	1.89	1.33	0.84

[1]Thermomechanical properties according to Peil and Wichers [PW05]

The propagation speed and thus the length of the thermal wave is further dependent on the material. According to Maldague [Mal01], wavelength λ_{th} and velocity c_{th} can be expressed as follows. Here, the thermal diffusion length introduces a dispersive behaviour.

$$\lambda_{\text{th}} = 2\pi\,\mu_{\text{th}} \tag{3.105}$$

$$c_{\text{th}} = f_{\text{mod}} \cdot \lambda_{\text{th}} = \sqrt{4\pi\, f_{\text{mod}}\, \alpha} \tag{3.106}$$

Assuming a periodically activated, planar heat source in the depth z, the temperature oscillation observed at the surface above will show a relative phase lag

$$\varphi_{\text{th}} = 2\pi \frac{z}{\lambda_{\text{th}}} = \frac{z}{\mu_{\text{th}}} \quad , \tag{3.107}$$

which is directly related to z and thus enables a depth resolved defect characterisation.

Similar as with mechanical waves also thermal waves are reflected at interfaces between two different materials. Considering homogeneous materials only, the thermal reflection coefficient of two media with subscripts 1 and 2 is defined as

$$R_{\text{th}} = \frac{(e_2/e_1) - 1}{(e_2/e_1) + 1} \quad , \tag{3.108}$$

where the incident thermal wave propagates in media 1 and gets reflected partly at the interface. A reflection coefficient of -1 indicates a perfect reflection, which is nearly the case when considering steel and the surrounding air [Mal01].

In order to clarify the fundamental relations between a lateral surface heat load and the local temperature evolution, a three-dimensional model of a 10 mm thick steel plate is used for parametric studies. Figure 3.19 shows the finite element model, meshed with 8-node brick thermal elements (ANSYS SOLID70) and symmetry conditions at one face. The coloured area indicates the element facets, which are subjected to an external heat load. All other boundary faces are assumed to be adiabatic. Note that the mesh is strongly refined near the heated zone to better resolve the resulting high temperature gradients. Here, the solid elements are cubic with an edge length of 1 mm. The material properties of low-carbon steel S355 measured by Peil and Wichers [PW05] are used.

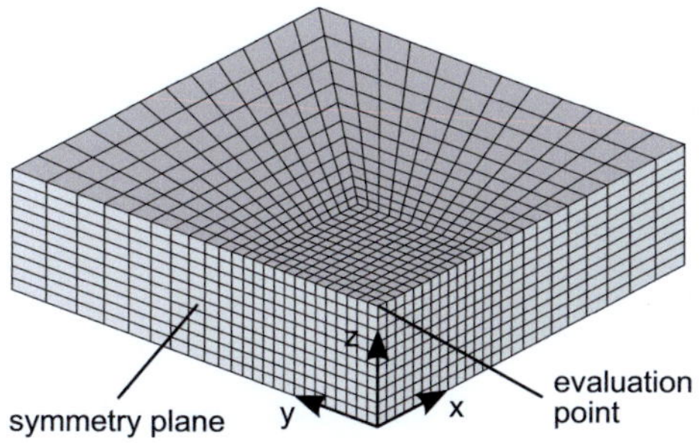

(a) Reduced model with symmetry plane

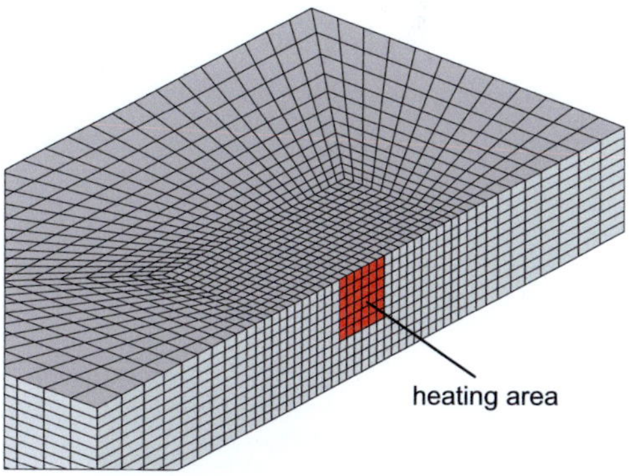

(b) Resulting expanded model

Figure 3.19: 3D finite element model of a $35 \times 35 \times 10\,\text{mm}^3$ plate used for parametric studies on conductive heat transfer.

First of all, a pulse-like heating of 150 ms duration acting in different depths with respect to the top plate surface is simulated. This transient thermal problem is solved for various heat loads ranging from 3 to 15 W. The total heated area of $4 \times 4\,\text{mm}^2$ is hold constant, so depths from 2 to 8 mm can be distinguished. Throughout the study only the nodal temperature at the top surface, next to and centred about the heating area is analysed.

Figure 3.20a illustrates the time histories of the temperature achieved for the different heat loads acting just below the plate surface. Obviously, the temperature evolution is linearly scaled in magnitude by the heating power. Figure 3.20b clarifies that if the depth of the source is hold constant, the nodal temperature is directly proportional to the applied heat flux. As can be seen from Figure 3.20c and 3.20d, the increasing depth of the source not only results in a dramatic decrease of the maximum temperature change ΔT at the plate's top surface but also affects the curve shape. Due to the three-dimensional heat conduction, the strong peak that was perceptible in the curve for 2 mm depth is disproportionately attenuated with the incrementally increasing depth. Furthermore, the occurrence of the maximum temperature at the evaluation point is more and more delayed in time. The peak values of ΔT would describe a hyperbolic function in the diagram presented in Figure 3.20c.

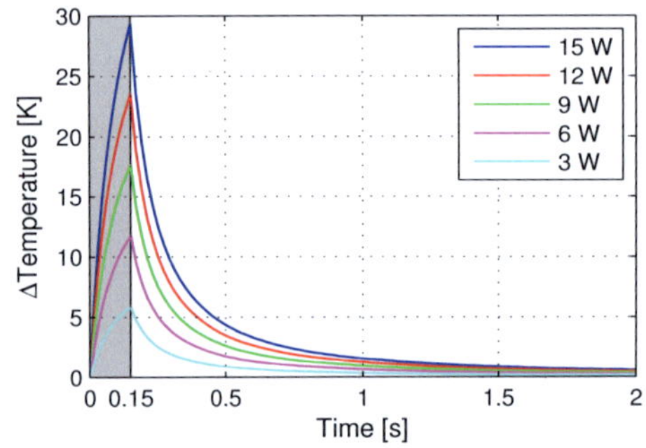

(a) Temperature increase at the evaluation point; heating power = var., depth = 2 mm

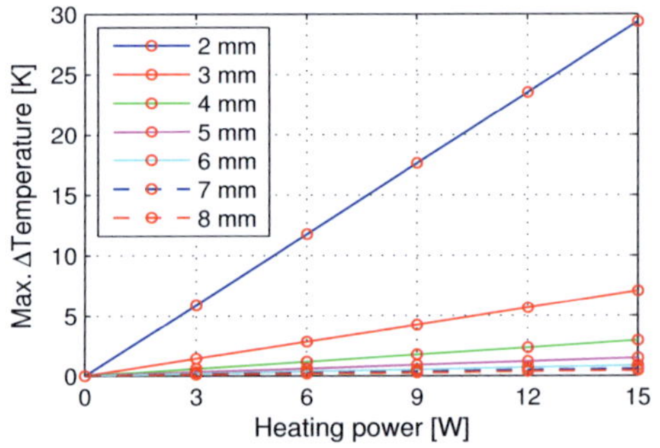

(b) Max. ΔT depending on heating power; depth = var.

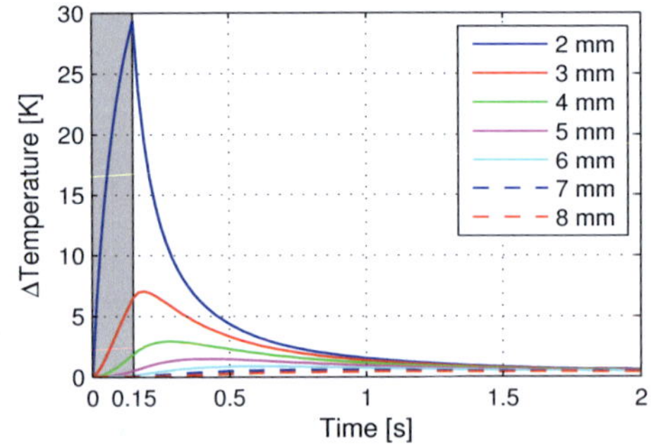

(c) Temperature increase at the evaluation point; heating power = 15 W, depth = var.

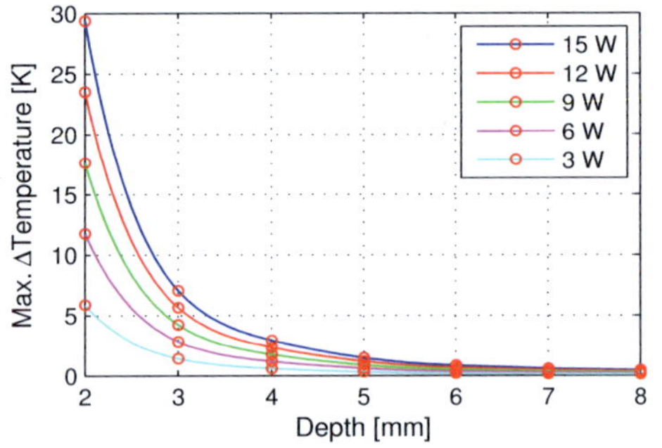

(d) Max. ΔT depending on depth; heating power = var.

Figure 3.20: Simulated temperature evolutions and maximum values found for varying power levels and depths of the heat load.

Secondly, in order to emphasise the influence of the material itself on the temperature increase, the thermomechanical properties are varied according to the values given in [BS06], see Table 3.1. Here, a three second heating and cooling phase was simulated, where a heating power of 10 W is applied to an area A of $6 \times 6\,\text{mm}^2$ beginning at the plate's top surface as indicated exemplarily in Figure 3.19b. The resulting temperature evolutions in Figure 3.21 are presented in terms of absolute changes and relative values normalised to the maximum temperature difference achieved.

A comparison of the ΔT curves obtained for five different metals reveals that titanium and chrome-nickel stainless steel heat up more than 200 % with respect to the low-carbon steel. The lower thermal diffusivity leads to a more pronounced accumulation of heat close to the heated area. Consequently, the cooling process monitored at the evaluation point is slowed down, too. In contrast, aluminium and copper heat up and cool down very quickly, but the maximum temperature increase is limited to values less than 25 % of that achieved for low-carbon steel.

As a last point, the effect of a change in the heating area A on the surface temperature should be pointed out. Similar computations have been conducted, whereas A was stepwise increased from $2 \times 2\,\text{mm}^2$, $3 \times 3\,\text{mm}^2$ up to $10 \times 10\,\text{mm}^2$ but always located next to the upper edge of the plate. The total heating power was hold constant at 10 W. As expected, the maximum temperature significantly decreases with the growing area because of the lower heat flux near the evaluation point and the increasing depth of the heating area's centre point. Figure 3.22 illustrates the results. The bigger the heating area is, the lower is the spatial temperature gradient around the monitoring node. Therefore, the heating and cooling process is relatively slow compared to a more concentrated heat load.

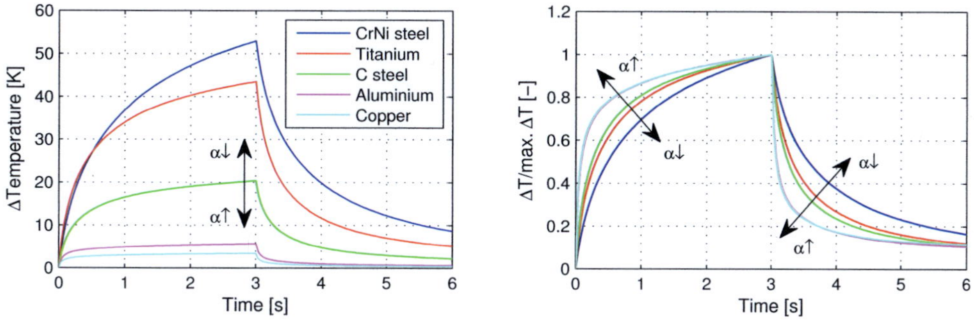

(a) Absolute temperature evolutions; α = var.

(b) Normalised temperature evolutions; α = var.

Figure 3.21: Simulated temperature evolutions for different metals due to a constant heat load of 10 W acting at a depth of 3 mm.

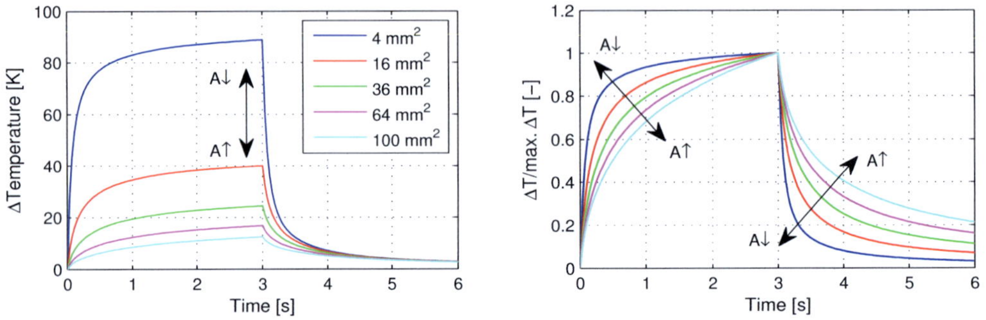

(a) Absolute temperature evolutions; A = var.

(b) Normalised temperature evolutions; A = var.

Figure 3.22: Simulated temperature evolutions for low-carbon steel due to a constant heat load of 10 W distributed over varying heating areas.

The described model calculations do not consider any additional layer such as a synthetic coating, which is commonly applied to the plate. According to ISO 12944-5, the typical thickness of a protective coating based on polyurethane and epoxy resins is 200 µm for steelworks classified in the corrosiveness category C4.

If one assumes that the plate is heated pulse-like or even periodically (see ultrasonic lock-in thermography in Section 3.5.2) just beneath the coating, the temperature response at the coating's outer surface will be damped to some extent. The attenuation depends on the coating thickness h_{coat} and its thermal diffusivity α but also on the pulse width or the heating frequency f_{mod}. McKelvie [McK87] proposes to adopt a symmetric 1D heat conduction model described by Carslaw and Jaeger [CJ59]:

$$\frac{\hat{T}_{\text{o}}}{\hat{T}_{\text{i}}} = \left| \frac{\cosh\left[\mu^* x (1+\text{i})\right]}{\cosh\left[\mu^* h_{\text{coat}} (1+\text{i})\right]} \right| \quad , \qquad \mu^* = \sqrt{\frac{\pi f_{\text{mod}}}{\alpha}} \tag{3.109}$$

For $x = 0$ McKelvie's formula gives the attenuation ratio, which is the relation between the temperature amplitudes at the outer and inner surface of the coating $\hat{T}_{\text{o}}$ and $\hat{T}_{\text{i}}$, respectively.

Domininghaus [Dom98] gives the thermal properties of most technically relevant plastics. In order to estimate the thermal lag caused by organic coatings based on epoxy resin, polyurethane or polyvinyl chloride, Table 3.3 presents a small choice of raw materials to clarify the range of the thermal diffusivity α. The computed values are consistent with the report of d'Almeida et al. [dCMM98], who measured values for an epoxy system ranging from 9×10^{-8} to 2.1×10^{-7} m^2/s depending on the hardener content. Therefore, equation (3.109) is evaluated for a coating thickness

Table 3.3: Thermomechanical properties of coating relevant plastics [Dom98].

Plastic	ρ [kg/m^3]	C_p [J/(kg K)]	k [W/(m K)]	α [10^{-6} m^2/s]
Epoxy moulding compound 871 (EP)	1900	800	0.88	0.58[2]
Polyurethane casting resin (PUR)	1050	1760	0.58	0.31[2]
Linear polyurethane U50 (PUR)	1210	400	1.8	3.72[2]
Polyvinyl chloride (PVC-U)	1465[1]	875[1]	0.16[1]	0.12[2]
Polyvinyl chloride/plasticiser (PVC-P)	1255[1]	1350[1]	0.15	0.09[2]

[1]Mean value of the range originally provided by Domininghaus [Dom98]
[2]Value computed according to the relation $\alpha = k/(\rho C_p)$, equation (3.101)

of 200 µm and diffusivities between 5×10^{-8} and 5×10^{-6} m^2/s. Figure 3.23a presents the resulting attenuation ratios depending on the heating pulse width or modulation frequency, respectively. In a steady state at $f_{\text{mod}} = 0$ Hz the outer coating temperature must be equal to the inner and no attenuation occurs. The thermal lag is negligible in case of a relatively high diffusivity even for such a thick coating, but it can have a significant impact on the outer surface temperature if the coating material effectively inhibits the heat flow as for example the low diffusive polyvinyl chloride.

Under laboratory conditions, where there is no need for corrosion control, the situation is not problematic at all. For the thermographic investigation of steel samples it is generally accepted to apply a thin layer of dark matt paint to the metallic surface, especially if small temperature changes are expected, e.g. in thermoelastic stress analysis [RDQB10]. Here, very thin coatings are realised that just exceed the opacity limit. McKelvie cites a report by Belgen [Bel68], in which the diffusivity of a carbon-filled acrylic-based formulation (35 % graphite, 65 % acrylic) was measured as 0.5×10^{-6} m^2/s. Figure 3.23b illustrates the frequency dependent attenuation ratios for a range of paint thicknesses from 20 to 100 µm. Obviously, the thermal lag is negligible at low modulation frequencies, which are of relevance in practice.

From a theoretical point of view the capability of detecting a buried heat source strongly decreases if one assumes reflectance of the thermal wave at the interface between steel and plastic coating. Evaluation of equation (3.108) for the material pairings of steel and the synthetic materials listed in Table 3.3 yields a thermal reflection coefficient R_{th} ranging from -0.93 to -0.83, which means that only a small fraction of the thermal energy would pass the interface at all. Here, the question arises whether equation (3.108) is applicable for such small coating thicknesses. The case of h_{coat} approaching zero requires further discussion, which is beyond the scope of this work.

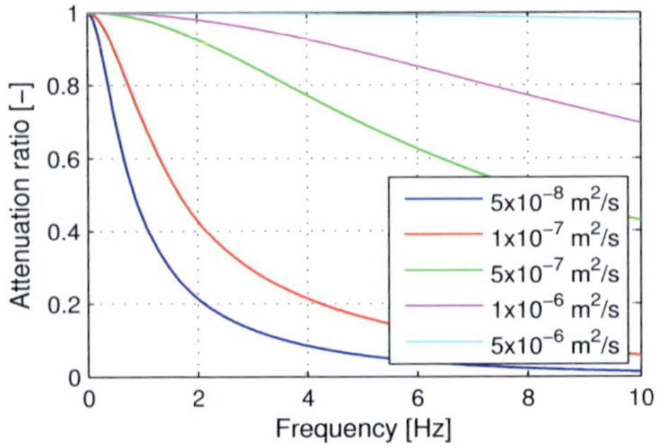

(a) Protective film with a thickness of 200 µm; $\alpha = \text{var.}$

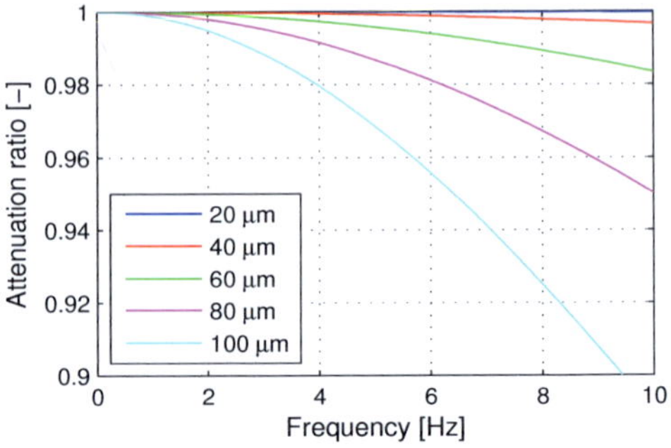

(b) Graphite/acrylic coating with a diffusivity of $0.5 \times 10^{-6}\,\mathrm{m^2/s}$; $h_{\text{coat}} = \text{var.}$

Figure 3.23: Attenuation of the surface temperature amplitude due to applied coatings.

3.3.2 Radiation

All objects with a thermodynamic temperature above zero Kelvin emit radiation energy. A body's thermal energy can be understood as the intensity of random movements of atoms or molecules and the oscillation of their charges. This in turn results in the conversion of kinetic energy into electromagnetic radiation. In most solid bodies and also liquids the radiation of molecules inside is absorbed by the neighbouring particles and does not reach the surface. The emitted radiation originates only from the molecules of a very thin layer just beneath the surface. Since this layer has a thickness of about 1 µm, radiation properties are to be attributed rather to the surface than to the body.

Thermal emission can be quantified by the photon emission of discrete quantities of energy or by electromagnetic wave emission. Both concepts are related by equation (3.110). For a given wavelength $\lambda = c/f$ the photonic energy released is

$$E_{\text{ph}} = \frac{hc}{\lambda} = hf \quad , \tag{3.110}$$

where $h = 6.6261 \times 10^{-34}\,\mathrm{J\,s}$ is the Planck constant, c the electromagnetic wave propagation velocity and f the frequency. Thus, the amount of energy transmitted by electromagnetic waves depends on the wavelength λ. The spectral distribution of the emitted radiation energy is characteristic for a body's material or rather for its surface.

The radiative heat exchange between two or more separated bodies can be attenuated by a surrounding medium such as a gas but is not depending on any "connecting" matter. The net heat flow essentially depends on the temperature differences, on the surface properties as well as on the view factors of the involved bodies. Since the emission, reflection and absorption of radiation energy is dependent on both the wavelength and direction in space, the exact quantitative characterisation becomes rather complex. Four different types of radiometric quantities, which are associated with each other are to be distinguished when describing emission, reflectance and absorbance [BS06]:

- Spectral-directional quantities: The distributions over wavelengths and directions in space are explicitly taken into account. These quantities are the most detailed. By integration over all wavelengths and/or all directions the following three integral quantities can be derived.
- Spectral quantities: These quantities summarise the emission, reflectance and absorbance within the hemisphere of a surface element and depend only on the wavelength.
- Directional quantities: Analogous to spectral quantities the total directional quantities describe integral values over all wavelengths and depend only on the direction.
- Hemispherical total quantities: These quantities represent the total emission, reflectance and absorbance over all wavelengths within the hemisphere of a surface element. No information is given about the directional distribution and the wavelength dependency.

The emitted radiant power Φ originated from a surface A is distributed over the wavelength spectrum and the directions of the covering hemisphere. Within the spherical coordinate system shown in Figure 3.24a a certain direction in space is given by the zenith angle β and the azimuth angle φ with an arbitrarily chosen zero position. The radiant power emitted from a differential surface patch $\mathrm{d}A$ passing a target element $\mathrm{d}A_n$, which is positioned normal to the beam direction and located in a distance r from the source, is proportional to the unit solid angle $\mathrm{d}\Omega$ [Mal01, BS06]. Figure 3.24b clarifies the size of $\mathrm{d}A_n$ generated by infinitesimal changes $\mathrm{d}\beta$ and $\mathrm{d}\varphi$ of the zenith angle β and the azimuth angle φ, respectively. With the assumption of small differential angles the size of the target element is

$$\mathrm{d}A_n = r\,\mathrm{d}\beta \cdot r\sin\beta\,\mathrm{d}\varphi = r^2\sin\beta\,\mathrm{d}\beta\,\mathrm{d}\varphi \tag{3.111}$$

and the unit solid angle is geometrically defined as:

$$\mathrm{d}\Omega = \frac{\mathrm{d}A_n}{r^2} = \sin\beta\,\mathrm{d}\beta\,\mathrm{d}\varphi \tag{3.112}$$

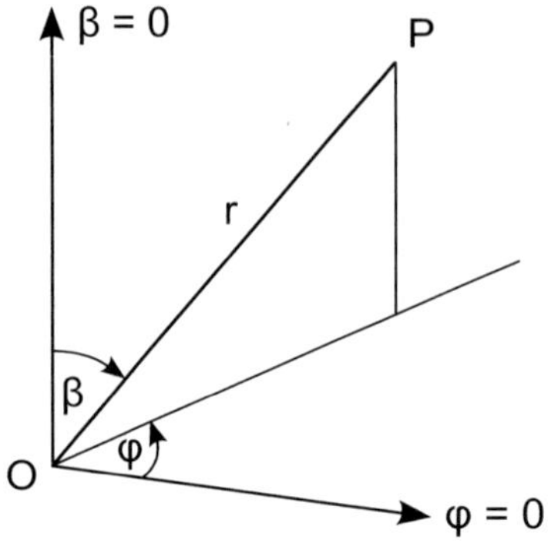

(a) Spherical coordinate system with origin O and a point P at (r,β,φ)

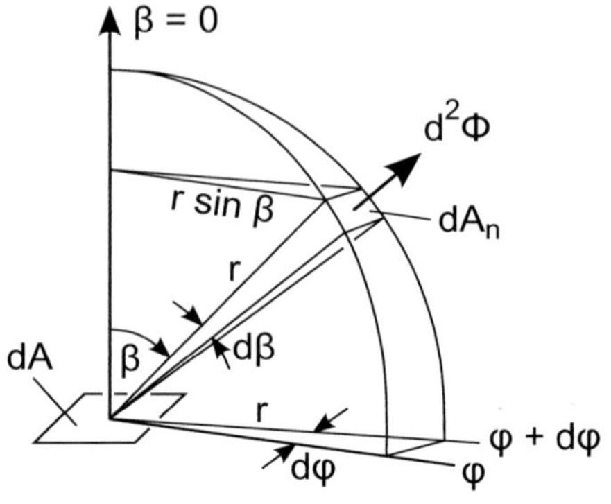

(b) Heat flux $d^2\Phi$ within a unit solid angle $d\Omega$ in direction of β and φ

Figure 3.24: Notations used for solid angle definition.

The spectral radiance $L_{\lambda\Omega}$, which is the basic spectral-directional quantity, specifies the radiant power $d^3\Phi$ emitted per surface element dA and solid angle $d\Omega$ and wavelength interval $d\lambda$. Here, the radiant flux is referred to the projected area $\cos\beta \cdot dA$. With the geometric view factor $\cos\beta$ the decrease of emitted power due to an increasing zenith angle β is considered.

$$L_{\lambda\Omega}(\lambda,\beta,\varphi,T) = \frac{d^3\Phi}{\cos\beta \, dA \, d\Omega \, d\lambda} \tag{3.113}$$

Making use of equation (3.112) the integration of the spectral radiance $L_{\lambda\Omega}$ over the whole hemisphere yields the spectral specific radiance $M_\lambda(\lambda,T)$:

$$M_\lambda(\lambda,T) = \frac{d^2\Phi}{dA \, d\lambda} = \int_{\varphi=0}^{2\pi} \int_{\beta=0}^{\pi/2} L_{\lambda\Omega}(\lambda,\beta,\varphi,T) \cos\beta \, \sin\beta \, d\beta \, d\varphi \tag{3.114}$$

For real surfaces the spectral radiance $L_{\lambda\Omega}$ depends on the direction in space. In case of a diffuse or Lambertian emitter the spectral radiance is not dependent on the direction. The assumption of a diffuse emitter can be accepted for many rough surfaces. With $L_{\lambda\Omega} = L_\lambda(\lambda,T)$ the expression (3.114) simplifies to ([BS06, BSMM97]):

$$M_\lambda(\lambda,T) = L_\lambda(\lambda,T)\int_{\varphi=0}^{2\pi}\int_{\beta=0}^{\pi/2}\cos\beta\,\sin\beta\,\mathrm{d}\beta\,\mathrm{d}\varphi = 2\pi L_\lambda(\lambda,T)\int_{\beta=0}^{\pi/2}\cos\beta\,\sin\beta\,\mathrm{d}\beta$$

$$= 2\pi L_\lambda(\lambda,T)\left[\frac{1}{2}\sin^2\beta\right]_0^{\pi/2} = \pi L_\lambda(\lambda,T) \tag{3.115}$$

Finally, the integration of the spectral quantity $M_\lambda(\lambda,T)$ over all wavelengths results in the total specific radiance $M(T)$, which is the corresponding hemispherical total quantity.

$$M(T) = \frac{\mathrm{d}\Phi}{\mathrm{d}A} = \int_0^\infty M_\lambda(\lambda,T)\,\mathrm{d}\lambda \tag{3.116}$$

The radiation capability of real bodies or surfaces is assessed by comparison with an ideal diffuse emitter. Gustav Robert Kirchhoff introduced the concept of the so called blackbody, an idealised object, which absorbs all incident radiation no matter which wavelength it has and no matter from which direction it falls onto. None of the electromagnetic radiation passes through the body and none is reflected. Such an emitter can be approximately realised by a cavity blackened with soot that is covered by an adiabatic enclosure [FR06]. Figure 3.25 shows a schematic blackbody and the path of incident rays. Because of the small aperture, almost 100 % of them are repeatedly reflected within the hollow space and finally absorbed. If the body's temperature does not change over time, such as in a steady state, the absorbed radiation must be equal to the emitted radiation. The spectral specific radiance of a blackbody was found by Max Planck in 1900:

$$M_{\lambda\mathrm{bb}}(\lambda,T) = \pi L_{\lambda\mathrm{bb}}(\lambda,T) = \frac{2\pi h c_0^2}{\lambda^5\left[\exp\left(\frac{h c_0}{\lambda k T}\right)-1\right]} = \frac{c_1}{\lambda^5\left[\exp\left(\frac{c_2}{\lambda T}\right)-1\right]} \tag{3.117}$$

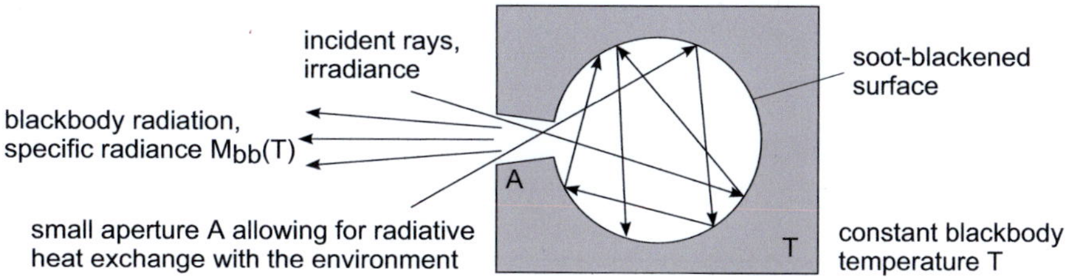

Figure 3.25: Ideal blackbody [FR06].

Two constants $c_1 = 2\pi h c_0^2 \approx 3.7418 \times 10^8\,\mathrm{W\,\mu m^4/m^2}$ and $c_2 = h c_0 / k \approx 1.4388 \times 10^4\,\mathrm{K\,\mu m}$ are used, which depend on the speed of light in vacuum $c_0 \approx 299\,792\,\mathrm{km/s}$, the Planck constant h and the Boltzmann constant $k \approx 1.3806 \times 10^{-23}\,\mathrm{J/K}$. Commonly, Planck's law is represented by a curve family. In Figure 3.26 the spectral specific radiance of a blackbody is plotted for several absolute temperatures. In the log-log graph the temperature dependent wavelengths at the peak radiance λ_{max} are lying on a straight line. The shift in wavelength is referred to as Wien's displacement law:

$$\lambda_{\mathrm{max}} = \frac{c_3}{T} \quad , \qquad c_3 \approx 2.8978 \times 10^3\,\mathrm{K\,\mu m} \tag{3.118}$$

Corresponding to equation (3.116) the total specific radiance of a blackbody at temperature T is found by integration of Planck's law over all wavelengths. The result is known as the Stefan-Boltzmann law that reads:

$$M_{\mathrm{bb}}(T) = \int_0^\infty M_{\lambda\mathrm{bb}}(\lambda,T)\,\mathrm{d}\lambda = \sigma T^4 \quad , \qquad \sigma = \frac{c_1}{c_2^4} \cdot \frac{\pi^4}{15} = \frac{2\pi^5 k^4}{15 c_0^2 h^3} \tag{3.119}$$

The constant of proportionality $\sigma \approx 5.6704 \times 10^{-8}\,\mathrm{W/(m^2\,K^4)}$ denotes the Stefan-Boltzmann constant. The thermal emission of real bodies is always lower than that of a blackbody having the same temperature. Therefore, emissivities $\epsilon < 1$ are introduced by comparing the different radiometric quantities of a real body with those of a blackbody. For example, the hemispherical spectral emissivity, which takes the temperature and wavelength dependencies into account, is formally defined as:

$$\epsilon_\lambda(\lambda,T) = \frac{M_\lambda(\lambda,T)}{M_{\lambda\mathrm{bb}}(\lambda,T)} \tag{3.120}$$

Thus, the spectral specific radiance $M_\lambda(\lambda,T)$ of an emitter is completely determined by its surface emissivity and temperature.

$$M(T) = \int_0^\infty M_\lambda(\lambda,T)\,\mathrm{d}\lambda = \int_0^\infty \epsilon_\lambda(\lambda,T)\,M_{\lambda\mathrm{bb}}(\lambda,T)\,\mathrm{d}\lambda \tag{3.121}$$

$$= \epsilon(T)\,M_{\mathrm{bb}}(T) = \epsilon(T)\,\sigma T^4 \tag{3.122}$$

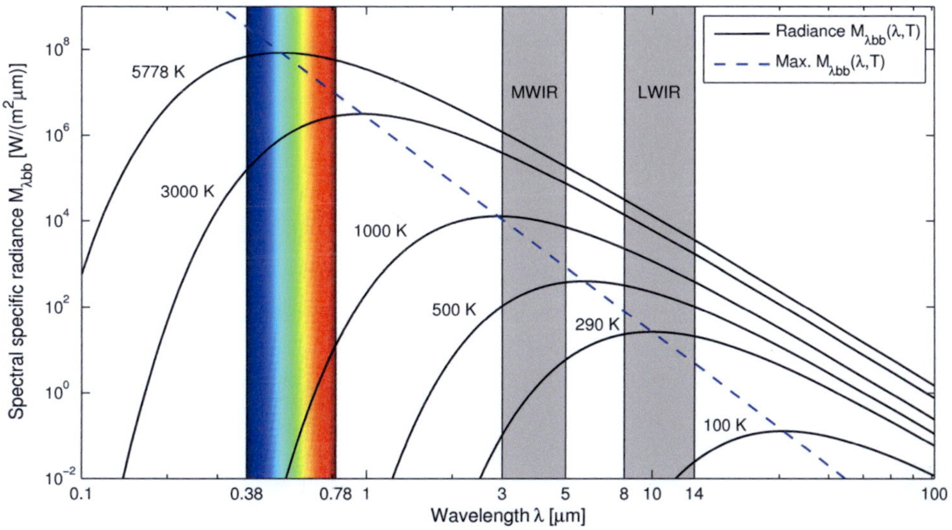

Figure 3.26: Spectral specific radiance of a blackbody at different temperatures according to Planck's law and Wien's displacement law; visible light spectrum and technically relevant infrared bands MWIR and LWIR.

A body that emits a certain fraction of the blackbody radiation at all wavelengths is called a grey body. The model of a grey emitter, which is fully characterised by its hemispherical total emissivity is preferably used as a viable approach for the true radiation capability of real bodies or surfaces:

$$\epsilon(T) = \frac{M(T)}{M_{\text{bb}}(T)} = \frac{M(T)}{\sigma T^4} \tag{3.123}$$

The interaction between two bodies, namely radiation exchange, is first explained for the case of ideal emitters. When two blackbodies with the surfaces A_1 and A_2 have different temperatures $T_1 > T_2$ a net heat flow Φ_{12}^* sets in depending on the individual directed flows from surface A_1 towards A_2 (Φ_{12}) and vice versa (Φ_{21}).

$$\Phi_{12}^* = \Phi_{12} - \Phi_{21} = A_1 F_{12} \sigma T_1^4 - A_2 F_{21} \sigma T_2^4 \tag{3.124}$$

Each single heat flow might be limited by the respective view factor

$$F_{ij} = \frac{\Phi_{ij}}{\Phi_i} \leq 1 \tag{3.125}$$

unless the whole hemisphere of A_i is covered by A_j. If there is no temperature difference ($T_1 = T_2$) the net heat flow vanishes ($\Phi^*_{12} = 0$) and the reciprocity rule $A_1 F_{12} = A_2 F_{21}$ is found [BS06]. Thus, equation (3.124) simplifies to:

$$\Phi^*_{12} = A_1 F_{12}\, \sigma \left(T_1^4 - T_2^4\right) = A_2 F_{21}\, \sigma \left(T_1^4 - T_2^4\right) \tag{3.126}$$

Regarding the practice of ultrasound excited thermography, the focus is mainly on a single body, which is placed in a certain environment, for example the laboratory or the outside. The radiative net heat flux between a grey body with constant values T_1, A_1, ϵ_1 and its surrounding with T_2, A_2 and ϵ_2 is derived by Stephan and Baehr [BS06]:

$$q_r = \frac{\Phi^*_{12}}{A_1} = \epsilon_{12}\, \sigma \left(T_1^4 - T_2^4\right) \quad , \qquad \frac{1}{\epsilon_{12}} = \frac{1}{F_{12}} + \frac{1}{\epsilon_1} - 1 + \frac{A_1}{A_2}\left(\frac{1}{\epsilon_2} - 1\right) \tag{3.127}$$

Assuming that the environment completely encompasses the body under investigation, the view factor F_{12} yields 1. Furthermore, the term $1/\epsilon_{12}$ clarifies that a far surrounding with $A_2 \gg A_1$ can always be thought of as a blackbody since A_1/A_2 approaches zero and therefore ϵ_2 has no influence on the heat flux q_r. Then, equation (3.127) reduces to ([BS06, GM01]):

$$q_r = \frac{\Phi^*_{12}}{A_1} = \epsilon_1\, \sigma \left(T_1^4 - T_2^4\right) \tag{3.128}$$

3.3.3 Convection

Heat convection is most often understood as a third mode of heat transfer, although it is based on heat conduction as mentioned by von Böckh [vB06]. Convective heat transfer typically describes the heat exchange between a solid body and a surrounding fluid, which is either initially at rest (free convection) or moving and circulating around the body due to an external driving force, e.g. a pressure gradient (forced convection). In both cases, the fluid adjacent to the body is slowed-down due to friction and forms a thin boundary layer [LILV03]. If a temperature difference between the solid body and the boundary layer exists, conductive heat transfer sets in. The resulting local change of the fluid density leads to buoyancy or downforce, depending on the initial situation. Free or natural convection is characterised by a (steady) fluid flow induced solely by the temperature difference itself. In the case of thermosonic experiments conducted under laboratory conditions, the object's energy loss to the surrounding air should be estimated in order to assess the influence on the defect heating.

A simplified approach, which is commonly used in technical applications, is known as Newton's law of cooling. It states that the convective heat flux is proportional to the temperature difference between the body and the ambient air.

$$q_c = \alpha_c \cdot (T_{obj} - T_{fluid}) \tag{3.129}$$

The unknown constant α_c is referred to as the film coefficient or convective heat transfer coefficient. Determination of α_c is a fairly complicated task. Depending on the situation, the film coefficient can take values ranging over several orders of magnitude [LILV03].

An estimation of α_c can be made by means of the similarity principle of heat transfer. This concept is based on the assumption that a temperature field and an associated velocity field can be characterised by dimensionless parameters. Thus, it is implied that a physical problem can be solved independently of its (arbitrarily chosen) scale.

As a first parameter, the Nusselt number Nu establishes a general relation between α_c, the length l, which is characteristic for the heat transfer problem, and the thermal conductivity k.

$$Nu = \alpha_c \cdot \frac{l}{k} \tag{3.130}$$

It is shown in [BS06] that in case of free convection, which is assumed here, Nu can be expressed as a function of two fundamental parameters Gr and Pr and the geometry characterising the heat transfer problem (K_{geom}).

$$Nu = f(Pr, Gr, K_{geom}) \tag{3.131}$$

The dimensionless Prandtl number Pr is based on material parameters only, relating thermal and dynamic properties of the fluid. The Prandtl number is defined as

$$Pr = \frac{\eta C_p}{k} = \frac{\upsilon}{\alpha} \tag{3.132}$$

with η being the dynamic viscosity, υ the kinematic viscosity, C_p the heat capacity at constant pressure and α the thermal diffusivity. Dynamic and kinematic viscosity are simply related through

$$\upsilon = \frac{\eta}{\rho} \quad . \tag{3.133}$$

In case of natural convection, the Grashof number Gr, which introduces the gravitational acceleration g and the volume expansion coefficient β, is required for an estimation of Nu. The Grashof number is given by Baehr and Stephan [BS06] as:

$$Gr = \frac{l^3 \cdot g \cdot \beta \cdot (T_{\text{obj}} - T_{\text{fluid}})}{\upsilon^2} \tag{3.134}$$

Here, the thermal expansion coefficient can be approximated by $\beta = 1/T_{\text{fluid}}$ if an ideal gas is assumed. For the sake of completeness the Rayleigh number Ra should be mentioned, which is simply defined as:

$$Ra = Pr \cdot Gr \tag{3.135}$$

In the literature quite a lot of empirical formulas for Nu based on Pr, Gr and Ra can be found, which consider different situations of convective heat transfer, e.g. horizontal, inclined and upright walls or cylinders. In Section 6.2.4 three of them are compared in order to estimate α_c.

3.4 Infrared thermography

This section gives a very brief introduction to infrared thermography focusing only on the basic approach of "measuring" object temperatures by means of a focal plane array (FPA) camera.

3.4.1 Electromagnetic spectrum and composition of detector irradiance

In his famous experiment the astronomer William Herschel split up sunlight into its spectrum by means of a glass prism and observed the strongest heating well beyond the visible red light. He correctly concluded that there has to be another kind of energetic radiation invisible to the human eye – the infrared radiation. Figure 3.27 shows a partition of the electromagnetic spectrum in terms of wavelength. The infrared band is commonly segmented into three or more divisions, e.g. according to DIN 5031-7: Near infrared (NIR, 780 nm to 3 µm), mid infrared (MIR, 3 to 5 µm) and far infrared (FIR, 5 µm to 1 mm). However, the used terms and wavelength bounds strongly depend on the technical application, standard and field of science.[3]

[3] See for example [HW90] for a variety of other subdivisions of the IR range.

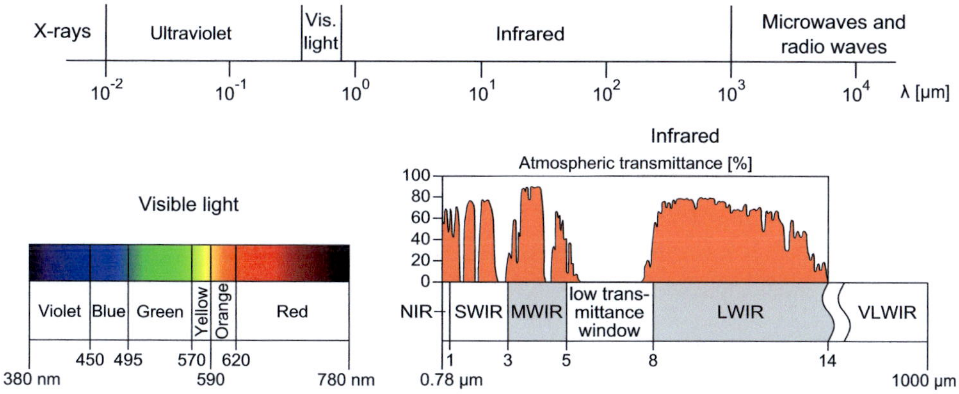

Figure 3.27: Ranges of optical and infrared radiation within the electromagnetic spectrum and atmospheric transmittance of technically relevant infrared wavelengths based on a 2000 m long measurement path at 15 °C and a relative humidity of 75 % [FR06, IC05].

Despite of the broad infrared range that covers wavelengths from 780 nm to 1 mm, mainly two discrete and relatively small wavelength intervals have established for thermal radiation measurements by means of infrared cameras: The ranges known as mid wave infrared (MWIR, 3 to 5 µm) and long wave infrared (LWIR, 8 to 14 µm). Besides of different detector materials and technologies, the choice of these narrow windows is primarily owed to the transmission properties of the surrounding atmosphere, namely the carbon dioxide and especially the water vapour content [FR06]. As a consequence, modern camera systems are optimised to operate in the MWIR or the LWIR band.

The absolute temperature "measurement" of a distant object by means of infrared thermography is a non-trivial task. Since only the irradiance (the radiant flux density) received at the detector is measured by the instrument, a radiation model is required to deduce the object's surface temperature from such measurements.

In order to trace and quantify infrared radiation originated from various sources reaching the detector, the energy balance of involved bodies and Kirchoff's law are regarded. At first, the radiation impinging on an object equals the sum of absorbed, transmitted and reflected radiation. Furthermore, the concept of an ideal blackbody radiator implies that absorption and emission must be equal in magnitude. This also applies to real bodies when in thermal equilibrium. Therefore, emissivity ϵ, transmittance τ and reflectance ρ are related through

$$\epsilon + \tau + \rho = 1 \quad . \tag{3.136}$$

Since only opaque bodies are inspected within this work, there is no radiation from background sources passing through the object under investigation. Assuming that any real object is a grey body with transmissivity $\tau_{obj} = 0$, its radiance depends only on the surface emissivity $\epsilon_{obj} < 1$ and the absolute temperature T_{obj}. Recalling equation (3.123) the object radiance is a fraction of the ideal blackbody radiance, namely $\epsilon_{obj} \cdot M_{bb}(T_{obj})$. Since τ_{obj} equals zero, the reflectance ρ of the body is $1 - \epsilon_{obj}$. As a consequence, a part of the ambient radiation originated from the environment, $(1 - \epsilon_{obj}) \cdot M_{bb}(T_{env})$, is reflected from the object. Note that the environment can be treated as a blackbody if it completely encloses the measurement scene (see Section 3.3.2).

Because of the very small distances between object and infrared detector realised in this work (typically 20 cm), it is assumed that there is no interference caused by the atmosphere ($\tau_{atm} = 1$). Instead, the transmissivity of the lens system and the heating of the camera chassis are taken into account. Typically, the lens transmission is about 90 %. That means both parts of the incoming radiant flux, the emitted and the reflected one, are to be multiplied by τ_{lens}. Neglecting any reflectance of the lens its theoretical emissivity $1 - \tau_{lens}$ is incorporated. This way, the actual lens temperature T_{lens}, which causes an additional radiance of $(1 - \tau_{lens}) \cdot M_{bb}(T_{lens})$, allows to represent the long-term heating of the camera chassis. In summary, the relation between detector irradiance E_{FPA} and the three infrared sources reads as follows:

$$E_{FPA} = \tau_{lens} \cdot \left[\left(1 - \epsilon_{obj}\right) \cdot M_{bb}(T_{env}) + \epsilon_{obj} \cdot M_{bb}(T_{obj}) \right] + (1 - \tau_{lens}) \cdot M_{bb}(T_{lens}) \qquad (3.137)$$

Obviously, the described radiation model, though being simplified, involves a number of parameters, which are not exactly known in general. Thus, absolute temperature measurements by means of an infrared camera should be critically reviewed or verified by alternative measurement techniques if possible. The remote measurement becomes fault-prone if the object's emissivity is very low, e.g. in case of polished metals. Then, ambient heat sources such as room illumination, moving people, the thermographer or the camera itself may lead to reflections, which prevent a flaw from being detected. Applying a matt black coating is an efficient measure against too high reflectance.

3.4.2 Non-uniformity correction and bad pixel replacement

In general, focal plane array based infrared cameras suffer from a detector non-uniformity. Each individual quantum detector of the array exhibits a slightly different response to the incident radiation. To achieve a homogeneous thermogram, a two-point non-uniformity correction (NUC) is performed. The dynamic response regarding a change of incoming flux as well as the stationary output can be equalised for each detector by applying an individual gain factor g and an offset value o to the raw detector reading. The corrected grey value image i^* with dimensions of $m \times n$ pixels is computed from the raw image i composed of values i_{ij} by applying equation (3.138) [FLI05b].

$$i^*_{ij} = i_{ij} \cdot g_{ij} + o_{ij} \quad , \qquad i = 1,2,\ldots,m \quad , \qquad j = 1,2,\ldots,n \tag{3.138}$$

Gain and offset for each detector can be found by means of a blackbody calibration source, which is set to two different temperatures $T_1 < T_2$. While the infrared camera is staring at the blackbody source, a "cold" image c_{ij} at T_1 and a "hot" image h_{ij} at T_2 is acquired. In order to reduce the noise level and to obtain representative detector readings, c and h are generated by averaging a sequence of images (e.g. 100 frames according to [Med08]). Then, gain and offset matrices are computed based on the normalised raw detector slope s_{ij}:

$$s_{ij} = \frac{(m \cdot n)\left(h_{ij} - c_{ij}\right)}{\sum_{i=1}^{m} \sum_{j=1}^{n} \left(h_{ij} - c_{ij}\right)} \quad , \qquad i = 1,2,\ldots,m \quad , \qquad j = 1,2,\ldots,n \tag{3.139}$$

$$g_{ij} = \frac{1}{s_{ij}} \tag{3.140}$$

$$o_{ij} = \frac{1}{m \cdot n} \sum_{i=1}^{m} \sum_{j=1}^{n} \left(c_{ij} \cdot g_{ij}\right) - c_{ij} \cdot g_{ij} \tag{3.141}$$

Some of the individual detectors exhibit an inappropriate response behaviour (e.g. always zero, always saturated, negative slope). According to [FLI05b], corrupted or "dead" pixels can be identified by applying an acceptance band AB to the normalised slopes s.

$$s_{\min} = \frac{\bar{s}}{1 + AB} \leq s_{ij} \leq s_{\max} = \frac{\bar{s}}{1 - AB} \quad , \tag{3.142}$$

$$\bar{s} = \frac{1}{m \cdot n} \sum_{i=1}^{m} \sum_{j=1}^{n} s_{ij} \quad , \qquad i = 1,2,\ldots,m \quad , \qquad j = 1,2,\ldots,n \tag{3.143}$$

If certain slopes s_{ij} fall outside this range, they are considered bad pixels. Replacement is achieved by median filtering based on a square window (3×3, 5×5, …) centred at the suspicious pixel of each image i^*. After collecting all pixel values within the window and sorting them in a list, the value of the centre pixel is replaced by the median of the list. Median filtering is known to be an edge-preserving smoothing filter that effectively suppresses speckle or salt-and-pepper noise.

3.4.3 Temperature calibration and measurement

Modern infrared cameras based on focal plane array quantum detectors typically provide an intensity or grey value output with a 14-bit or 16-bit quantisation. In order to translate the digital counts into temperatures, a calibration is required that establishes the relation between the detector irradiance and the grey value reading for a certain temperature range of interest. For this purpose the nominal temperature of a blackbody source is varied stepwise and its radiation is "measured" by the camera. In a first step the internal camera settings, such as integration time, gain factors and offsets, are adjusted so that the minimum and the maximum temperature result in camera readings in the lower and the upper range of available digits. Using the radiation model described before with T_{obj} and ϵ_{obj} now being the blackbody temperature T_{bb} and the true blackbody emissivity ϵ_{bb}, respectively, the detector irradiance E_{FPA} is given by:

$$E_{\mathrm{FPA}} = \tau_{\mathrm{lens}} \cdot [(1-\epsilon_{\mathrm{bb}}) \cdot M_{\mathrm{bb}}(T_{\mathrm{env}}) + \epsilon_{\mathrm{bb}} \cdot M_{\mathrm{bb}}(T_{\mathrm{bb}})] + (1-\tau_{\mathrm{lens}}) \cdot M_{\mathrm{bb}}(T_{\mathrm{lens}}) \tag{3.144}$$

Here, τ_{lens}, ϵ_{bb}, T_{env}, T_{lens} and of course T_{bb} are known. In the following, all quantities $M_{\mathrm{bb}}(T)$ are related to the camera's sensitive range of wavelengths λ_{min} to λ_{max} and can be computed by integration of Planck's law:

$$M_{\mathrm{bb}}(T) = \int_{\lambda_{\mathrm{min}}}^{\lambda_{\mathrm{max}}} M_{\lambda\mathrm{bb}}(\lambda, T)\,\mathrm{d}\lambda = \int_{\lambda_{\mathrm{min}}}^{\lambda_{\mathrm{max}}} \frac{c_1}{\lambda^5\left[\exp\left(\frac{c_2}{\lambda T}\right)-1\right]}\,\mathrm{d}\lambda \tag{3.145}$$

Thus, equation (3.145) is evaluated for the ambient temperature T_{env}, the lens temperature T_{lens} and a series of blackbody temperatures T_{bb}. Plugging the values M_{bb} into Equation (3.144) yields the relation between detector irradiance and grey value reading in the form $E_{\mathrm{FPA}}(i^*)$. Here, the noise level can be significantly reduced by spatial and temporal averaging of the camera output.

Figure 3.28a shows how M_{bb} develops with increasing temperature if different spectral ranges are considered. Obviously, thermal imagers working in the range of 3 to 5 µm, which is typical for indium antimonide (InSb) detectors, are more sensitive at high temperatures. In contrast, mercury cadmium telluride (HgCdTe) detectors working in the 8 to 14 µm range are more sensitive at lower temperatures. Both curves intersect at about 350 °C. Throughout this work a camera with an InSb detector is used, see Section 4.2.4 for details on the hardware. The MWIR excitance for relevant temperatures is exemplarily depicted in Figure 3.28b.

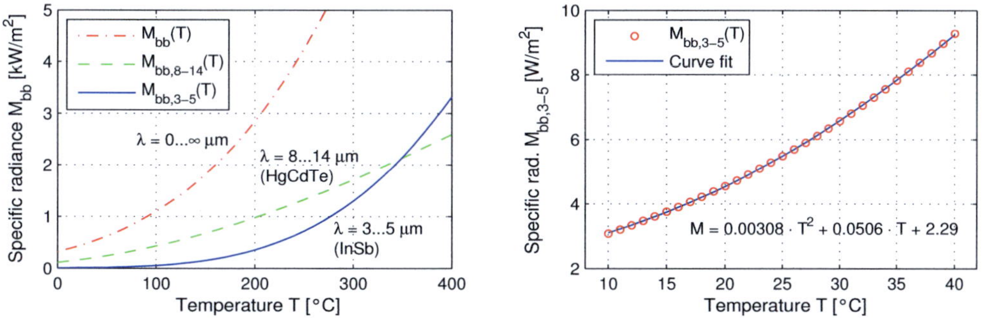

(a) MWIR, LWIR and total specific radiance (b) Detail of MWIR radiance and curve fit

Figure 3.28: Specific radiance of a blackbody depending on the spectral range.

If absolute temperatures are to be deduced based on the presented radiation model, equation (3.137) is solved for $M_{\mathrm{bb}}(T_{\mathrm{obj}})$. Again, τ_{lens}, ϵ_{obj}, T_{env}, T_{lens} need to be known or estimated. The unknown detector irradiance E_{FPA} is now assessed by using the calibration curve $E_{\mathrm{FPA}}(i^*)$. Finally, the object temperature is estimated through the inverse function of $M_{\mathrm{bb}}(T_{\mathrm{obj}})$.

Figure 3.29 presents an experimentally achieved relation $E_{\mathrm{FPA}}(i^*)$ using actual values $\epsilon_{\mathrm{bb}} = 0.9713$ [Mik07], $\tau_{\mathrm{lens}} = 0.92$ [Jan07], $T_{\mathrm{env}} = 20\,^{\circ}\mathrm{C}$ and $T_{\mathrm{lens}} = 26\,^{\circ}\mathrm{C}$ as well as the inverted curve $T(M_{\mathrm{bb}})$. Both functional relations $E_{\mathrm{FPA}}(i^*)$ and $T(M_{\mathrm{bb}})$ can be approximated by fitted polynomial functions in order to provide computationally simple conversions from grey values via detector irradiance and object radiance to absolute object temperatures.

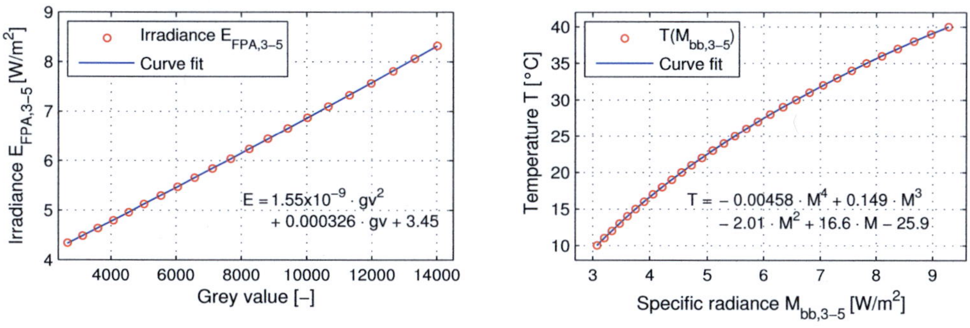

(a) Experimentally found calibration cure $E_{\mathrm{FPA}}(i^*)$ (b) Inverse function of $M_{\mathrm{bb}}(T_{\mathrm{obj}})$ valid for MWIR

Figure 3.29: Deduction of object temperatures through a calibration curve and Planck's law.

3.5 Excitation methods and data processing techniques

In addition to the mono-frequency and constant amplitude excitation of a structure, more advanced methods of excitation and corresponding data processing are available. In this section the most common methods of ultrasound excited thermography are described, namely ultrasonic sweep thermography, ultrasonic lock-in thermography and ultrasonic burst-phase thermography. The focus of this work is on the sweep and the lock-in approach.

3.5.1 Ultrasound sweep thermography

According to the experimental findings reported by Rothenfusser and Homma [RH05], Gleiter et al. [GSB07] and Holland [Hol07], vibrothermographic detection of crack-like flaws in metal parts is strongly dependent on the excitation frequency. Thus, raising the transducer frequency gradually or stepwise is useful to identify the inspected component's thermal response to ultrasonic excitation at various frequencies within a relatively short time. During this procedure the nominal oscillation amplitude is usually kept at a constant level.

The essential control parameters are the excitation power, the start and stop frequency and the sweep duration. Assuming that the frequency range of interest should be swept linearly in time within a certain duration d, the question arises whether a continuous or a stepped sweep is more advantageous. In the latter case an appropriate frequency step size Δf needs to be chosen, which is small enough to accurately resolve the "acousto-thermal spectrum", the thermal response with respect to the excitation frequency. In case of a large Δf value, there is a risk of skipping efficient frequencies that would lead to (a stronger) crack heating. On the other hand, the oscillating system needs some time to turn from a transient to a steady state vibration, which will enhance the crack detection in case of resonance [Hom07]. With a continuous sweep the duration of resonant excitation theoretically tends to zero if the resonant peaks were ideally sharp.

Considering a digital control of the exciter, the continuous sweep function needs to be reproduced by a finite number N of data samples n. The discretised equation describing the swept sine waveform with linearly increasing frequency over time is given by Rowe [Row00]:

$$y(n) = \hat{A} \cdot \sin\left(\frac{a \cdot n^2}{2} + b \cdot n\right) \quad , \qquad n = 1, 2, \ldots, N \tag{3.146}$$

Here, $\hat{A}$ denotes the signal amplitude and the samples n refer to the discrete integer steps. The constants a and b are defined as

$$a = \frac{2\pi}{N} \cdot \left(\frac{f_{\text{stop}}}{f_{\text{s}}} - \frac{f_{\text{start}}}{f_{\text{s}}} \right) \quad , \qquad b = 2\pi \cdot \frac{f_{\text{start}}}{f_{\text{s}}} \quad , \tag{3.147}$$

where f_{s} is the sampling rate and f_{start} and f_{stop} are the start and stop frequency, respectively. Equation (3.146) can be reformulated to yield the continuous waveform depending on time t and the sweep duration d by making use of the basic relations $f_{\text{s}} = N/d$ and $t = n/N \cdot d$:

$$y(t) = \hat{A} \cdot \sin\left(\pi \cdot \left(f_{\text{stop}} - f_{\text{start}} \right) \frac{t^2}{d} + 2\pi \cdot f_{\text{start}} \cdot t \right) \tag{3.148}$$

In contrast to the described continuous waveform, the stepped frequency sweep results in unavoidable discontinuities and phase jumps in the time domain. Even if this is true for the digital control signal, in practice any mechanical system will not be able to reproduce such a jumpy signal but smooth out those discontinuities. An illustrative comparison of a continuous and a stepped waveform with a massively exaggerated rate of frequency increase is shown in Figure 3.30.

Another important aspect is related to the requirements of accompanying measurement techniques e.g. laser vibrometry. If steady state vibrations are to be analysed, a stepped frequency sweep is more convenient than a continuous one since the data for each frequency level can be processed individually. Therefore, only stepped frequency sweeps are performed throughout this work. Figure 3.30b indicates a further subdivision of the time step width Δt. The time intervals Δt_1, Δt_2 and Δt_3 correspond to a waiting, a data acquisition and a buffering phase, which will be used and defined in Chapter 5.

Based on the infrared data stream acquired during the sweep, a result image could be simply generated by plotting the maximum absolute temperature achieved for each pixel. However, such an absolute value thermogram is directly affected by static reflections or an inhomogeneous surface emissivity. Therefore, differential images are usually computed by evaluating the maximum temperature increase per pixel. Gleiter et al. [GSB07] proposed to smooth the pixel signal and to calculate the maximum value of its first derivative with respect to time or rather frequency. This approach clearly leads to an improved contrast and robustness compared to the mapping of absolute and differential temperatures, especially if the thermal response of the defect arises faster than a possible long-term heating due to bulk material damping. Ultrasound sweep thermography provides a defect-selective imaging NDT method comparable to the lock-in method described in the next section.

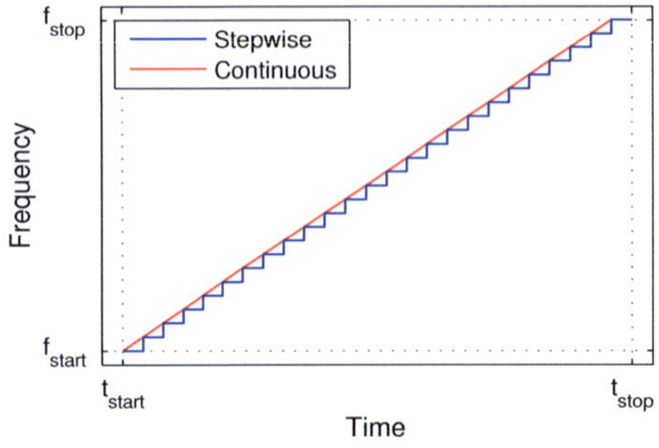

(a) Smooth and gradual frequency increase

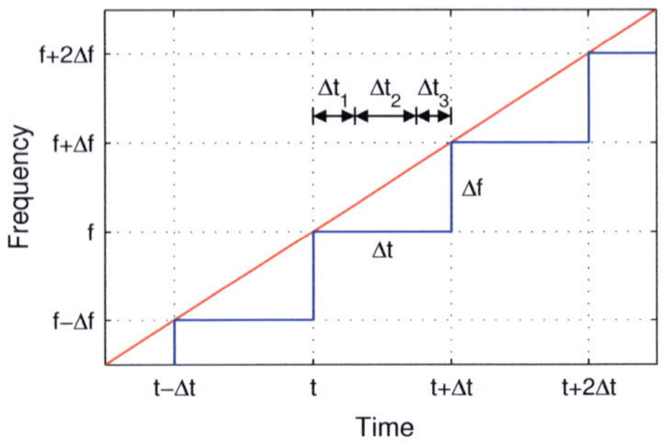

(b) Frequency step width and temporal subdivision

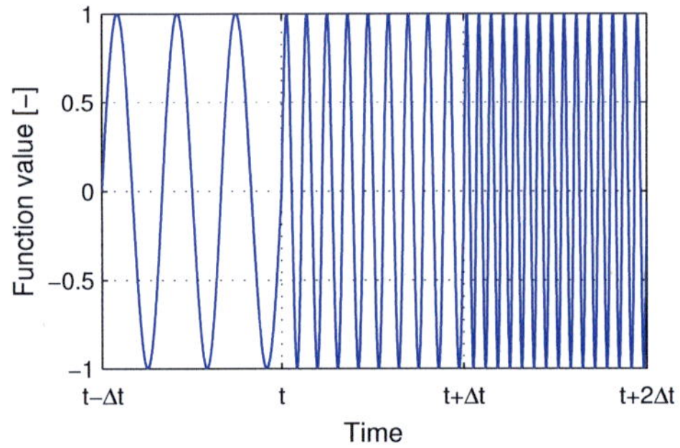

(c) Stepped sweep signal

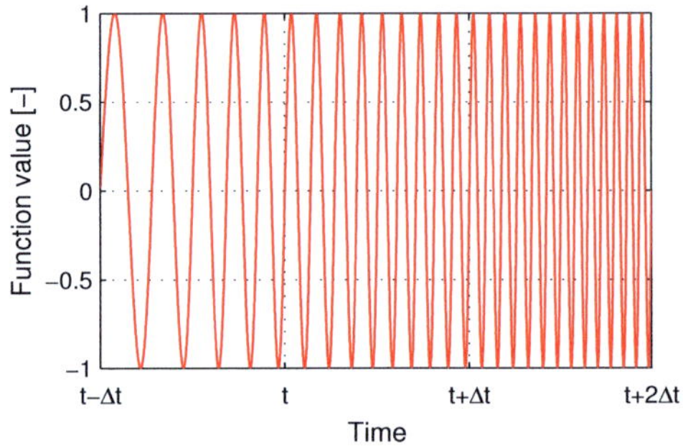

(d) Continuous sweep signal

Figure 3.30: Schematic comparison of stepped and continuous sine sweep signals.

3.5.2 Ultrasound lock-in thermography

In contrast to ultrasound sweep thermography, which has been recently developed owing to the pronounced frequency dependence of crack heating observed in metals, the lock-in method in general has a much longer history and established in the field of active thermography several decades ago [CB76, KFA$^+$88, BWK92].

The idea of lock-in excitation and data processing is to modulate the energy input harmonically at a fixed lock-in frequency and to evaluate the thermal response corresponding to the modulation frequency only. This way, a narrow-band filtering is realised, by which noise that is distributed over a wide spectrum is efficiently suppressed. Also, any other disturbing reflections that do not vary in accordance with the modulation are eliminated. Since the modulation history serves as a temporal reference signal, the thermal response over time can be characterised by the temperature amplitude and its phase shift with respect to the modulated excitation. The phase information is of great interest because it provides access to the spatio-temporal evolution of the thermal response and the propagation of thermal waves.

The well-known principle of power modulated excitation was soon adapted to ultrasound excited thermography [RWB96, RWSB96, DBW99]. While the ultrasound frequency f is kept constant, the oscillation amplitude is being sinusoidally modulated at a low lock-in frequency, e.g. $f_{\text{mod}} = 0.2\,\text{Hz}$.[4] The output waveform $y(t)$ can be constructed in the form

$$y(t) = \left[\frac{\hat{A}_{\text{max}} - \hat{A}_{\text{min}}}{2} \cdot \sin(2\pi \cdot f_{\text{mod}} \cdot t + \varphi) + \frac{\hat{A}_{\text{max}} + \hat{A}_{\text{min}}}{2} \right] \cdot \sin(2\pi \cdot f \cdot t) \quad , \tag{3.149}$$

where $\hat{A}_{\text{max}}$ and $\hat{A}_{\text{min}}$ denote the maximum and the minimum oscillation amplitude. The phase angle φ can be freely adjusted, for example to obtain a smooth starting of the modulated waveform. Figure 3.31 exemplarily shows two waveforms with different modulation depths. For better perceptibility the high-frequency ultrasound oscillation is represented by a sine wave with frequency $f = 15 \cdot f_{\text{mod}}$ in these graphs.

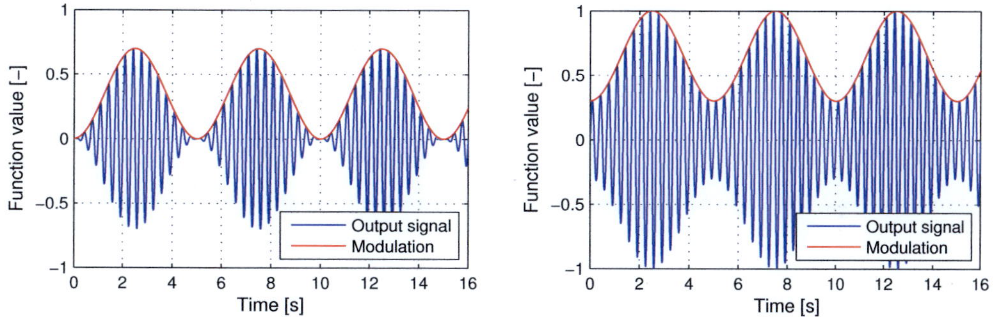

(a) Amplitude modulated waveform with $\hat{A}_{\text{max}} = 0.7$, $\hat{A}_{\text{min}} = 0$, $f_{\text{mod}} = 0.2\,\text{Hz}$ and $\varphi = -\pi/2$

(b) Amplitude modulated waveform with $\hat{A}_{\text{max}} = 1$, $\hat{A}_{\text{min}} = 0.3$, $f_{\text{mod}} = 0.2\,\text{Hz}$ and $\varphi = -\pi/2$

Figure 3.31: Amplitude modulation of a sine signal with different modulation depths.

An energy dissipating crack is expected to generate heat correspondingly to the intensity-modulated excitation. After several modulation cycles have been performed, a lock-in data processing is applied to the recorded sequence of infrared images. For each pixel of the stream the temperature

[4] Zweschper et al. propose a permanent variation of the ultrasound frequency (wobbling) during lock-in excitation, in order to avoid any standing wave patterns [ZDR+03, ZRDB04].

amplitude at the lock-in frequency and the phase shift related to the modulation history or to an arbitrary reference are calculated and composed as amplitude and phase maps.

Today, the discrete Fourier transform (DFT) is commonly used for this task. Alternative methods include the wavelet transform [OWG+07] and the four bucket method [BWK92]. A performance comparison of four different algorithms, which were introduced in the early works on lock-in thermography, was conducted by Krapez [Kra98]. Presuming that both the infrared stream and the modulation or reference signal are synchronously acquired, the lock-in post-processing typically comprises the following steps:

- DFT of the modulation signal, optional detection of the actual lock-in frequency f_{mod} and storing of the phase at f_{mod}
- Trend removal from each infrared pixel history, e.g. by fitting a polynomial function to the long-term temperature evolution
- Pixelwise DFT of the infrared sequence, mapping of the temperature amplitude and the phase difference at f_{mod} with respect to the phasing of the modulation signal

Since the infrared sequence is analysed regarding only temperature changes at the modulation frequency, the signal-to-noise ratio of the amplitude map is greatly improved compared to a single thermogram taken from the stream. High temperature amplitudes directly indicate the location of dissipative effects. The phase image illustrates the temporal lag of the thermal waves originated from heat sources. Contrary to the amplitude distribution, which could still be affected by inhomogeneous surface emissivity, the phase image is much less susceptible to such an interference.

Both the amplitude and phase signature observed at the object surface are strongly influenced by the actual depth of the heat source and the modulation frequency. In order to illustrate the influence of both parameters on the surface temperature variation, the FE model representing a corner of a 10 mm thick plate (Fig. 3.19a) is deployed again without applying symmetry conditions at $x = 0$.

A modulated heat flux is applied to a $4 \times 4\,\mathrm{mm}^2$ square area located next to the edge at $x = 0\,\mathrm{mm}$ to simulate a highly localised heat source acting in different depths. The lock-in frequency ranges from 0.01 to 5 Hz, where eight modulation periods are considered using a time step size of $1/(40\,f_{\mathrm{mod}})$. Here, the lateral faces at $x, y = 35\,\mathrm{mm}$ are assumed to keep their initial temperature to allow for an outward heat flow. This way, the long-term temperature increase monitored along the surface edge at $y = 0\,\mathrm{mm}$ asymptotically approaches a steady state. A typical temperature evolution composed of the transient heating and the superimposed oscillation is depicted in Figure 3.32a. Correspondingly,

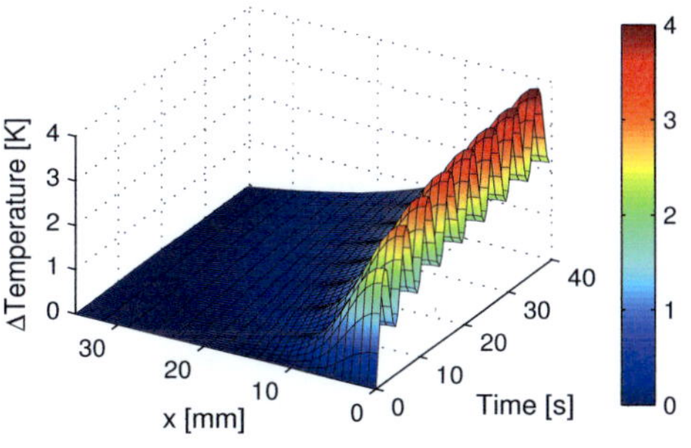

(a) Surface temperature evolution along $y = 0\,\text{mm}$; $P_{\text{max}} = 1.5\,\text{W}$, $f_{\text{mod}} = 0.2\,\text{Hz}$, depth $= 4\,\text{mm}$

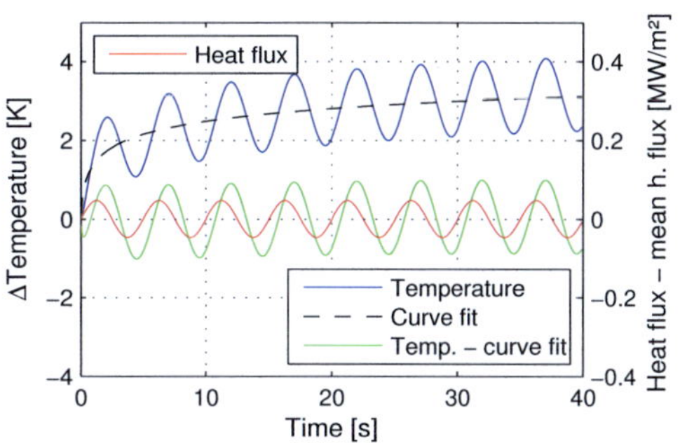

(b) Variation of heat flux and corner temperature; $P_{\text{max}} = 1.5\,\text{W}$, $f_{\text{mod}} = 0.2\,\text{Hz}$, depth $= 4\,\text{mm}$

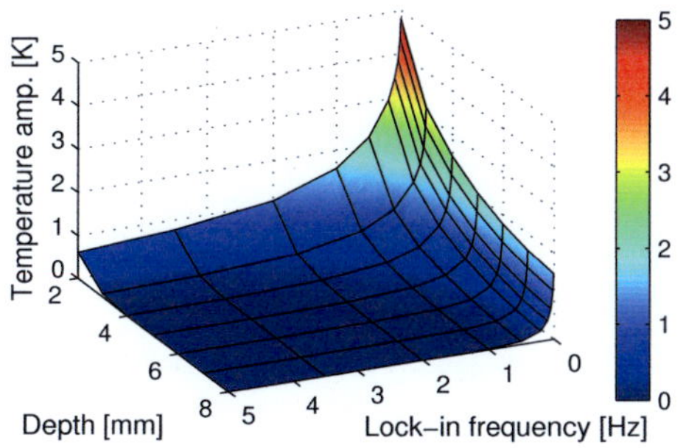

(c) Temperature amplitude at the corner node; $P_{\text{max}} = 1.5\,\text{W}$

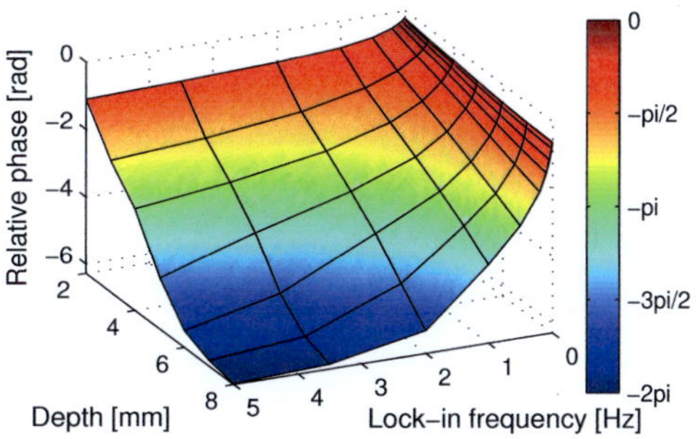

(d) Relative phase shift at the corner node; $P_{\text{max}} = 1.5\,\text{W}$

Figure 3.32: Simulated temperature evolution along the surface edge; amplitude and relative phase values found for different modulation frequencies and depths of the heat source.

Figure 3.32b shows the time histories of the applied heat flux, the nodal temperature at the surface corner ($x = 0\,\text{mm}$, $y = 0\,\text{mm}$, $z = 10\,\text{mm}$) as well as the remaining temperature pulsation after trend removal. The presented results have been achieved for a heat load of 1.5 W, a moderate depth of 4 mm and a lock-in frequency of 0.2 Hz. With these parameters the temperature oscillation as well as the phase shift with respect to the heating is clearly detectable. The situation dramatically changes with increasing depth and modulation frequency. As a result of the performed parametric study Figures 3.32c and 3.32d display the temperature amplitude and the unwrapped relative phase at the corner node, respectively.

Recalling equation (3.107) one finds that the relative phase varies linearly with the depth at any constant modulation frequency. The phase data obtained from Fourier analyses of the noise free synthetic time signals appears as a smooth surface with almost constant slope at lock-in frequencies up to 1 Hz. Nevertheless, it has to be pointed out clearly that the 1D thermal diffusion length is exceeded by far for many combinations of depth and modulation frequency. In the considered three-dimensional problem any lateral heat flow from the buried, point-like hot spot as well as the flow towards the opposite plate surface additionally lower the magnitude of temperature oscillation at the evaluation node. Manual judgement on the temperature variation revealed that the oscillatory part of the signal is almost completely damped out and thus practically undetectable. Interfering noise will render the phase to be random in case of a vanishingly low amplitude at f_{mod}, which is typical for vibrothermographic phase images. Only in a certain range around the pulsating heat source, which strongly depends on the diffusion length, meaningful phase values are achieved. Any inactive regions such as the background and zones far away from the defect appear noisy in the phase image.

The discrete Fourier transform of an acquired data stream represents the essential step within the lock-in data processing. It will be used in Chapter 5 in the fashion of a standard method. Therefore, its practical use is described in more detail at this point. In general, the Fourier transform (FT) converts a continuous time signal $s(t)$ from the time domain to the frequency domain. The frequency domain representation $\underline{S}(f)$ of the original signal as a complex function of frequency f is given by

$$\underline{S}(f) = \int_{-\infty}^{\infty} s(t) \cdot \mathrm{e}^{-2\pi \mathrm{i} f t} \, \mathrm{d}t \quad . \tag{3.150}$$

The complex Fourier transform can be decomposed in its real part $\mathrm{Re}\{\underline{S}(f)\}$ and imaginary part $\mathrm{Im}\{\underline{S}(f)\}$. Alternatively, $\underline{S}(f)$ is represented by its amplitude spectrum $|\underline{S}(f)|$ and the associated phase spectrum $\varphi(f)$ using the following relations:

$$\underline{S}(f) = \mathrm{Re}\{\underline{S}(f)\} + \mathrm{i} \cdot \mathrm{Im}\{\underline{S}(f)\} = |\underline{S}(f)| \cdot \mathrm{e}^{\mathrm{i}\varphi(f)} \tag{3.151}$$

$$|\underline{S}(f)| = \sqrt{\mathrm{Re}^2\{\underline{S}(f)\} + \mathrm{Im}^2\{\underline{S}(f)\}} \tag{3.152}$$

$$\varphi(f) = \arctan\left(\frac{\mathrm{Im}\{\underline{S}(f)\}}{\mathrm{Re}\{\underline{S}(f)\}}\right) \tag{3.153}$$

Numerical integration of equation (3.150) yields the discrete Fourier transform (DFT). Presuming a constant sampling rate f_s, the DFT is applicable in the following form [Bri97, OSB99][5]:

$$\underline{S}(n) = \sum_{k=1}^{N} s(k) \cdot \mathrm{e}^{-2\pi\mathrm{i}(k-1)(n-1)/N} \quad , \qquad n = 1,2,\ldots,N \tag{3.154}$$

Here, n and k represent the vector indices of the discrete Fourier transformed function $\underline{S}$ and the discrete signal samples s, respectively. The frequency resolution of the DFT is given as the reciprocal of the measurement time or f_s/N.

Referring to Shannon's sampling theorem only spectral components below the Nyquist frequency $f_N = f_s/2$ can be reconstructed correctly. The single-sided spectrum providing physical amplitude values is therefore found through the following normalisation

$$\hat{S}(n) = \begin{cases} |\underline{S}(n)|/N & \text{for} \quad n = 1 \\ |\underline{S}(n)| \cdot 2/N & \text{for} \quad n = 2,3,\ldots,N/2 \quad , \end{cases} \tag{3.155}$$

whereas phase values are computed analogously to the continuous transform using equation (3.153). It should be noted that frequencies higher than f_N still affect the analysis since such spectral components are unintentionally detected as alias frequencies (mirrored about the Nyquist frequency). Aliasing can only be avoided completely by means of an analogue low-pass filter, which is applied prior to any digital signal processing.

Any acquired data block is of finite length and its start and end point in the time domain can be arbitrary. In case of lock-in thermography, the infrared record might not exactly contain multiples of the modulation period length. As a result, the accuracy of the calculated amplitude spectrum using the DFT approach suffers from leakage effects. Apart from the strongest frequency components, which are actually contained in the time signal, additional side lobes smear the amplitude spectrum due to the violation of the signal periodicity prerequisite. If unknown, the true frequency might be difficult to identify from the spectrum. Moreover, the achieved peak values are lower than the actual amplitude since the energy of the signal is spread over several neighbouring frequencies.

A common countermeasure that can improve the accuracy of an amplitude spectrum computed by Fourier transform is the use of weighting functions, which artificially enforce the signal to satisfy

[5] In the cited books by Brigham [Bri97] and Oppenheim et al. [OSB99] the vector indices count from 0 to N-1. Against this convention the following equations directly refer to the practical computation and data processing using MATLAB. Here, the first index is 1 instead of 0 (compare [The08]).

the assumption of periodicity by fading the actual signal in and out smoothly. Several standard window functions, e.g. Hamming, Blackman, Chebyshev, Hann or Bartlett, are known and widely used in signal processing [Bri97]. Harris [Har78] compiled a concise review of windows and their effects on the detection of harmonic signals buried in noise. Without going into details one of the main intention is to achieve a greatly accelerated roll-off rate so the first side lobes in the frequency domain drop significantly in contrast to the use of no weighting function, which corresponds to a rectangular window. Figure 3.33 shows the Hann window function (named after Julius von Hann) and a so called Flat Top window function in the time domain.

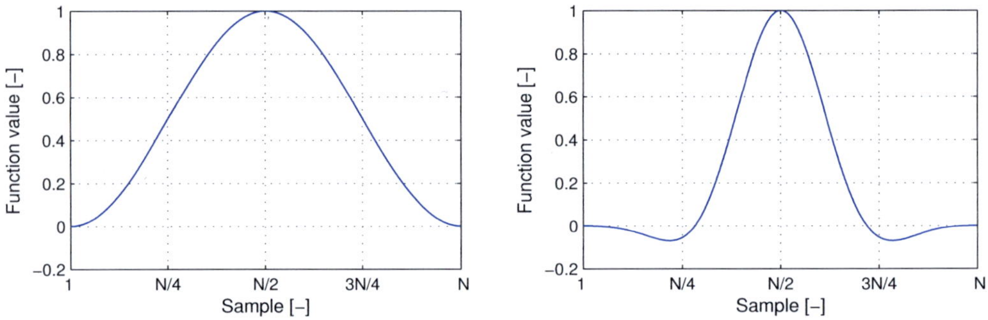

(a) Hann window

(b) Flat Top window (max. value normalised to 1.0)

Figure 3.33: Hann and Flat Top window functions used in spectral analysis.

The Hann window represents a versatile weighting function, which gives a good trade-off between frequency resolution, spectral leakage and amplitude accuracy, especially when analysing random signals. In contrast, the Flat Top window is perfectly suited to detect the actual amplitude of sinusoidal signals buried in noise. On the downside, it results in a poor frequency resolution.

The sampled functions $w(n)$ of both the periodic Hann and Flat Top windows with data point index n and length N are computed through

$$\begin{aligned} w(n) = a_0 - a_1 \cos\left(\frac{2\pi(n-1)}{N}\right) + a_2 \cos\left(\frac{4\pi(n-1)}{N}\right) \\ - a_3 \cos\left(\frac{6\pi(n-1)}{N}\right) + a_4 \cos\left(\frac{8\pi(n-1)}{N}\right) \quad , \qquad n = 1,2,\ldots,N \end{aligned} \tag{3.156}$$

The coefficients a_0 to a_4 are given in Table 3.4. Note that different sets of coefficients exist to construct the Flat Top window. Since the discrete signal data points are multiplied by the weighting function $w(n)$, the resulting amplitude values computed by means of the DFT are lower compared to the use of a rectangular window function. To compensate for this artificial amplitude decrease, the function values $w(n)$ are normalised by multiplication with a factor

$$a = \frac{N}{\sum\limits_{n=1}^{N} w(n)} \quad , \tag{3.157}$$

which finally yields the amplitude corrected window function:

$$w_{\text{corr}}(n) = a \cdot w(n) \quad , \qquad n = 1, 2, \ldots, N \tag{3.158}$$

Table 3.4: Coefficients of the Hann and Flat Top window functions.

Window	a_0	a_1	a_2	a_3	a_4
Hann [Har78, OSB99]	0.5	0.5	0	0	0
Flat Top [ISO 18431-1]	1.0	1.9330	1.2860	0.3880	0.0322

Since the DFT is excessively used within this work, it should be mentioned that the computational cost can be greatly decreased by use of the fast Fourier transform (FFT). With the declaration $\underline{c} = \mathrm{e}^{-2\pi \mathrm{i}/N}$ equation (3.154) can be written in matrix form:

$$\begin{bmatrix} \underline{S}(1) \\ \underline{S}(2) \\ \underline{S}(3) \\ \vdots \\ \underline{S}(N) \end{bmatrix} = \begin{bmatrix} \underline{c}^0 & \underline{c}^0 & \underline{c}^0 & \cdots & \underline{c}^0 \\ \underline{c}^0 & \underline{c}^1 & \underline{c}^2 & \cdots & \underline{c}^{(N-1)} \\ \underline{c}^0 & \underline{c}^2 & \underline{c}^4 & \cdots & \underline{c}^{2\cdot(N-1)} \\ \vdots & \vdots & \vdots & \ddots & \vdots \\ \underline{c}^0 & \underline{c}^{(N-1)} & \underline{c}^{(N-1)\cdot 2} & \cdots & \underline{c}^{(N-1)\cdot(N-1)} \end{bmatrix} \cdot \begin{bmatrix} s(1) \\ s(2) \\ s(3) \\ \vdots \\ s(N) \end{bmatrix} \tag{3.159}$$

This notation clarifies that for a direct evaluation of the DFT N^2 complex multiplications and $N \cdot (N-1)$ additions are necessary. FFT algorithms reduce the number of multiplications to $N \cdot \log_2(N)$. Since the calculation time is proportional to the number of multiplications, this results in an enormous time saving for large values of N. The most popular algorithm for the efficient computation of the DFT was developed by Cooley and Tukey in 1965 [CT65] and is commonly implemented in today's computer algebra software packages. Comprehensive discussions as well as derivations of alternatives can be found in textbooks by Brigham [Bri97] and Oppenheim et al. [OSB99].

3.5.3 Ultrasound burst-phase thermography

Both methods described in the preceding two sections, ultrasound sweep thermography and ultrasound lock-in thermography, represent approaches that require the excitation system to be controlled precisely regarding frequency and relative power, respectively. There is a third technique available, which is derived from lock-in thermography representing the counterpart to the established optically excited pulse-phase thermography [MM96, GMVC99]. It is based on an excitation pulse, which means that the ultrasound generator is just turned on for a short duration typically ranging from several hundreds of a second up to a few seconds [DZRB03]. Consequently, a short heating pulse is initiated by a dissipating defect so that a thermal wave is generated that spreads into the structure and reaches the surface.

In ultrasound burst-phase thermography the heating period and the subsequent cooling phase are thermographically acquired [DZB00]. The infrared stream is then pixelwise transformed to the frequency domain via Fourier analysis similar as with the lock-in processing. However, there is no dominant frequency at which one can expect the strongest thermal response. Instead, amplitude and phase maps are computed for a variety of frequencies, which is meaningful since a short heating pulse in the time domain corresponds to a broad spectrum in the frequency domain.[6] In contrast to the lock-in approach, where a stationary temperature variation at a single modulation frequency is induced, pulse heating corresponds to the simultaneous launching of thermal waves of various amplitudes and phases in a transient mode [Mal01].

It depends on the size, orientation and depth of the defect at which evaluation frequency the greatest amplitude and phase contrasts are achieved. In particular, the phase information is of interest since it enables a qualitative deduction of the defect depth [Dil01]. However, in case of crack faces oriented perpendicular to the surface, any kind of depth probing will become difficult, especially if the active heat source is smeared and non-uniformly distributed along a certain depth range.

Although the signal-to-noise ratio of the result images will be lower compared to the ultrasound lock-in processing, burst-phase thermography combines the advantages of robust phase maps, less interference-prone defect detection similar to the lock-in approach and a short inspection time known from pulse-phase thermography. The general issue related to the interference of crack detection due to standing wave patterns can be addressed by applying a certain frequency modulation to the burst excitation [ZRDB03, ZRDB04]. This measure might also help to overcome the pronounced frequency dependence of crack heating in metals.

[6]The Dirac delta function in the time domain corresponds to a flat frequency spectrum and vice versa [Bri97].

4 Experimental analysis – Vibrothermography

In this central chapter it is attempted to give an encompassing overview of ultrasound excited thermography applied to typical plate-like components used in steel construction. After a detailed description of the specimen manufacture with an emphasis on the generation of fatigue cracks, the vibrothermography test equipment is introduced. In the following, the main part of the chapter deals with the manifold influences of the setup, the excitation parameters, the initial temperature, the crack depth, the plate thickness and the prestress state on the thermographic crack detection. In addition, the chapter covers various phenomena such as standing wave patterns on flawless plates, the effects of long-term excitation on the thermal response and the fracture surface as well as a simulation of in-situ crack detection.

4.1 Manufacture of test specimens

Thanks to modern CNC machining, non-standard test pieces can be reproducibly manufactured, which fulfil the requirements of vibrothermographic testing. After describing the used specimen type and preparation, the focus of this section is on fatigue testing to produce through-thickness cracks that are to be detected later. The section includes an introduction to the local strain approach, which is used to predict the necessary fatigue loading to achieve the desired load cycles to crack initiation.

4.1.1 Specimen geometry and preparation

Systematic experimental investigation of the thermal response due to high-power ultrasound excitation requires appropriate test specimens, which need to fulfil several functions regarding the planned fatigue testing but also vibrational properties and various measurement approaches. The following requirements need to be met by the specimen:

- Size large enough to resolve a sufficient number of sound wave lengths
- Plate thickness large enough to simulate heavy load bearing members
- Size small enough to allow for series production by CNC machining and manual handling
- Strong localisation of fatigue damage to ensure a reproducible crack location
- Simple load application and quick installation when mounting in a testing machine
- Capability to withstand high static loads allowing for prestressing the crack region

All of the above listed demands can be satisfied by a rather simple plate geometry, which has been developed by researchers at the Institute of Steel Structures, Braunschweig University, Germany [Meh03]. The shape and detailed dimensions of the specimen are depicted in Figure 4.1. Two holes with a diameter of 50 mm allow to mount the plate via pin connections into customised adapters, which in turn can be clamped by the hydraulic grips of available testing machines.

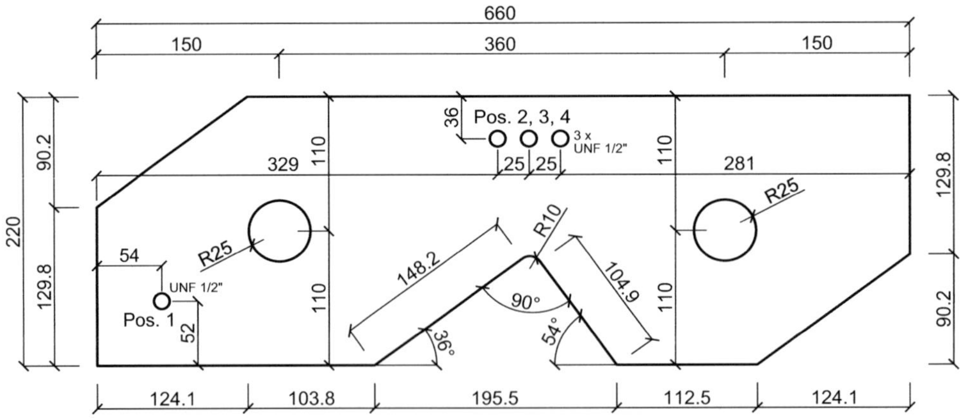

Figure 4.1: Specimen geometry with coupling positions (plate thickness $h = 6 \ldots 14\,\text{mm}$).

The deep necking of the cross section leads to a geometrical stress concentration at the circular fillet. Fatigue crack initiation will occur at this location, and the path of crack propagation is expected to develop approximately perpendicular to the plate's longitudinal axis. It is unlikely that the macroscopic crack initiation occurs along the whole plate thickness at once. Typically, the crack initiates at one of the fillet edges first and propagates with a curved crack front. At this stage the defect is characterised by the crack depth a and the crack length c (compare Fig. 2.6). It is not known a priori at which value of a the edge flaw turns into a through-thickness crack with depths a being approximately the same at both sides of the plate. The presence of an ideal through-thickness crack with $c = h$ is essential for efficient prestressing of the crack faces and advantageous for a clear assignment of mode shapes by means of laser vibrometry.

In order to investigate the influences of the plate thickness h and the crack depth a, both parameters are varied. Five different plate thicknesses $h = 6, 8, 10, 12, 14\,\text{mm}$ are investigated, and the nominal crack depths $a_{\text{nom}} = 1, 3, 6, 12, 24, 40\,\text{mm}$ are considered for each thickness. Additional plates without cracks are produced as reference samples. Focusing on the medium thickness of 10 mm, another ten plates have been manufactured in order to study the natural scattering of the cracks and the thermal response. Furthermore, several specimens might be damaged by prestressing or long-term excitation.

The plate specimens should represent steel components as they can be found in real structures. Metal fatigue is generally a problem with light-weight steel structures, which are dynamically stressed by wind, traffic loads or complex phenomena such as (rain-)wind induced oscillations, etc. Usually, the components are not mis-dimensioned so that low cycle fatigue (LCF) failure can be excluded in the majority of cases. Instead, the focus is clearly on fatigue damages that show up as macroscopic cracks after more than 10^4 load cycles, which is commonly regarded as the lower limit of the high cycle fatigue (HCF) regime. However, three specimens are intended to be excessively loaded in order to generate significantly rougher crack faces due to severe plastic straining.

All specimens are made from European steel grade S355J2+N according to EN 10025-2. The chemical compositions of the five different steel slabs according to manufacturer ladle analyses are summarised in Table A.2. Mechanical properties such as yield stresses and tensile strengths are given in Table A.3, which presents the results from tensile tests besides of the values taken from the inspection certificates. Both tables, together with representative plots of the monotonic stress-strain curve can be found in Appendix A.2.

After CNC machining of the plates, enamelled copper wires with a diameter of 0.08 mm have been applied to the fillet, close to the edges, and to both sides of the specimen according to Figure 4.2a. The loop in the fillet is intended to act as a crack detection wire, which is electrically monitored during fatigue loading. In case of rupture due to macro crack initiation, the current number of cycles is documented by the controller software of the testing machine. The straight wires on both sides, which are glued at the nominal crack depth, are connected in series so that a machine stop is triggered as soon as the desired crack depth is achieved on one side of the specimen. Prepared like this, the raw specimens are ready to be fatigue loaded in the next step.

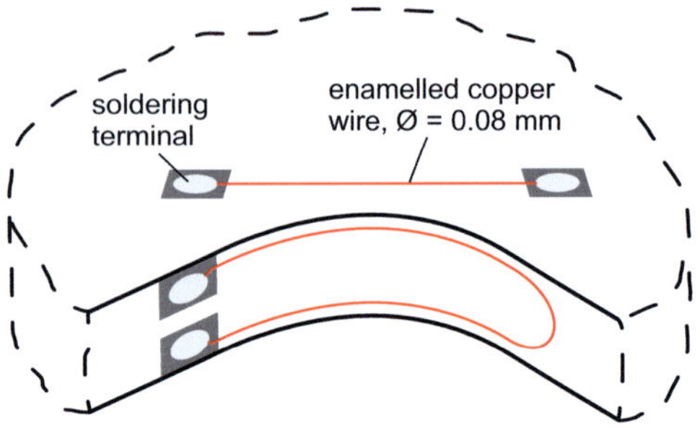

(a) Scheme of wire installation, crack depth detection wires applied to both sides

(b) Realised application of copper wires, 6 mm plate thickness, 3 mm distance to groove base

Figure 4.2: Specimen notch detail with attached crack detection wires and soldering terminals.

4.1.2 Fatigue loading of steel plate specimens

The aim of fatigue loading is to provoke crack initiation at the fillet radius of each plate specimen within a certain number of load cycles. Since the crack face morphology and especially its roughness will be influenced by the extent of cyclic plastic deformation at the notch and the crack front, the load level is understood as an important parameter that should be carefully chosen. Of course, the fatigue life is characterised by a significant natural scatter that cannot be overcome. For three of 39 specimens a target number of cycles to macroscopic crack initiation $N_c = 2500$ is chosen. All other plates are intended to show their cracks after 250000 cycles. To prevent any buckling of the plates, a fixed load ratio $R = F_{min}/F_{max} = 0.1$ is selected for high cycle fatigue as well as for low cycle fatigue loading throughout this work.

Since the present steel sample with the specific notch geometry cannot be categorised according to any catalogue of standard notch details, the nominal stress approach is not suitable for establishing a relation between loading and sustained cycles. In order to estimate the appropriate LCF and HCF load ranges defined by the minimum and the maximum machine forces F_{min} and F_{max}, the local strain approach is used.

The underlying idea of the local strain approach is the assumption that a small material element located at the notch of a complex specimen shows the same cyclic material behaviour as a smooth uniaxially loaded tensile specimen as long as the local elastic-plastic loading is the same. The requisites for a lifetime estimation with the local strain approach according to Naubereit and Weihert [NW99] are:

- Load function
- Cyclic material parameters
- Damage accumulation

Considering a simple load spectrum with $R = 0.1$, the purely elastic notch stresses under maximum and minimum machine force can be determined by means of a static finite element simulation of the specimen. A solid model of the 10 mm thick plate is shown in Figure 4.3. Note that this model consists of only one volume that was automatically meshed using tetrahedron elements with quadratic shape functions (ANSYS element SOLID92). A two-step mesh refinement of the notch region ensures a high resolution of the maximum notch stresses.

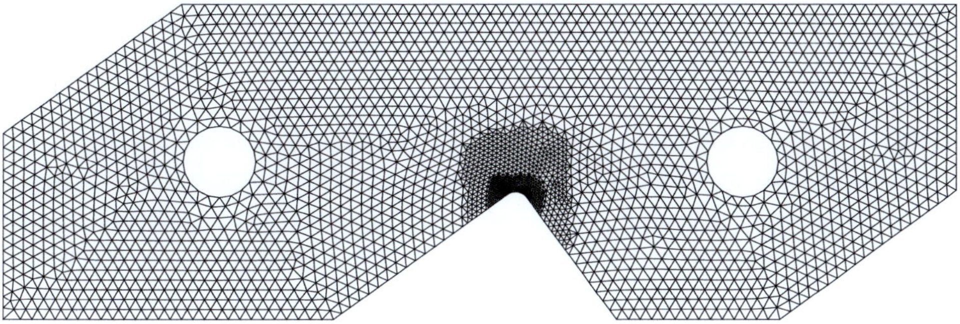

Figure 4.3: Finite element discretisation of the steel plate specimen for notch stress analysis.

The plate modelling was complemented by two short bolts equipped with surface-to-surface contact elements with friction coefficients set to 0.1 (SOLID92 and CONTA174, not shown). The corresponding faces of the holes were defined as target surfaces (TARGE170) so that a machine force of 1 kN can be realistically applied by distributing the unit force over the outer nodes of one of the bolt snag while the other one is restrained. Figure 4.4a shows the refined mesh of the fillet region, and Figure 4.4b depicts a contour plot of the first principal stress σ_1. Obviously, the maximum notch stress $\sigma_1 = 6.63\,\text{MPa/kN})$ occurs at half the thickness due to a stronger restraint of the transversal strain near the mid-surface of the plate. The maximum stress at the fillet edges is slightly lower ($\sigma_1 = 6.32\,\text{MPa/kN}$). The stress difference caused by the transition from plane stress conditions at the free plate surface to plane strain conditions at the inner plate regions tends to be more pronounced in case of thicker plates.

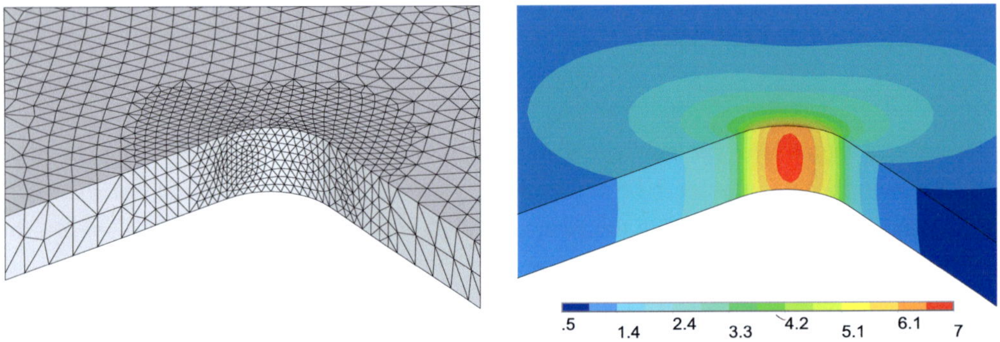

(a) Mesh refinement of the notch region (b) Notch stress σ_1 per 1 kN force

Figure 4.4: Finite element mesh detail of the notch region and distribution of the principal stress σ_1 due to a tensile force of 1 kN.

The stress concentration factor K_t is defined as the ratio of notch stress and nominal stress, which is calculated using elementary bending theory, considering a point load acting orthogonally on the reduced plate cross section indicated by the path A-B in Figure 4.5a. Analysis of the numerically calculated longitudinal stress σ_x, the principal stress σ_1 and the nominal stress along the path, as plotted in Figure 4.5b, clarifies the effect of the cross-sectional tapering on the local notch stress. K_t is found to be 2.72, based on the slightly higher notch stress at half thickness.

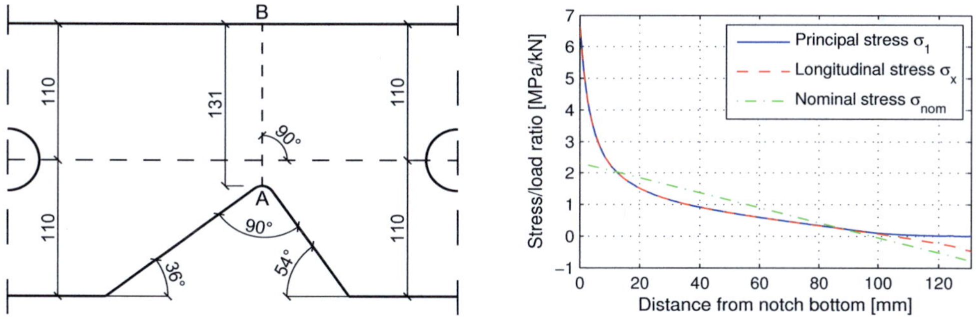

(a) Shortest path A-B orthogonal to the specimen's longitudinal axis

(b) Stresses σ_1 and σ_x at half thickness along the path A-B with respect to the applied load

Figure 4.5: Path definition and related stresses along the tapered plate cross section.

The previous analysis is based on a purely elastic calculation. In case of fatigue loading, cyclic elastic-plastic strains will occur at stress concentrations, even if the static yield stress is not exceeded. The local strain approach is based on the assessment of the actual strain amplitude, which is essential for estimation of the fatigue life.

Commonly, the stress-strain hysteresis shown in Figure 4.6a is decomposed in its elastic and plastic components. The total strain amplitude ε_a is then given by:

$$\varepsilon_a = \varepsilon_{a,e} + \varepsilon_{a,p} = \frac{\Delta\varepsilon}{2} = \frac{\Delta\varepsilon_e}{2} + \frac{\Delta\varepsilon_p}{2} = \frac{\Delta\sigma}{2E} + \frac{\Delta\varepsilon_p}{2} \tag{4.1}$$

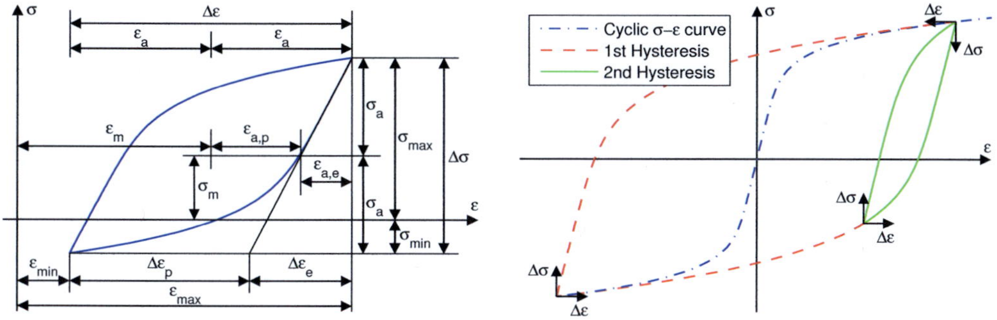

(a) Parameters of a hysteresis loop

(b) Masing behaviour of low-carbon steel

Figure 4.6: Parameters of a closed stress-strain hysteresis loop and Masing behaviour.

In the case of constant load amplitudes with zero mean stress the turning points of any possible stabilised hysteresis loop constitute the cyclic stress-strain curve. It provides the general relationship between stress amplitude σ_a and strain amplitude ε_a. Mathematically, the cyclic stress-strain curve is commonly approximated by a Ramberg-Osgood power law that introduces two material constants K' and n' [Hai02]:

$$\varepsilon_\mathrm{a} = \varepsilon_\mathrm{a,e} + \varepsilon_\mathrm{a,p} = \frac{\sigma_\mathrm{a}}{E} + \left(\frac{\sigma_\mathrm{a}}{K'}\right)^{1/n'} \tag{4.2}$$

Furthermore, the cyclic material behaviour of low-carbon steel is characterised by the validity of Masing's hypothesis. It states that any stabilised hysteresis loop located in the stress-strain space follows the shape of the cyclic stress-strain curve scaled by a factor of two. Thus, according to [Rad03], a single branch of a hysteresis loop is described through:

$$\Delta\varepsilon = 2\,\varepsilon_\mathrm{a} = \frac{\Delta\sigma}{E} + 2\left(\frac{\Delta\sigma}{2K'}\right)^{1/n'} \tag{4.3}$$

For historical reasons the concept of strain-life curves in its original form provides the relation between the total strain amplitude and the number of reversals to failure $2N_\mathrm{f}$ instead of cycles to failure N_f. The fundamental works of Coffin, Manson and Morrow prove that the strain-life curves for metals can be approximated by an additive composition of two straight lines in a log-log graph, which can be mathematically described by four material parameters [Hai02, Rad03]. The first line represents the material's resistance in the high cycle fatigue regime. Equation 4.4 is known as the Basquin law, which introduces the fatigue strength coefficient σ'_f and the fatigue strength exponent or Basquin exponent b.

$$\frac{\Delta\sigma}{2} = \sigma'_\mathrm{f}\,(2N_\mathrm{f})^{b} \tag{4.4}$$

A second equation relates the plastic strain amplitudes and the number of reversals to failure. The Manson-Coffin law accurately describes the material's cyclic resistance in the low cycle fatigue regime. Again, two material parameters are introduced: The fatigue ductility coefficient ε'_f and the fatigue ductility exponent or Coffin exponent c.

$$\frac{\Delta\varepsilon_\mathrm{p}}{2} = \varepsilon'_\mathrm{f}\,(2N_\mathrm{f})^{c} \tag{4.5}$$

Plugging Equations (4.4) and (4.5) into (4.1) yields the equation of the strain-life curve in its original form, which can be rearranged according to [Hai02] so that ε is a function of N_f.

$$\varepsilon_\mathrm{a} = \frac{\sigma_\mathrm{f}'}{E}(2N_\mathrm{f})^b + \varepsilon_\mathrm{f}'(2N_\mathrm{f})^c = 2^b \cdot \frac{\sigma_\mathrm{f}'}{E} \cdot N_\mathrm{f}^b + 2^c \cdot \varepsilon_\mathrm{f}' \cdot N_\mathrm{f}^c \tag{4.6}$$

Due to the compatibility of the cyclic stress-strain curve and the strain-life curve, the coefficients K' and n' can be calculated if σ_f', ε_f', b and c are known. The following relations can be derived by comparison of the elastic and plastic parts of equations (4.2) and (4.6) assuming the cyclic stress-strain curve being independent of N_f [NW99]:

$$n' = \frac{b}{c} \quad , \qquad K' = \frac{\sigma_\mathrm{f}'}{\left(\varepsilon_\mathrm{f}'\right)^{n'}} \tag{4.7}$$

Since the introduced parameters are not known for the delivered steel slabs, the cyclic stress-strain curve and the strain-life curve need to be estimated. Meggiolaro and Castro presented an extensive statistical evaluation of ten different Coffin-Manson parameter estimates based on experimental results of more than 700 steels. The overall accuracy of all models was found to be rather poor. Good predictions could be achieved using the methods proposed by Roessle-Fatemi, Bäumel-Seeger and Muralidharan-Manson [MC04a]. Another comparison was conducted by Haibach [Hai02] based on the database of Boller and Seeger [BS87]. According to this work, the uniform material law (UML) derived by Bäumel and Seeger results in a more accurate prediction than Manson's method of universal slopes and the modified universal slopes equation by Muralidharan and Manson. Thus, strain-life curve parameters are estimated using the uniform material law for low-carbon steels according to Table 4.1.

Table 4.1: Uniform material law for unalloyed and low-alloyed steels [Rad03].

Type of metal	σ_f' [MPa]	b [−]	ε_f' [−]	c [−]	n' [−]	K' [MPa]
Non/low-alloy steels	$1.50 \cdot R_\mathrm{m}$	-0.087	$0.59 \cdot \Psi$	-0.58	0.15	$1.65 \cdot R_\mathrm{m}$
Aluminium/titanium alloys	$1.67 \cdot R_\mathrm{m}$	-0.095	0.35	-0.69	0.11	$1.61 \cdot R_\mathrm{m}$

$\Psi = 1$ for $R_\mathrm{m}/E \leq 3 \times 10^{-3}$, $\Psi = 1.375 - 125 \cdot R_\mathrm{m}/E$ for $R_\mathrm{m}/E > 3 \times 10^{-3}$

Besides of evaluating the uniform material law (UML) based on the actual material strengths, independent values of all six parameters according to Manson, Coffin and Morrow can be found in the literature. Boller and Seeger published a huge compilation of processed fatigue test results including the model parameters [BS87]. The database contains eight different parameter sets, which apply to the present steel grade S355J2+N (former designations S355J2G3, St 52-3). Four of them are included in Table 4.2 to get an idea of the natural scattering.

Table 4.2: Comparison of Manson-Coffin parameters based on UML estimates and test results.

Data source	R_{eH} [MPa]	R_m [MPa]	σ'_f [MPa]	b [−]	ε'_f [−]	c [−]	n' [−]	K' [MPa]
Measured, averaged	384	534	802	-0.087	0.590	-0.580	0.150	882
Certificates, averaged	394	558	821	-0.087	0.590	-0.580	0.150	904
Cert., avg. 6 and 10 mm	425	568	851	-0.087	0.590	-0.580	0.150	936
Cert., avg. 8, 12, 14 mm	373	534	802	-0.087	0.590	-0.580	0.150	882
St 52-3 [BS87], p. 158	400	597	1193	-0.110	0.660	-0.553	0.185	1228
St 52-3 [BS87], p. 162	400	597	1293	-0.115	0.495	-0.515	0.207	1337
St 52 [BS87], p. 166	420	549	980	-0.100	0.495	-0.541	0.185	1109
St 52-3 [BS87], p. 169	378	–	733	-0.078	0.305	-0.569	0.139	871

Figure 4.7a depicts the cyclic stress-strain curves derived for the present steel melts corresponding to the five different slab thicknesses. The additive composition of the strain-life curve based on Basquin's law and the Manson-Coffin law is exemplarily shown in Figure 4.7b for the 10 mm thick plates. Here, the parameter estimations are based on R_m taken from the inspection certificates.

The relationship between the actual elasto-plastic stress-strain state, specified by σ and ε, and the hypothetical elastic stress level can be established by using Neuber's rule. Either the stress concentration factor K_t and the nominal stress amplitude $\sigma_{a,nom}$ are known or the Hookean notch stress amplitude $\sigma_{a,H} = \sigma_{a,1}$ is calculated by means of a purely elastic finite element model as it was done for the plate specimen [NW99]:

$$\frac{\sigma_{a,H}^2}{E} = \frac{(\sigma_{a,nom} \cdot K_t)^2}{E} = \sigma_a \cdot \varepsilon_a \tag{4.8}$$

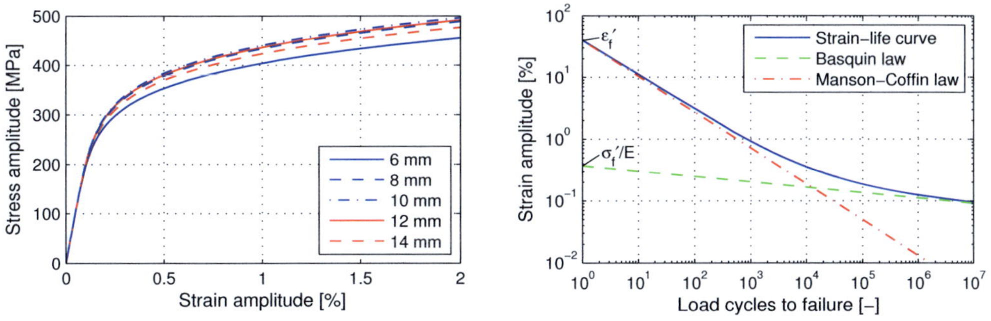

(a) Stabilised cyclic stress-strain curves; h = var.

(b) Strain-life curve for 10 mm thick base material

Figure 4.7: Stabilised cyclic stress-strain curves and composition of a strain-life curve based on the uniform material law.

Substituting ε_a with the elastic and plastic strain amplitude in equation (4.2) leads to the following relation, which can only be solved iteratively for σ_a.

$$\sigma_{a,H}^2 = \sigma_a^2 + E\,\sigma_a \left(\frac{\sigma_a}{K'}\right)^{1/n'} \tag{4.9}$$

Inserting σ_a into the expression for the cyclic stress-strain curve (equation (4.2)) yields an approximation of the elasto-plastic strain amplitude ε_a. In case of alternating loads ($R = -1$) the strain-life curve (4.6) could be used directly to predict the number of cycles to failure. Mean stresses might be taken into account by the modified strain-life curve

$$\varepsilon_a = \frac{\sigma_f' - \sigma_m}{E} (2N_f)^b + \varepsilon_f' (2N_f)^c \tag{4.10}$$

or by evaluation of a damage parameter life curve [Hai02]. The parameter according to Smith, Watson and Topper [SWT70] incorporates mean stresses if greater than zero.

$$P_{SWT} = \sqrt{\sigma_{max} \cdot \varepsilon_a \cdot E} \quad , \qquad \sigma_{max} = \begin{cases} \sigma_a & \text{for } \sigma_m \leq 0 \\ \sigma_a + \sigma_m & \text{for } \sigma_m > 0 \end{cases} \tag{4.11}$$

Its definition is to be interpreted as a new measure of stress but it also gives the damage parameter life curve for $R = -1$ at the same time. Inserting Hooke's law $\sigma_a = E \cdot \varepsilon_{a,e}$, making use of the Basquin law and the expression for the total strain amplitude ε_a given by equations (4.1), (4.4) and (4.6) finally yields the P_{SWT}-life curve [Hai02, Rad03].

$$P_{\mathrm{SWT}} = \sqrt{\sigma_{\mathrm{f}}^{\prime 2} \cdot (2N_{\mathrm{f}})^{2b} + \sigma_{\mathrm{f}}' \cdot \varepsilon_{\mathrm{f}}' \cdot E \cdot (2N_{\mathrm{f}})^{(b+c)}} \quad (4.12)$$

Figure 4.8a depicts the P_{SWT}-life curve regarding the different R_{m} values of the delivered base materials. In case of high cycle fatigue loading the target number of cycles to crack initiation N_{c} is set to 250000 and for low cycle fatigue 2500 cycles are chosen. In order to apply the local strain approach, it has to be assumed that N_{c} corresponds to the failure load cycles N_{f} found for the fatigue tests the UML is based on. This is a conservative assumption since the load cycles spent for crack propagation are neglected. The required P_{SWT} load is found from the damage parameter Wöhler curve, which is solely based on the previously introduced Coffin-Manson parameters. To estimate the appropriate machine forces, the P_{SWT} loading was calculated for varying maximum forces assuming $R = 0.1$ and considering the elastic stress/load ratio that applies to the fillet edge. The maximum force F_{max} that leads to the desired number of cycles is then looked up from curves like those shown in Figure 4.8b. Based on the six parameters σ_{f}', b, $\varepsilon_{\mathrm{f}}'$, c, K' and n', the machine force F_{max} is calculated for both LCF and HCF target values of N_{f}. One of the major uncertainties involved is the mean stress. Depending on the actual load control mode, which is not clear in case of a real structure's notch, either cyclic creep in load control or mean stress relaxation in displacement control may occur. The limits of P_{SWT} for both cases, taking the mean stress into account ($\sigma_{\mathrm{max}} = \sigma_{\mathrm{a}} + \sigma_{\mathrm{m}}$) and neglecting it ($\sigma_{\mathrm{max}} = \sigma_{\mathrm{a}}$), are considered. Finally, Table 4.3 presents the predicted machine forces F_{max}.

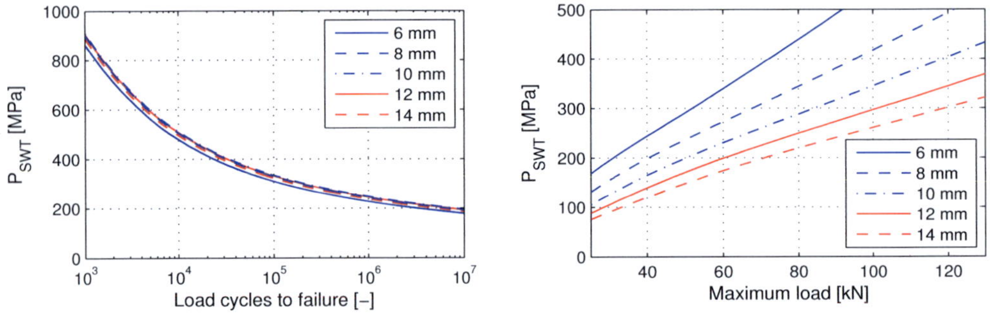

(a) P_{SWT}-life curve; $h = \text{var.}$ (b) P_{SWT} with respect to the maximum load; $h = \text{var.}$

Figure 4.8: Damage parameter Wöhler curve and relation between P_{SWT} and the maximum force.

Table 4.3: Comparison of the predicted maximum machine forces per unit thickness based on different Manson-Coffin parameters and the load ratio $R = 0.1$.

Data source	LCF loading: $N_f = 2.5 \times 10^3$			HCF loading: $N_f = 0.25 \times 10^6$		
	P_{SWT}	F_{max} [kN/mm]		P_{SWT}	F_{max} [kN/mm]	
	[MPa]	$\sigma_a + \sigma_m$	σ_a	[MPa]	$\sigma_a + \sigma_m$	σ_a
Measured, averaged	696	21.6	24.5	285	8.0	10.0
Certificates, averaged	707	22.0	24.9	291	8.2	10.2
Cert., avg. 6 and 10 mm	724	22.4	25.5	301	8.4	10.6
Cert., avg. 8, 12, 14 mm	696	21.6	24.5	285	8.0	10.0
St 52-3 [BS87], p. 158	896	27.2	31.5	327	9.0	11.5
St 52-3 [BS87], p. 162	930	27.9	32.7	341	9.5	12.0
St 52 [BS87], p. 166	780	23.7	27.4	304	8.5	10.7
St 52-3 [BS87], p. 169	576	17.8	19.6	281	7.7	9.9

Based on these results, HCF loading was started between 8 and 9 kN per millimetre thickness, which resulted in a runout. Later on, after several plates were cracked, 12 kN/mm have been applied to most of the specimens in order to statistically approach the target number of cycles. Figure 4.9 shows one of the plate specimens mounted in a resonant testing machine, which is perfectly suited for HCF loading since load frequencies of about 40 to 45 Hz could be achieved. The bolted joints and the two independent electric circuits connected to the crack detection wires are clearly visible. LCF loading was performed using a servo-hydraulic testing machine with an identical setup regarding load application and electric crack detection. Here, 28.8 kN/mm was applied after one specimen has been loaded with stepwise increasing forces.

Tables 4.4 to 4.8 summarise the applied loading for all specimens and the number of cycles to macroscopic crack initiation N_c. For the sake of completeness, the total number of cycles reached until the nominal crack depth was achieved at one side of the plate, which is denoted as N_a, is also given. Furthermore, the exact plate thickness and all relevant crack dimensions are listed. These include the actual crack depths a_A and a_B measured on both sides A and B of the plate, the crack lengths c in thickness direction as well as the crack mouth opening displacements CMOD. The values $CMOD_{50}$ are based on additional measurements at half the crack depth. Note that the mid crack opening was only determined in the case of a CMOD greater than 25 µm.

Table 4.4: Loading and crack geometry of 6 mm thick plate specimens.

Specimen	a_{nom} [mm]	Mode [–]	F_{max} [kN]	F_{min} [kN]	R [–]	N_c [–]	N_a [–]	h_{act} [mm]	a_A [mm]	a_B [mm]	c [mm]	CMOD [µm]	$CMOD_{50}$ [µm]
06-01	1	HCF	50	5	0.1	runout[1]	–	5.93	–	–	–	–	–
06-01	1	HCF	60	6	0.1	890 394	907 504	5.93	0	1.8	2.3	2…5	–[2]
06-03	3	HCF	58	5.8	0.1	194 748	218 375	5.93	0	3	4	5	–[2]
06-06	6	HCF	58	5.8	0.1	366 113	405 245	5.95	4	6	6 $(=h)$	2…6	–[2]
06-12	12	HCF	58	5.8	0.1	183 612	223 494	5.95	12.2	11.9	6 $(=h)$	10	–[2]
06-24	24	HCF	60	6	0.1	110 948	150 231	5.93	24	23.6	6 $(=h)$	42	25
06-40	40	HCF	65	6.5	0.1	78 348	147 166	5.96	39.4	39.6	6 $(=h)$	690	500
06-00	–	–	–	–	–	–	–	6.01	–	–	–	–	–

[1]Specimen passed 2×10^6 cycles
[2]$CMOD_{50}$ not measured

Table 4.5: Loading and crack geometry of 8 mm thick plate specimens.

Specimen	a_{nom} [mm]	Mode [–]	F_{max} [kN]	F_{min} [kN]	R [–]	N_c [–]	N_a [–]	h_{act} [mm]	a_A [mm]	a_B [mm]	c [mm]	CMOD [µm]	$CMOD_{50}$ [µm]
08-01	1	HCF	96	9.6	0.1	64 871	64 993	7.75	0	1	5.7	2…4	–[1]
08-03	3	HCF	90	9	0.1	259 186	285 090	7.79	3.6	0	4.7	1…3	–[1]
08-06	6	HCF	90	9	0.1	148 929	182 152	7.79	6.2	2.9	8 $(=h)$	5	–[1]
08-12	12	HCF	90	9	0.1	139 900	158 005	7.77	12	12.1	8 $(=h)$	5	–[1]
08-24	24	HCF	86	8.6	0.1	168 218	200 324	7.7	25.1	24.7	8 $(=h)$	25	8
08-40	40	HCF	86	8.6	0.1	81 997	127 924	7.75	40.2	40.7	8 $(=h)$	920	710
08-00	–	–	–	–	–	–	–	7.79	–	–	–	–	–

[1]$CMOD_{50}$ not measured

Table 4.6: Loading and crack geometry of 10 mm thick plate specimens.

Specimen	a_{nom} [mm]	Mode [–]	F_{max} [kN]	F_{min} [kN]	R [–]	N_c [–]	N_a [–]	h_{act} [mm]	a_A [mm]	a_B [mm]	c [mm]	CMOD [µm]	$CMOD_{50}$ [µm]
10-01-I	1	HCF	120	12	0.1	1 031 474	1 195 751	9.98	1.2	0	7.7	< 2	–[2]
10-01-II	1	HCF	120	12	0.1	290 139	291 255	9.91	1.2	0	6.2	< 2	–[2]
10-03	3	HCF	110	11	0.1	runout[1]	–	10.23	–	–	–	–	–[2]
10-03	3	HCF	125	12.5	0.1	374 877	378 901	10.23	0	3	7.1	3	–[2]
10-06-I	6	HCF	120	12	0.1	866 410	905 520	9.97	0	6.1	7.8	8	–[2]
10-06-II	6	HCF	120	12	0.1	223 576	240 451	9.94	6.2	2.4	10 (= h)	3	–[2]
10-12-I	12	HCF	120	12	0.1	287 001	304 811	9.98	12.1	12.2	10 (= h)	8	–[2]
10-12-II	12	HCF	120	12	0.1	230 436	283 296	9.99	13.4	12.9	10 (= h)	8	–[2]
10-12-III	12	HCF	120	12	0.1	176 668	220 191	9.96	10.7	12.7	10 (= h)	14	–[2]
10-12-IV	12	LCF	288	28.8	0.1	2435	3660	9.96	12	12	10 (= h)	360	320
10-12-V	12	LCF	288	28.8	0.1	3100	4660	9.97	12.2	9.6	10 (= h)	440	300
10-24-I	24	HCF	120	12	0.1	286 628	348 268	9.96	24.1	23.9	10 (= h)	35	18
10-24-II	24	HCF	120	12	0.1	405 515	426 292	9.95	24.2	24.3	10 (= h)	25	10
10-24-III	24	HCF	120	12	0.1	505 789	554 038	9.96	24.3	24.3	10 (= h)	40	15
10-24-IV	24	LCF	var[3]	var[3]	0.1	12 000	13 500	10.00	23.8	24.1	10 (= h)	1900	1700
10-40	40	HCF	120	12	0.1	246 180	278 677	9.98	42	41.1	10 (= h)	780	620
10-00	–	–	–	–	–	–	–	10.11	–	–	–	–	–

[1]Specimen passed 2×10^6 cycles
[2]$CMOD_{50}$ not measured
[3]Loading was increased stepwise after 2000 cycles, F_{max}: 133, 155, 177, 200, 222, 244, 266 kN, $F_{min} = 0.1\,F_{max}$

Table 4.7: Loading and crack geometry of 12 mm thick plate specimens.

Specimen	a_{nom} [mm]	Mode [–]	F_{max} [kN]	F_{min} [kN]	R [–]	N_c [–]	N_a [–]	h_{act} [mm]	a_A [mm]	a_B [mm]	c [mm]	CMOD [μm]	$CMOD_{50}$ [μm]
12-01	1	HCF	145	14.5	0.1	266173	267312	12.18	0	1	8.8	< 2	–[1]
12-03	3	HCF	145	14.5	0.1	190335	192368	12	0	3	11.4	< 2	–[1]
12-06	6	HCF	145	14.5	0.1	273965	285217	12.08	6.9	2.1	12 ($=h$)	2...3	–[1]
12-12	12	HCF	145	14.5	0.1	226928	274216	12.09	12.1	12	12 ($=h$)	2...5	–[1]
12-24	24	HCF	145	14.5	0.1	109726	310111	12.14	24.2	24.1	12 ($=h$)	23	6
12-40	40	HCF	145	14.5	0.1	120956	157187	12.05	39.5	39.5	12 ($=h$)	580	430
12-00	–	–	–	–	–	–	–	12.25	–	–	–	–	–

[1]$CMOD_{50}$ not measured

Table 4.8: Loading and crack geometry of 14 mm thick plate specimens.

Specimen	a_{nom} [mm]	Mode [–]	F_{max} [kN]	F_{min} [kN]	R [–]	N_c [–]	N_a [–]	h_{act} [mm]	a_A [mm]	a_B [mm]	c [mm]	CMOD [μm]	$CMOD_{50}$ [μm]
14-01	1	HCF	168	16.8	0.1	216006	217775	13.75	1	0	5.5	< 2	–[1]
14-03	3	HCF	160	16	0.1	438915	465029	13.72	0	3.3	5.5	< 2	–[1]
14-06	6	HCF	160	16	0.1	189407	211427	13.7	0	8.3	12.6	1...3	–[1]
14-12	12	HCF	168	16.8	0.1	163939	173577	13.72	12.1	12.1	14 ($=h$)	3...4	–[1]
14-24	24	HCF	160	16	0.1	405452	430221	13.72	23.7	24.1	14 ($=h$)	34	15
14-40	40	HCF	160	16	0.1	282620	306251	13.74	34.9	35.4[2]	14 ($=h$)	410	360
14-00	–	–	–	–	–	–	–	13.86	–	–	–	–	–

[1]$CMOD_{50}$ not measured
[2]a_{nom} possibly not achieved due to premature rupture of the crack detection wire

(a) Testing machine, hydraulic grips, clamped mounting forks and plate specimen

(b) Plate specimen constrained by pin-joints, connection cables for crack dectection wires

Figure 4.9: Test setup for fatigue loading of steel plate specimens using a resonant testing machine SincoTec Power Swing 1000 MOT.

After all specimens have been fatigue loaded, the mean value of N_c for the majority of specimens (20 of 36) loaded with 12 kN/mm is found to be 309 000 cycles. As stated in [Rad03], only logarithmised failure cycles approximately follow a normal distribution. The expected specimen life $N_{c,50\,\%}$ that corresponds to a 50 % probability of survival P_s is given by Issler [IRH97]:

$$\log N_{c,50\,\%} = \frac{1}{m} \sum_{i=1}^{m} \log N_{c,i} \tag{4.13}$$

Here, $N_{c,i}$ are the m individual numbers of cycles to macroscopic cracking. Note that equation (4.13) holds only for $P_s = 50\,\%$ presuming a Gaussian distribution of $\log N_{c,i}$. The resulting value of $N_{c,50\,\%}$ is 250 430, which is (rather accidentally) exactly the target number of cycles. The overall mean HCF load F_{max} was 11.5 kN/mm. Here, the mean value of N_c is 291 500 and $N_{c,50\,\%}$ is found to be 235 700 load cycles.

It can be concluded by comparison of the applied loading and the predicted loading in Table 4.3 that all of the parameter sets given in Table 4.2 and especially the ones based on the measured material strengths lead to a conservative estimation of the fatigue life. Reasons for that can be found in the approach itself. The predicted load or fatigue life is based on fatigue tests of smooth specimens with a uniform distribution of stress and strain. The estimates are transferred to notched specimens with a high stress gradient. These exhibit the same strain only in a thin layer near the surface of the notch root. According to Bílý [Bil93] this is one possible reason for a delayed crack initiation. Furthermore, in larger bodies a certain crack growth must be included in the evaluation of the fatigue life. Otherwise the dimensions of the smooth specimens (which are tested until rupture) and the notch region of the actual specimens do not have a similar scale and thus cannot be compared.

The best prediction of the necessary HCF loading was achieved when using the Manson-Coffin parameters given by Boller and Seeger ([BS87], p. 162) and neglecting the mean stress for evaluation of the damage parameter. This load exactly matched the applied maximum machine force of 12 kN/mm that led to the target number of cycles. In case of LCF loading, the same set of parameters performed best in prediction of the actual value $F_{\max} = 28.8\,\mathrm{kN/mm}$.

4.1.3 High-emissivity coating

Prior to any thermographic testing, the inspected surface of each specimen was coated with a matt grey graphite layer (conductive paint Graphit 33, CRC Industries/Kontakt Chemie) to equalise any surface inhomogeneities. The graphite spray was used earlier by Medgenberg [Med08], who found that it provides the best trade-off between emissivity, thermal conductivity and applicability compared to different other sprays and lacquers. With an emissivity $\epsilon \approx 0.87$ the paint is well suited for detecting smallest temperature changes and therefore applied to all specimens tested throughout this work. According to the manufacturer, a typical value for the coating thickness is 20 µm [CRC04]. Although the actual coating thickness was not measured, a possible influence of the coating on the thermal response is considered negligible since extremely low attenuation ratios are estimated in Section 3.3.1 based on a thermal diffusivity value found in the literature.

4.2 Test equipment and setup for vibrothermography

This section introduces the used equipment for thermosonic testing including the ultrasound generators and the infrared camera. The stationary test setup for standardised investigation of plate specimens is described in detail. Moreover, different portable excitation devices are presented.

4.2.1 Test stand for plate specimens

The most experiments conducted within this work are carried out on cracked plate specimens whose manufacture has been described in Section 4.1. For basic thermographic investigations a test stand has been developed that allows for secure and reproducible retaining of the specimens. The setup is based on an adjustable frame system with fork-like aluminium adapters mounted onto a horizontal transom. One of those is shown in Figure 4.10a. Two holders support the specimen and keep it in place by means of threaded rods, which can be adjusted to fit various plate thicknesses. To prevent a point-like clamping of the specimens, the rods are equipped with circular plates of 20 mm diameter. A very smooth support without any metal to metal contacts is realised by placing soft foam material between the steel specimen and its holders at each supporting/clamping location (Fig. 4.10b).

(a) Fork-like aluminium holder with threaded rods

(b) Test stand for vibrothermography with the Branson ultrasound transducer screw-coupled to a plate specimen at coupling position 1

Figure 4.10: Stationary test setup for vibrothermographic inspection of steel plates.

4.2.2 Ultrasound generators

Two distinct systems were available for the generation/amplification of ultrasound and insonification of steel components. The one that is used for pure thermosonic testing consists of a modified Branson 2.2 kW ultrasound generator and transducer. In the past, ultrasonic welders have been utilised by numerous researchers since they provide high vibrational power and are easily computerised, which is essential for reproducible and automated excitation of structures. A major drawback is the narrow frequency bandwidth. Weld generators typically work at fixed frequencies of 20, 30, 35 or 40 kHz and have limited tuning capabilities. This is sufficient for welding since all parts of the system such as converter, booster and sonotrodes are designed to match a specified resonance frequency.

In order to meet the requirements of ultrasound excited thermography, system suppliers offer specialised devices. Figure 4.11 shows the Branson 2000b generator and the piezoelectric transducer. This system was modified by Automation Technology GmbH/edevis GmbH enabling an extended frequency bandwidth of 15 to 25 kHz. Furthermore, the transducer oscillation amplitude can be adjusted from 0 to 100 % in terms of relative power. Both features are essential if metal components are to be tested. In contrast to amplifier systems, the ultrasound generator is controlled by constant voltage signals, which are proportional to the relative oscillation amplitude and to the ultrasound frequency [Bra04]. The (sine) waveform itself cannot be specified in any way. All input signals are read internally with a sampling rate of 1 kS/s so that short burst excitations and smooth frequency sweeps are possible. In case of sudden jumps in frequency or a too high oscillation set point, the generator abortively stops operating and requires a reset prior to restart.

(a) Ultrasound generator Branson 2000b and converter Branson CR20 with customised National Instruments I/O connector block SCB-68

(b) Ultrasound transducer Branson CR20 with setscrew (1/2″ - 20 UNF)

Figure 4.11: Modified 2.2 kW Branson ultrasonic plastic welding system.

An alternative ultrasound system was provided by Siemens Corporate Technology München, Germany. The linear amplifier S250K, manufactured by TIB GmbH, offers analogue waveform amplification with an extremely low distortion [TIB10]. Its electrical power is only 250 W, about a tenth of the Branson system. Again, an ultrasound converter based on stacked piezo actuators is used to excite the structure under investigation. Figure 4.12 shows the equipment containing the amplifier, a transformer and the converter with a custom threaded adapter that allows for using the setscrews suitable for the Branson transducer.

Both ultrasound systems are controlled by means of a National Instruments data acquisition (DAQ) card PCI-6251 and customised LabVIEW programmes. Connections to the available generator/amplifier interfaces were realised using shielded I/O connector blocks SCB-68. Custom wiring included several optocouplers and voltage converters to enable relevant digital controls such as remote on/off and reset after overloading. Table 4.9 summarises the main features of the different ultrasound systems.

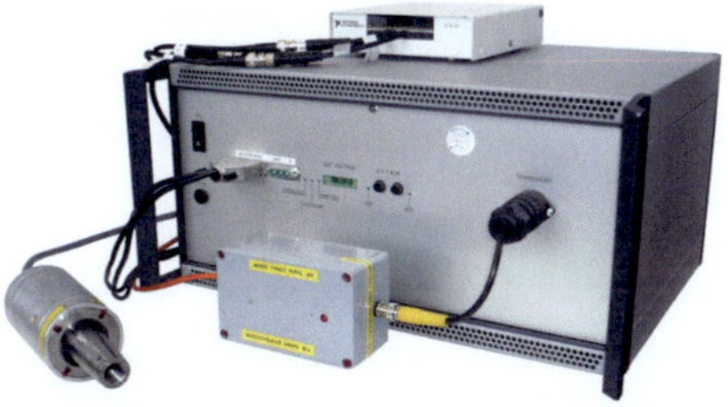

(a) Amplifier TIB S250K, transformer and converter TIB HS USW 1000 with customised National Instruments I/O connector block SCB-68

(b) Broadband transducer TIB HS USW 1000 with threaded adapter and setscrew (1/2″ - 20 UNF)

Figure 4.12: 250 W TIB linear amplifier and ultrasound transducer.

Table 4.9: Specifications of the used ultrasound systems.

Device	Electric power [W]	Amplitude range [%]	Frequency range [kHz]	Input signal [−]
Branson 2000b	2200	10...100	19.5...20.5	constant
Branson 2000b mod.	2200	0...100	15...25	constant
TIB S250K	250	0...100	15...100	waveform

4.2.3 Coupling methods

Direct screw-coupling of the ultrasound transducer (as shown in Fig. 4.10b) probably provides the best reproducibility of excitation since both compressive and tensile forces can be transmitted from the converter tip to the steel structure under investigation. Excessive hammering of the transducer should be eliminated in this way (compare [HHW$^+$07]). However, drilling holes and cutting threads are not acceptable measures when it comes to in-situ testing. In this case, portable excitation methods are required that are fully removable and do not cause any damage to the structure. Surely, this is a non-trivial task since any high-power transducer, which is hammering on the surface coating, is likely to leave signs of wear. Therefore, a soft coupling material such as aluminium duct tape placed between specimen and horn was proposed by Zweschper et al. [ZRDB04]. This might not only prevent wearing but also allows for impedance matching (compare Section 3.1.1).

The use of stationary excitation devices was demonstrated by numerous researchers working in the field. Commonly, the converter tip or the complete welding tool (including booster and sonotrode) is pressed against the sample by means of a pneumatic piston, where the system pressure is understood as a tuning parameter ranging from 68 kPa [CKR06] to 345 kPa [PBML08]. Very few publications contain detailed information on the effective compression force. Polimeno, Barden and Almond, for example, specify the preload between 50 N and 80 N [PA09, BACM06].

The author built up several prototypes of portable excitation devices, which have been tested on a cracked girder. Thermographic results are presented in Section 4.6.2. In the left and middle column, Figure 4.13 depicts three devices, which allow for pressing the whole weld tool, the converter and booster or only the plain converter against the specimen. The compression force is balanced in different ways and the tensile retention force transmitted back to the sample is carried by one or two electromagnets. On the right hand side, two adjustable mounting clamps are shown that fit a wide range of plate thicknesses.

The most complex ultrasound exciter consists of two tubular studs connected to magnets, a robust transom and the assembled system of converter, booster and sonotrode. The heavy steel transom clutches the flange of the booster and slides up and down within the tubes. However, the frame is flexurally rigid. In order to press the sonotrode against the steel member, a tie rod is screwed to each magnet inside the stud passing through the transom. The system can be prestressed in a controlled manner by means of a simple nut on a spring. Each electromagnet (IBS Magnet EM8, 24 V, 15 W) with a diameter of 80 mm is capable of bearing the force of 1.8 kN in normal direction, whereas this value is strongly dependent on the plate thickness. The manufacturer recommends a minimum plate thickness of 9 mm (ferritic steel), in order to achieve the nominal retention force [IBS09]. Concerning the ultrasound transmission through the resonance unit, it should be noted that the booster does not amplify or attenuate the vibration amplitude. Instead, the employed catenary sonotrode having an input diameter D_{in} of 40 mm and a tip diameter D_{tip} of 30 mm causes an amplitude magnification M_c, which ranges between the magnifications of a stepped and an exponential horn M_s and M_e, respectively [Bra05] (compare Fig. 3.1):

$$M_e = \frac{D_{\text{in}}}{D_{\text{tip}}} = 1.33 < M_c < M_s = \left(\frac{D_{\text{in}}}{D_{\text{tip}}}\right)^2 = 1.78 \tag{4.14}$$

Two alternative devices based on the same principle are shown in the middle of Figure 4.13. Either the booster or the converter is clamped and hold by a steel transom. A simple steel post acts as a counterpart at the opposite end. Thus, sufficient levelling is achieved to facilitate a flat contact between booster or converter and the structure. Again, a tie rod is screwed to an electromagnet and passes through the transom. If the screw is tightened against the spring, the transom acts as a two-span beam and preloads the booster or converter. Secure fixation is realised by a strong magnet (IBS Magnet EM15, 24 V, 37 W) with a diameter of 150 mm that carries up to 9 kN if the subjacent steel plate has a thickness greater than 17 mm [IBS09].

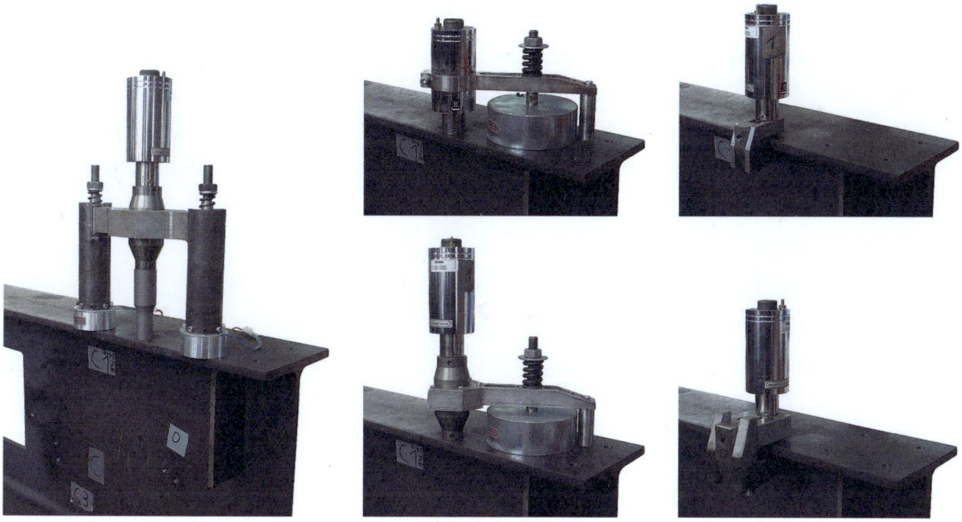

Figure 4.13: Portable coupling devices for in-situ ultrasound excitation.

The clamps are based on two parts, which are only held in place by fastening the bolt. In order to increase the local contact pressure and to enhance grip, both the jaws are milled to a smaller footprint and have a zig-zag contour. A thread hole in one part of the clamp allows to firmly attach the transducer.

4.2.4 Infrared camera

All thermographic investigations reported in this work have been carried out using a Flir Phoenix DTS infrared camera that features a 640×512 InSb focal plane array detector, which is sensitive in the wavelength range of 1.5 to 5 µm. The effective wavelength band is limited by a cold filter bandpass to the common mid wave infrared range of 3 to 5 µm. According to the manufacturer, the camera head enables thermal imaging with a noise-equivalent temperature difference (NETD) below 25 mK. To achieve this low noise level, the FPA quantum detector is cooled down to the boiling point of liquid nitrogen (≈ 77 K) by means of a Stirling closed cycle cooler [FLI05a]. Equipped with a Janos Technology MWIR Asio lens with a focal length of 50 mm and four interchangeable extension rings, a variety of camera distances can be realised. The typical lens transmission declared by the manufacturer is 92 % in the relevant 3 to 5 µm band [Jan07].

The camera control software enables access to advanced settings such as the detector integration time (typical values used: 0.4 to 2.0 µs) and the integration mode. Internal gain and offset values allow for exact adjustment of the detector response with respect to the incident radiation. These settings are essential in order to create customised temperature calibrations as described in Section 3.4.3. Digital camera counts (intensity or grey values) are output within a dynamic range of 14 bit. The camera provides an Ethernet interface in compliance to the GigE Vision standard and offers built-in synchronisation capabilities. Image capturing is done via Pleora's iPORT Universal IP Filter Driver using an Intel Pro/1000 gigabit Ethernet adapter. According to the camera specifications, frame rates up to 100 Hz can be achieved at full window size. Most of the infrared sequences have been recorded in a sub-window mode (192×300 px), enabling frame rates greater than 400 Hz.

Image capturing was done using proprietary software, which allows for complex capturing procedures (arbitrarily scheduled, looped, externally triggered, using varying frame rates and record durations, etc.). Acquired infrared streams can be comfortably handled and various image processing filters are applicable. Furthermore, the software enables to control an available blackbody radiator Mikron M245X6 so that the non-uniformity correction and temperature calibration can be performed without any user interaction. A total of four different temperature ranges have been

Figure 4.14: Infrared camera head Flir Phoenix DTS (separate digitalisation box not shown) and Janos MWIR Asio 50 mm lens with four extension rings.

calibrated, namely 16 to 33 °C, 18 to 55 °C, 18 to 95 °C and 20 to 150 °C, whereas the absolute accuracy of the latter one is rather poor at lower temperatures. This is mainly due to the short integration time used in this case. The other setups have been verified occasionally using the blackbody source and revealed good overall accuracy with absolute deviations below 0.2 K. Within the chosen sub-window a total of 203 "dead" pixels have been identified, which corresponds to a detector operability of 99.7 %. Analysis of a 1024 image stream recorded at 412 Hz revealed a uniformly distributed standard deviation of about 15 mK as shown in Figure 4.15a, which can be regarded as an approximate measure of the camera noise. Fourier transform of the spatial mean signal indicates a low-frequency temperature fluctuation below 1 Hz, probably caused by the blackbody control loop (Fig. 4.15b). The peak amplitude of about 1.7 mK at 60 Hz and the lower peak at the second harmonic (120 Hz) have been attributed to the detector cooler in an earlier work [Med08].

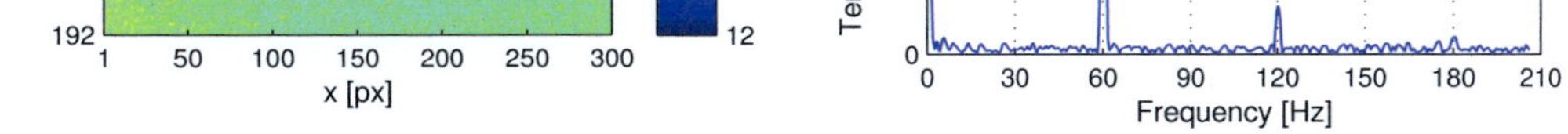

(a) Spatial distribution of the standard deviation

(b) Amplitude spectrum of the spatial mean signal

Figure 4.15: Noise characteristics of the used infrared camera found at 24 °C.

4.3 Standing wave patterns due to material damping

As several researches reported, standing wave patterns at resonance frequencies can interfere with the crack heating or even prevent the crack from being detected. In order to determine whether the effect is of relevance when testing thick-walled steel structures, the flawless reference plate 10-00 is first subjected to ultrasound excitation.

Having the Branson transducer screw-coupled at excitation position 1 (compare Fig. 4.1), the plate was inserted into the sample holders as shown in Figure 4.10b. All clamping screws were gently fastened so that the specimen is securely held in place but not excessively restrained.

While focusing the infrared camera on the notch region, a stepwise frequency sweep is performed within the full generator band ranging from 15 to 25 kHz. The frequency increment Δf was chosen to be 10 Hz and the relative excitation power P was set to 5 % since this value was the maximum possible allowing to complete the sweep within a duration d of 55 s without provoking a generator shut-off. Figure 4.16a shows the pixelwise achieved maximum temperature increase ΔT. At first sight, the notch region does not indicate any conspicuousness since a noisy but uniform thermal response was found. The situation changes significantly if the sweep rate is decreased from 182 Hz/s to 80 Hz/s and a limited frequency range of 19 to 23 kHz is analysed. It can be found from Figure 4.16b illustrating the result of the slowed down sweep at 7 % relative power that there is an overall heating of 0.1 K. Furthermore, the edge on the left side has reached a slightly higher temperature than the rest of the inspected region. Finally, a single frequency could be identified by a third run with a sweep rate of 56 Hz/s that leads to localised heating next to the fillet shown in Figure 4.16c. All infrared sequences were recorded using a frame rate of 32 Hz.

The heating which was obviously caused by exciting a resonance frequency was found to be quite low. Hence, the lock-in approach was employed to better detect the standing wave pattern along the whole specimen. The excitation level P was sinusoidally modulated at 0.5 Hz ranging from 0 to 13 % relative power, whereas the ultrasound frequency was kept constant at 21.62 kHz. Figure 4.17a shows the temperature amplitude distribution at $f_{\text{mod}} = 0.5\,\text{Hz}$ calculated by lock-in post-processing of the acquired infrared sequence. Although a flawless plate was tested, the notch region periodically heated up with a temperature amplitude of about 0.25 K. For better perceptibility of the heating pattern, the plate contour is highlighted and the upper limit of the colour scale is strongly decreased. Obviously, standing wave patterns can lead to misinterpretations of an amplitude image. The presented mapping is a good example of false crack detection since the notch region heated up by far the most, right next to the presumed location of a fatigue crack.

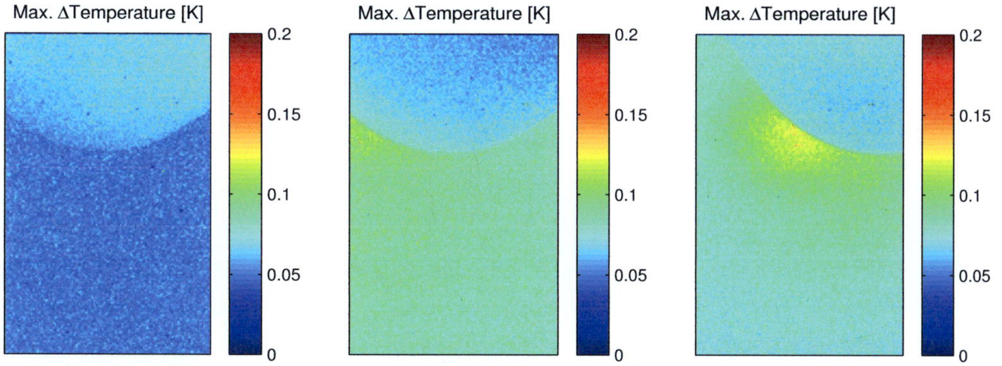

(a) $f = 15\ldots25\,\text{kHz}$, $d = 55\,\text{s}$, $\Delta f = 10\,\text{Hz}$, $P = 5\,\%$
(b) $f = 19\ldots23\,\text{kHz}$, $d = 50\,\text{s}$, $\Delta f = 10\,\text{Hz}$, $P = 7\,\%$
(c) $f = 21\ldots22\,\text{kHz}$, $d = 18\,\text{s}$, $\Delta f = 10\,\text{Hz}$, $P = 7\,\%$

Figure 4.16: Local heating of a 10 mm thick flawless plate due to bulk material damping.

The hot spot temperature evolution exemplarily depicted in Figure 4.17b shows the typical pulsation, which is superimposed by a long-term increase that asymptotically approaches a steady state. Trend removal by subtraction of a fitted curve based on a $\sqrt{t}$-law results in the adjusted ΔT history. A phase lag can be identified when plotting the modulated excitation history in the same graph after subtraction of the mean value.

According to the theory of viscoelastic material damping, the energy dissipated in the bulk per cycle is proportional to the square of the strain amplitude. Such a behaviour of PVC plates was experimentally demonstrated by Homma [Hom07]. If one assumes a linear relation between ultrasound excitation level and vibration amplitude, the local temperature increase of the steel plate should also satisfy a quadratic function of the relative power. An attempt to verify this relation is undertaken by means of ultrasound lock-in thermography with varying maximum power levels and modulation frequencies ranging from 0.1 to 5 Hz.

The achieved hot spot temperature amplitudes are illustrated in Figure 4.17c and 4.17d grouped by lock-in frequency and excitation level, respectively. To enhance the accuracy regarding the absolute value of temperature amplitudes, a Flat Top window was applied to the infrared data stream prior to the computation of the pixelwise discrete Fourier transform. It is found that f_{mod} has a large impact on the surface temperature amplitude. Although the heat source is not a discrete but a distributed one, a very similar behaviour to that predicted by the numerical simulations in Section 3.5.2 is observed. Since the thermal diffusion length μ_{th} decreases with increasing modulation frequency, the temperature amplitudes are more and more limited when approaching $f_{\text{mod}} = 5\,\text{Hz}$.

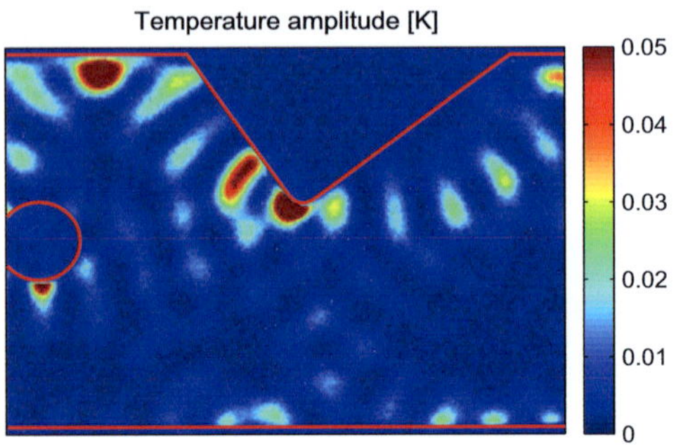

(a) Distribution of the temperature amplitude; $P_{max} = 13\,\%$, $f_{mod} = 0.5\,\text{Hz}$

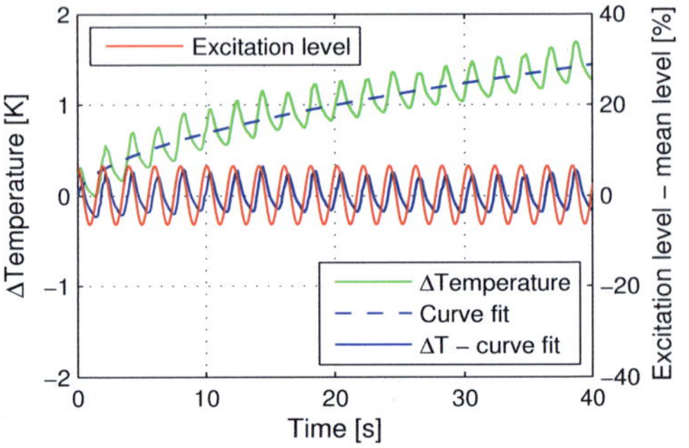

(b) Notch temperature increase and modulated excitation; $P_{max} = 13\,\%$, $f_{mod} = 0.5\,\text{Hz}$

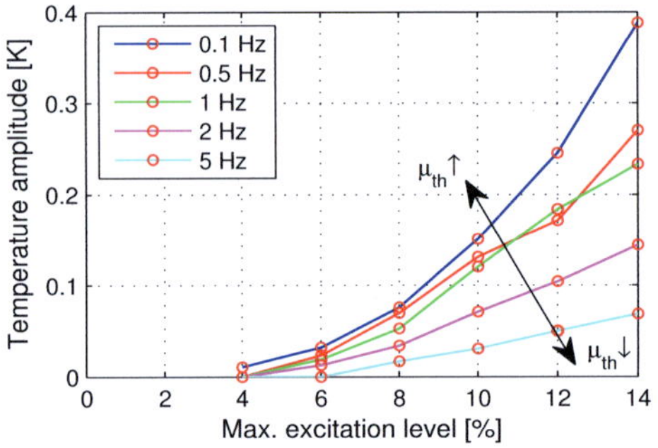

(c) Hot spot temperature amplitude; $f_{mod} = \text{var.}$

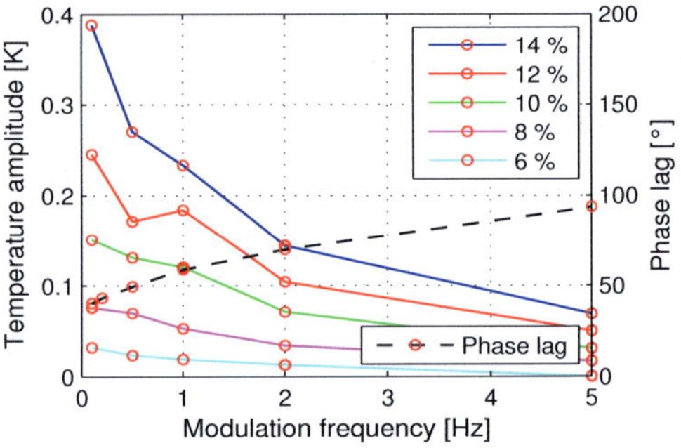

(d) Temperature amplitude and phase lag; $P_{max} = \text{var.}$

Figure 4.17: Thermographic visualisation of a standing wave pattern by means of ultrasound lock-in thermography (a); instant temperature response and long-term heating (b); hot spot temperature amplitude and phase lag achieved at different maximum excitation levels and modulation frequencies (c, d).

Obviously, such a high modulation frequency is rather impractical regarding the detection of buried heat sources. As expected, the phase lag of the hot spot temperature variation with respect to the modulation history increases with f_{mod} in the same manner as seen in Figure 3.32d. The predicted quadratic dependency of the temperature amplitude on the excitation level seems to be verified, at least in case of low f_{mod} values.

4.4 Detection of cracks in fatigue loaded steel plates

In the following, various aspects of vibrothermographic crack detection are being discussed in detail. The section covers all essential phenomena which affect the defect heating regarding excitation parameters, the vibrational system, initial bulk temperature and of course the crack size including the prestressing state. All presented results are based on experimental work using the high-power Branson ultrasound equipment and the previously described specimen type.

4.4.1 Influence of the excitation frequency and test reproducibility

One of the fundamental properties of ultrasound excited thermography applied to metal components is the pronounced frequency dependence of crack detection. From the experiments published in [GSB07, Hom07] and [PU09] it is well known that localised heating of cracks in metals can only be expected at narrow and sharply restricted frequency bands. Prior to any further investigation of the phenomena in thermosonic testing, the reproducibility of crack detection should be proven. Therefore, plate specimen 10-12-I was repeatedly excited by means of the Branson converter screwed at coupling position 1. Five consecutive frequency sweeps were performed at three different excitation levels. The sweep frequency range of 19 to 23 kHz, the frequency step size of 10 Hz and the total sweep duration of 45 s were not changed. It will be shown later on, that the used frequency range and increment is most suitable for the vibrothermographic tests conducted throughout this work. Between the individual tests, the plate, its support and the transducer coupling were not touched. To ensure reproducible test runs, infrared data acquisition was automatically triggered by digital signals with a pulse width of 30 ms, sent at the beginning and at the end of each sweep. The camera staring at the fillet region was recording at a frame rate of 41.2 Hz, a tenth of the maximum frame rate at the specified sub-window of 192×300 px.

During the frequency sweeps the crack flashed up several times in the infrared sequence. Once the thermal hot spot was identified, the average temperature evolution was extracted for each test based on a 7×7 px neighbourhood. Figure 4.18 shows all 15 temperature histories after subtraction of the minimum value. The exemplary curve plots of ΔT against frequency represent very typical results of ultrasound sweep thermography applied to metals. Within a range of 4 kHz only a few discrete frequencies are found that lead to a temperature increase of more than 1 K. Regarding the deviations between the three sets of curves, it can be concluded that the peak frequencies related to crack heating are in perfect agreement. Analysis of the achieved maximum temperature increases results in standard deviations of 0.04, 0.30 and 0.55 K for the excitation levels 3, 5 and 7 %, respectively. The peak temperatures reveal divergences of 10 to 12 % based on the coefficients of variation. In summary, the overall repeatability of the experiment seems to be fairly good, so the Branson ultrasound system clearly qualifies to be used for further investigations.

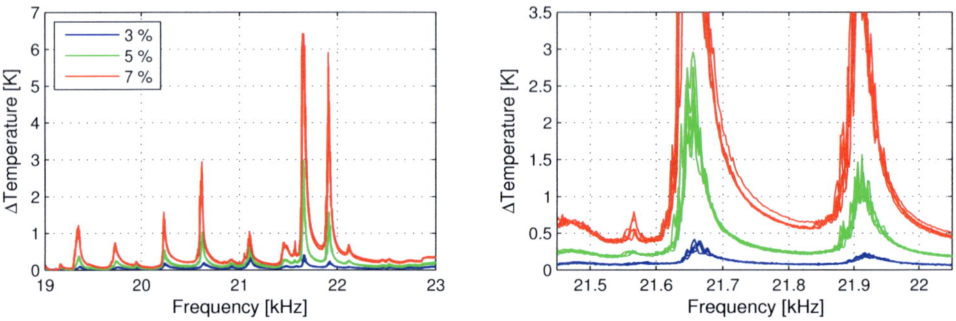

(a) Hot spot temperature increase; $P = \text{var.}$ (b) Hot spot temperature increase (detail); $P = \text{var.}$

Figure 4.18: Temperature increase achieved on plate 10-12-I-B with fixed support conditions.

4.4.2 Influence of the coupling position and the support conditions

Even if the ultrasound excitation itself is perfectly reproducible, substantial discrepancies in the thermal response might occur if a plate is accidentally moved or the clamping screws have been loosened. In order to assess the influence of variable support conditions on the thermal response, the plate 10-12-I was tested again, whereas the specimen was slightly moved in longitudinal direction. A centred position and relocations of 15 mm to the left and to the right have been considered. Furthermore, the threaded rods have been thoroughly tightened once and then loosened so that they just prevent the plate from tilting. The six combinations are repeated three times. Using the same frequency sweep parameters as before and a relative power level of 7 % it is found that those changing support boundary conditions have little effect on the temperature increase of the crack region. In contrast to the preceding section, the plate has been relocated between tests. Care has been taken to select exactly the thermal hot spot in the infrared streams for each combination so that the different records can be compared to each other. Figure 4.19 shows the averaged curves of the hot spot temperature evolution for each group. The coefficient of variation indicates a low deviation of 7 % when regarding all peak values of ΔT. Separation of the curves achieved with tightened and loosened screws reveals that looser fixture results in a slightly higher average temperature increase

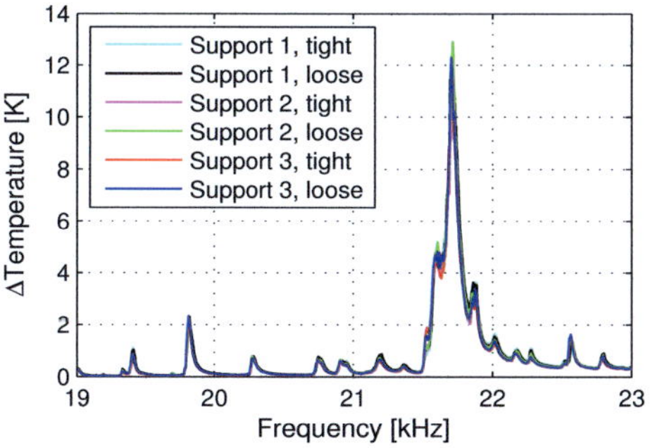

(a) Mean hot spot temperature increase

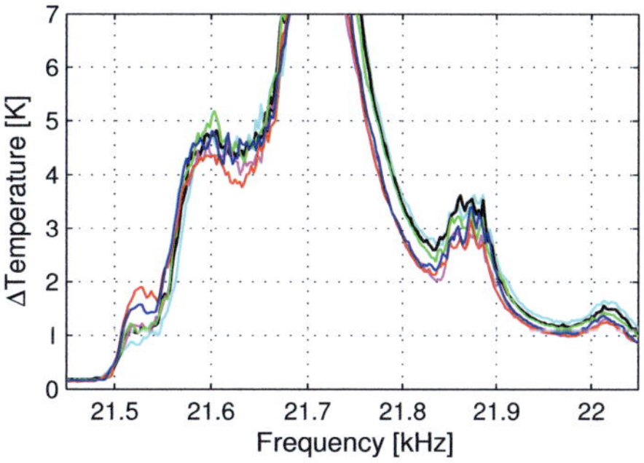

(b) Mean hot spot temperature increase (detail)

Figure 4.19: Temperature increase achieved on plate 10-12-I-B with varying support conditions.

of about 0.9 K compared to the situations with fastened screws. Grouping the results with respect to the mounting position of the plate does not indicate any tendency.

In contrast to the previous finding, the defect heating changes significantly if the coupling position is altered. While the excitation parameters were not changed, the thermal responses due to the frequency sweeps do not match any longer regarding their characteristic "acousto-thermal" spectra and the achieved maximum values of ΔT vary dramatically. Figure 4.20 manifests only a rough correlation of the peak frequencies. Comparing coupling positions 1 and 2, the max. ΔT values differ by a factor greater than 3.5. This finding is not quite unexpected since the oscillatory system comprising plate and transducer fundamentally changes. Homma reported similar ratios of achieved maximum and minimum infrared intensities in the range of 2 to 4 when inspecting turbine blades using different sonotrode positions [Hom07]. It should be mentioned that the maximum applicable relative power may also have changed when switching coupling positions. Experience showed that the maximum excitation level could only be raised to 9 or 10 % when applying the frequency sweep to a different plate of the same thickness or when changing the transducer position. However, such a dramatic change of defect heating seems to be impossible to compensate just by increasing the excitation power.

Besides of the present work, Hassan and colleagues achieved quite similar results on aero-engine fan blades. Stepwise relocation of the ultrasonic horn revealed a strongly varying crack heating. Although uncalibrated infrared measurements are shown in [HHW$^+$07], a difference by a factor of 12 can be estimated from the published graph, whereas plotting the thermal signal against contact position results in a relatively smooth curve. Furthermore, velocity data of a point next to the crack is presented that correlates with the infrared intensity. Considering the results of Hassan et al., the structure under investigation can be clearly understood as an influence field. Therefore, significant variations of occurring vibration mode shapes and amplitudes are most likely to have a large impact on the local crack heating. At the current stage the author considers this as a major weak point of the method, which seems to be widely ignored in the literature.

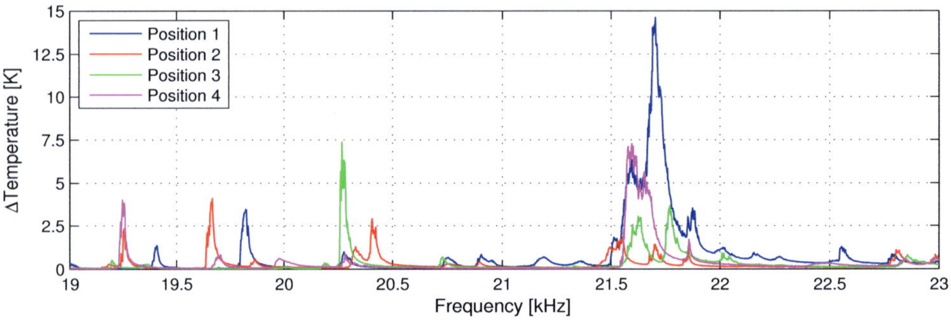

Figure 4.20: Influence of the coupling position on the thermal response of plate 10-12-I-B.

4.4.3 Influence of the frequency increment and the sweep duration

Although ultrasound sweep thermography represents an adequate approach if a pronounced frequency dependence of flaw detection exists, a systematic identification of optimal sweep parameters is generally not reported in the literature. However, a certain optimisation is surely required, especially in case of industrial series testing.

In order to determine standard settings which are used later on for routine testing of the plate specimens, a preliminary test was performed focusing on the influences of the sweep parameters, namely the frequency step width Δf and the total sweep duration d. Specimen 10-12-I was investigated again, where Δf values of 5, 10, 20, 50, 100, 200 and 500 Hz and durations of 15, 30, 60 and 120 s were chosen. The different durations correspond to sweep rates between 33 and 267 Hz/s considering the standard frequency range of 19 to 23 kHz. Accounting for the different sweep rates and also for the infrared record size, the camera frame rate was adjusted between 20 and 70 Hz. All infrared sequences of the 28 combinations have been processed as before focusing on the hot spot pixel history and the achieved maximum temperature increase. Figure 4.21a exemplarily illustrates the thermal responses achieved with a sweep duration of 60 s, and Figure 4.21b illustrates the maximum crack heating obtained from all records, depending on d and Δf.

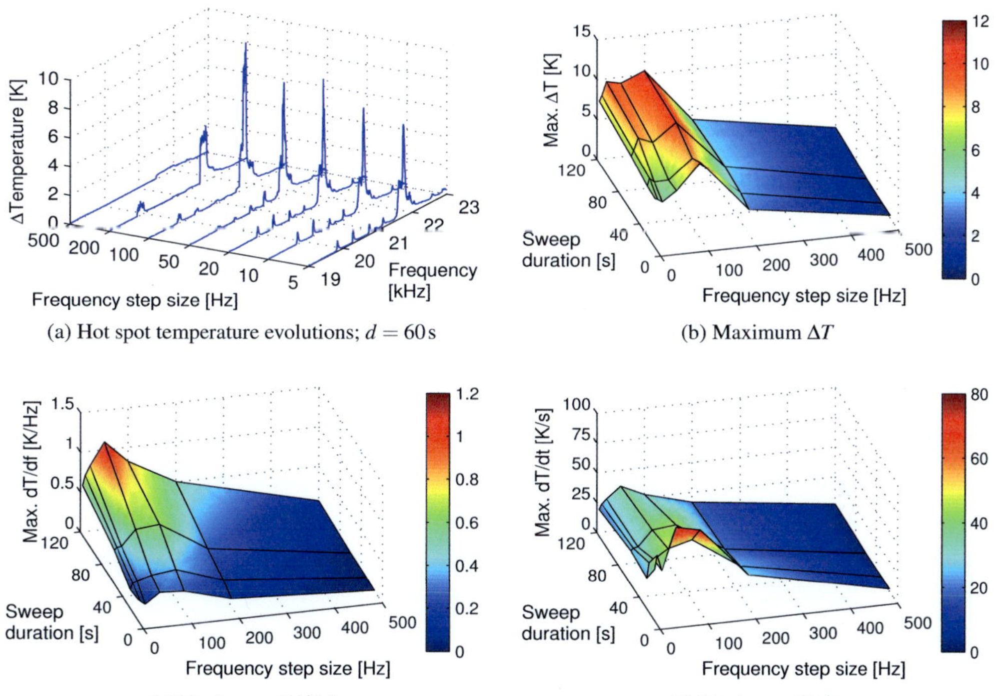

(a) Hot spot temperature evolutions; $d = 60\,\mathrm{s}$

(b) Maximum ΔT

(c) Maximum $\mathrm{d}T/\mathrm{d}f$

(d) Maximum $\mathrm{d}T/\mathrm{d}t$

Figure 4.21: Influence of the frequency step size and the sweep duration on the defect heating; specimen 10-12-I; $P = 7\,\%$.

The choice of the frequency step size is basically controlled by the typical width of the heating peak found in the "acousto-thermal" response curves but also determined by the requirements of associated investigations made in parallel. A continuous sweep with Δf approaching zero may be unfavourable if longer vibration measurements are required for which steady state oscillations are claimed. As illustrated in Figure 4.21a, it can be seen that a small frequency increment leads to a very high resolution of single heating peaks but does not necessarily bring the maximum defect heating. If Δf is excessively increased, many heating relevant frequencies are skipped. Crack detection finally turns out to be random since it just depends on if a potential heating frequency is matched by the coarsely resolved frequency sweep. The effect can be perfectly seen in the present example. While the peak at about 21.61 kHz is still detected using $\Delta f = 200\,\mathrm{Hz}$, the stronger heating at 21.74 kHz is not resolved. Crack detection by means of a sweep with $\Delta f = 500\,\mathrm{Hz}$ completely fails since no heating frequencies are incidentally multiples of 500 Hz. Lots of test runs proved that a frequency increment of 10 Hz suffices to accurately resolve the thermal response of the plate specimens. Most analysed heating peaks revealed widths greater than 30 Hz based on the accuracy of the Branson system.

As expected, increasing the total sweep duration leads to stronger crack heating since "effective" frequency bands are longer excited when the sweep pass is slowed down. Regarding the maximum heating, an excessively long duration is less efficient because the heat diffusion is gaining influence. Clearly, the total dissipated energy increases and causes a stronger net heating of the specimen, which might be required to detect a small defect. In search of a compromise between total energy input and time expenditure it is found from Figure 4.21b that a sweep duration of about 60 s is sufficient.

To complement this study, Figure 4.21c and 4.21d illustrate the derived quantities max. $\mathrm{d}T/\mathrm{d}f$ and max. $\mathrm{d}T/\mathrm{d}t$, respectively. These values characterise the achieved slope of the temperature increase and can also be used for imaging cracks [GSB07] while the inactive background is fully suppressed. If one relates the temperature increase to the frequency increment, a behaviour similar to max. ΔT is found. High values of the time derivative $\mathrm{d}T/\mathrm{d}t$ achieved by the fastest sweep reveal an instantaneous thermal response of the inspected plate.

4.4.4 Influence of the excitation level and the burst duration

In the previous sections some basic properties of ultrasound sweep thermography have been discussed. Similar to the preceding study the influence of the excitation duration in case of a mono-frequency burst should be examined prior to any routine testing. Variation of the burst duration will have a strong impact on the defect imaging concerning the maximum heating as well as the local defect contrast. Apart from this aspect, which is essentially governed by the heat diffusion, the loadability of the generator, which also depends on the burst duration, is of great importance regarding the practical use of the method.

In order to clarify the thermal response due to varying excitation durations, specimen 10-12-I was repeatedly excited by tone-bursts lasting 20 ms up to 5 s. The frequency of 21.69 kHz[1], which leads to strong heating, was chosen and held constant during the tests. It could be observed that the maximum power of the Branson system is more and more limited if the burst duration is increased.

[1] Note the small difference in the peak frequency comparing to the previous section. This is presumably caused by slightly varying mounting conditions or a change of the overall specimen temperature.

Accordingly, the relative power was carefully selected so that a strong heating is achieved but at the same time a long excitation period is still possible. Finally, each test was repeated three times using an excitation level of 12 %. Figure 4.22 depicts the maximum hot spot temperature increases for each burst duration as well as the averaged time histories of ΔT. For better perceptibility only a few curves are shown. Based on the experimental data of the 5 s excitation, an extrapolation was done using a curve fitting function of the form

$$\Delta T_{\text{fit},1}(t_{\text{exc}}) = c_1 \cdot \sqrt[c_2]{t_{\text{exc}}} \tag{4.15}$$

with c_1 and c_2 being the fit coefficients. As a first attempt a $\sqrt{t}$-law seems most natural since heat conduction states a diffusion problem. In case of $c_2 = 1$, equation (4.15) represents the exact one-dimensional solution for the surface temperature of a semi-infinite body subjected to a constant heat load [BS06]. However, a better approximation was found by adjusting the model equation

$$\Delta T_{\text{fit},2}(t_{\text{exc}}) = c_1 \cdot \frac{1}{(t_{\text{exc}} + c_2)^{c_3}} + c_4 \tag{4.16}$$

to fit the measured values. The curve labelled $\Delta T_{\text{fit},3}$ represents the version using $c_2 = 0$. Presuming that the inspected specimen showed a representative thermal response with respect to the shape of the heating curve, it is concluded from Figure 4.22 that a burst duration greater than 300 to 500 ms becomes rather inefficient. This is surely not a general rule since the depth of the heat source and the material properties will greatly influence the slope of the thermal response as shown in Chapter 3.

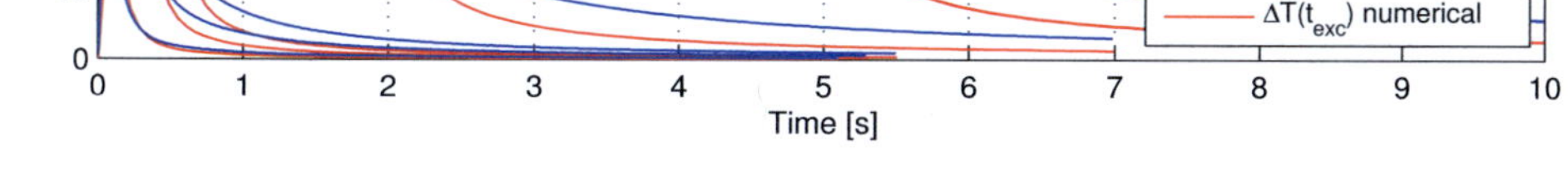

Figure 4.22: Experimental hot spot temperature evolutions measured on specimen 10-12-I-B using different burst durations, analytical approximations and numerical simulation results after manual fitting.

The present experimental study was complemented by a finite element analysis deploying the basic model of a steel plate introduced in Section 3.3.1. Considering the corner model in Figure 3.19a without symmetry conditions, a uniform heat flux was applied to the lateral face $y = 0$ and then cleared to simulate the experimental study. Adjustment of the heating power, the loaded surface area and its location on the simulated crack face finally leads to a perfect agreement between experimental and numerical heating curve. Even so, the decay behaviour still manifests differences, which are more pronounced at longer heating durations. These discrepancies might result from mismatched boundary conditions (radiation and convection have been fully neglected), material properties or from the simplified model geometry with its limited spatial resolution owing to the chosen finite element mesh. However, this study proves the experimental temperature increase to be the result of a continuous and sufficiently stable ultrasound excitation, which can be thought of as a steady heat flux applied to the crack faces. In this case, heating was applied to a $2 \times 2\,\text{mm}^2$ square, which is horizontally offset 3 mm and positioned in thickness direction just below the edge. Nodal temperature data was retrieved directly from the edge monitoring the greatest heating of the plate surface. The total applied heating power was 4.9 W. Even if the opposite side of the crack, which is not modelled, is taken into account, a total heat power of about 10 W is just a 4 % fraction of the nominal generator power that was theoretically available at the used excitation level and less than 0.5 % regarding the electrical peak power of 2.2 kW. Clearly, this rough estimation fully neglects the question of the system effectiveness and how much power is spent for keeping up the plate vibration, but it indicates that the overall efficiency of the used equipment is rather poor in terms of defect heating.

Another important aspect of thermosonic crack detection applied to metal components is of course the relationship between ultrasound power and the resulting heating. Although the Branson generator allows to readout the welding power of the last operation via an optional software tool, the author doubts that a conclusion can be directly drawn concerning the actual mechanical power induced into the specimen. Thus, to determine the influence of the excitation level on the thermal response, the relative power is varied while the ultrasound frequency is kept constant at the most "effective" one that is found by a preceding sweep thermography. A 150 ms tone-burst is performed, where the Branson nominal power level is stepwise increased until a repeated generator shut-off occurs. Each test is repeated five times in order to account for scattering of the excitation and heating. The test interval was large enough to allow for the specimens to cool down after each burst excitation. To enhance the significance of the analysis, four plate specimens 10-03, 10-12-I, 10-12-III and 10-24-III have been investigated in the same way. Figure 4.23 illustrates the achieved diagrams plotting max. ΔT against the excitation level. All inspected specimens essentially revealed a consistent thermal response. It is found that a certain level of the Branson power exists in each case which needs to be passed to achieve significant heating. Exceeding the threshold value finally reveals a linear relation between relative power and heating for the inspected cracks with depth $a \geq 12\,\mathrm{mm}$. The smooth transition between zero temperature increase and constant slope might be described by means of a power law. Here, the individual defect size and the temperature measurement range seem to be most critical regarding the significance of the non-linear part of the complete functional relation. In case of the 3 mm deep crack of plate 10-03 only the power law approximation of the measured data points can be found.

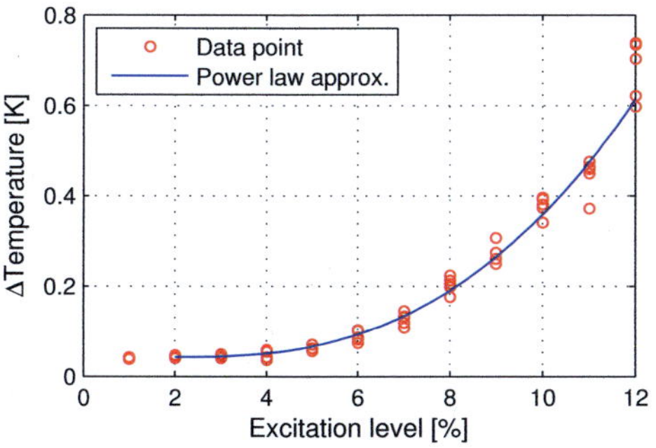

(a) Specimen 10-03-B

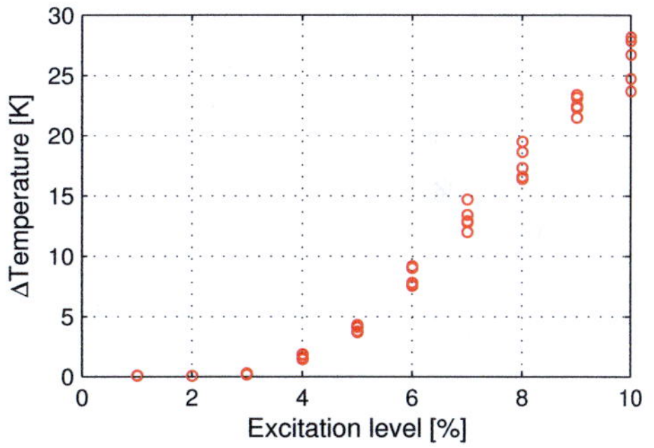

(b) Specimen 10-12-I-B

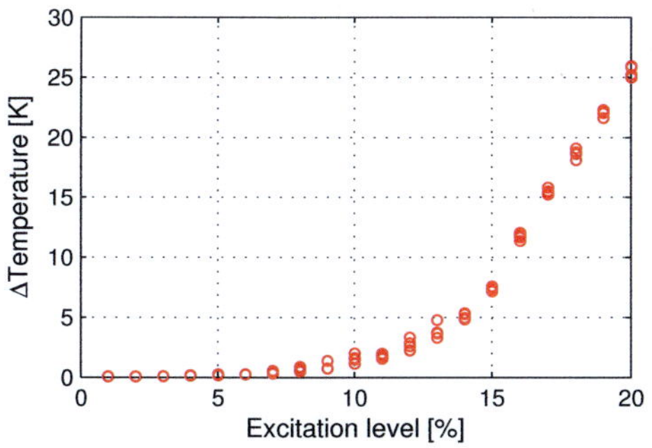

(c) Specimen 10-12-III-B

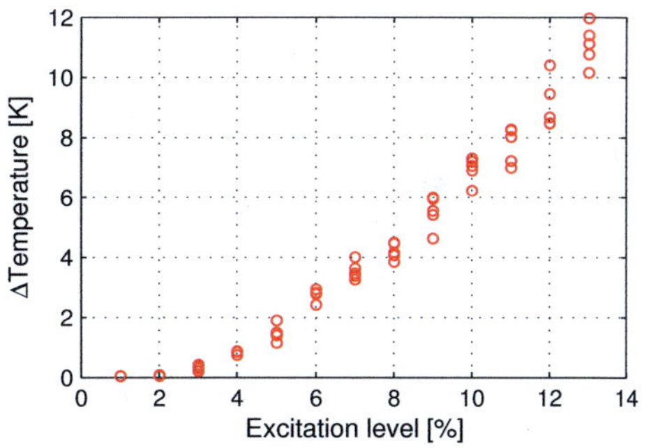

(d) Specimen 10-24-III-B

Figure 4.23: Temperature response due to mono-frequency excitation at varying power levels.

Until this point it is unclear whether the observed threshold behaviour results from a non-linear relation between input signal and output power or if the heating mechanism itself causes the threshold. Note that the used generator does not support power levels below 10 % in its original configuration. Nevertheless, it can be assumed that the Branson system is not the only reason for the threshold level since it varies between 4 and 12 % in the presented graphs. As the plate thicknesses are exactly the same, so that the mechanical impedance should be similar for all plates, it is unlikely that the difference is due to the internal control system of the generator. Additional investigations focusing on the vibration amplitude response are therefore carried out and described in Chapter 5 to allow for a discussion of the heating mechanism.

4.4.5 Influence of the initial temperature

During further preliminary testing an interesting phenomenon was observed when surveying the thermal response due to sweep excitations at varying nominal power levels. The effect is best illustrated on the thin plate specimen 06-06. Focusing on a narrow frequency band of 400 Hz, centred at the most efficient heating frequency, refined sweeps using $\Delta f = 5\,\text{Hz}$ have been performed successively. Figure 4.24a shows the hot spot temperature increases found using excitation levels up to 50 %. The frequency at which the maximum heating occurs reveals a significant shift towards lower values with increasing excitation power. When plotting only the maximum values of ΔT in Figure 4.24b, this trend is more clearly recognisable. A second peak frequency at about 19.55 kHz is vaguely perceptible in Figure 4.24a, showing a similar frequency change.

Furthermore, when exceeding the power level of 20 % a sudden shift of about -50 Hz was observed. This might be explained based on the results found in Section 4.4.1 and the conclusion made from the previous study. When having a closer look at the temperature histories shown in Figure 4.18, it is clearly evident that the maximum heating, observed at the two most efficient frequencies, does not proportionally increase with the excitation power. Although Figure 4.23 depicts results achieved from various plate specimens, it cannot be expected that different heating relevant frequencies manifest the same threshold levels and slopes of the functional relation between ΔT and the excitation level. Therefore, a change of the most efficient frequency seems reasonable. If a physical threshold level is exceeded, a neighbouring frequency might become the decisive one, concerning crack heating. However, the observed "jump" of 50 Hz is quite small, and the original peak frequency found at low power does not protrude at all when considering higher excitation levels. Therefore, a fundamental change of the inherent crack face interaction, which is possibly caused by a different vibration mode shape can only be suspected at this point.

For the sake of completeness the maximum values of the derived temperature histories $\mathrm{d}T/\mathrm{d}f$, plotted against the nominal excitation frequency, are shown in Figure 4.24c. Here, a smooth transition of the most responsive frequency from 19.5 to 19.35 kHz is found.

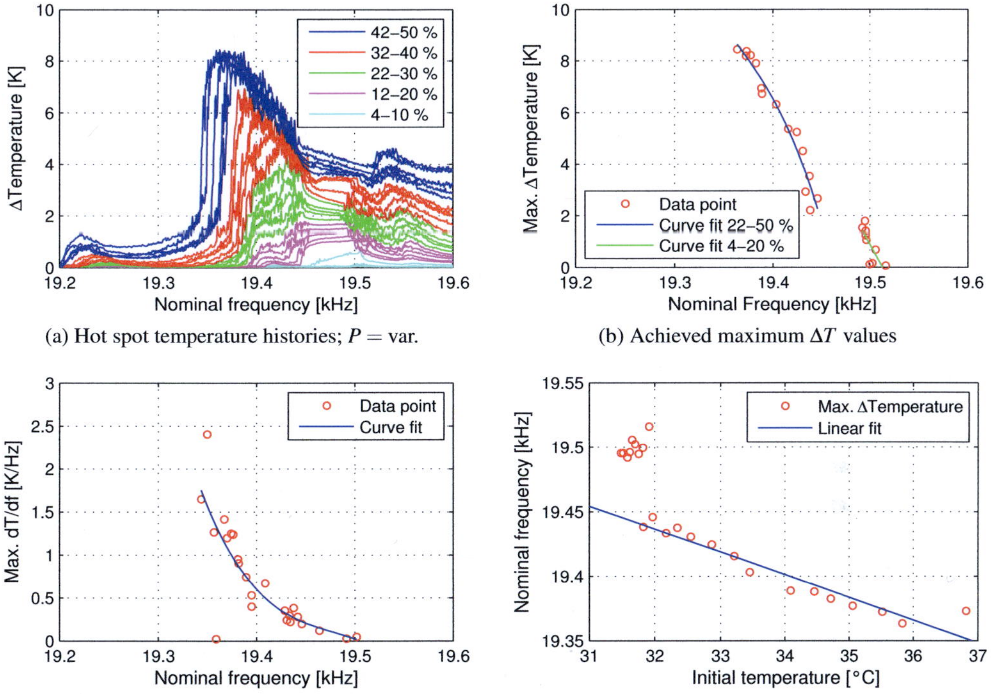

(a) Hot spot temperature histories; $P =$ var.

(b) Achieved maximum ΔT values

(c) Achieved maximum $\mathrm{d}T/\mathrm{d}f$ values

(d) ΔT peak frequencies against initial temperature

Figure 4.24: Shift of the most efficient heating relevant frequency and relation to the initial plate temperature; specimen 06-06.

Unless the ultrasound system is fully characterised in terms of frequency accuracy, the reason for the peak frequency shift cannot be determined without any doubt. At least four different effects need to be considered when assuming that the max. ΔT peak frequency corresponds to a plate eigenfrequency:

- The plate's natural frequencies decrease with increasing overall plate temperature.
- The transducer's actual frequency drifts towards higher values with increasing power.
- The temperature of the transducer chassis/piezo stack increases and detunes the system.
- The plate itself exhibits a softening non-linear dynamic behaviour induced by the interacting crack faces.

It is well known that eigenfrequencies of a metallic body vary with temperature. Heating-up results in an enlargement given by the coefficient of thermal expansion. In addition, the mechanical properties such as the modulus of elasticity, the Poisson's ratio and the density change at the same time. Homma [Hom07] found a linear decrease of the natural frequency with increasing temperature when exciting a copper block, which was heated from 25 to 40 °C in a controlled way.

The ultrasound sweeps have been conducted in a row, starting at low power, with a cooling pause of less than one minute after each run. Although the transducer was permanently cooled by pressurised air flowing through its chassis, the transducer tip, the internal piezo stack and of course the plate specimen heated up during the tests. Extraction of the initial hot spot temperature from the first infrared image of each record clearly reveals a long-term temperature increase of more than 5 K. When plotting the max. ΔT peak frequencies against the corresponding initial temperature in Figure 4.24d, a linear fit gives a good approximation of the data. Here, the first data points achieved at low excitation power have been ignored for the curve fitting procedure because of the frequency jump. Based on this finding, it is very likely that the max. ΔT peak frequency corresponds to a plate resonance and that the frequency shift results from an increasing initial plate temperature.

However, at this point it cannot be excluded that the excitation system itself causes an overall drift in frequency either due to the rising oscillation amplitude or due the increasing piezo stack temperature. A complete characterisation of the Branson generator/transducer system is carried out and reported in Section 5.2.

Quite a different effect might also influence the thermal response of the plate with respect to the relevant frequencies. As described earlier in Chapter 3, the micro-slip friction model introduces a shear stiffness of the mating crack faces that reduces with increasing relative displacement and approaches zero if all asperities are sliding (compare Fig. 3.12). Consequently, the resistance against sliding is greater at low excitation levels and smaller at high-power excitation. Therefore, a softening behaviour is conceivable, where the plate's resonant frequency depends on the oscillation amplitude. Based on vibration measurements, further discussion of the plate reaction to increasing excitation power can be found in Section 5.3.2.

Regarding the maximum achievable temperature increase, which is to be determined in the fashion of a routine testing of all available plate specimens, the present study clarifies that it might be insufficient to focus only on the frequencies found by means of a rather low-power sweep excitation. A possible shift of the most efficient excitation frequency requires manual tuning and repeated adjustment or a closed-loop control if highest generator power levels and maximum defect heating are desired.

4.4.6 Influence of the crack depth and the plate thickness

It is to be expected that the extent of dissipated heat inside the crack region varies with the actual crack depth a. Clearly, an increasing crack depth goes along with an enlarged crack mouth opening displacement as long as the specimens have been fatigue loaded by a pulsating tensile force. Therefore, a prediction of the effective crack face areas, which are able to contact themselves under mechanical excitation, is hardly possible. In case of 40 mm deep HCF cracks and for the three plates loaded in the LCF regime one can presume that no heating will occur along the crack path since the gap between both sides of the crack is too large (360 to 1900 µm). Here, the question arises whether the crack tip is detectable.

In order to investigate the influence of the crack depth and the plate thickness, all available specimens were mounted and clamped into the aluminium holders of the stationary test stand. The Branson transducer was screw-coupled to each plate as shown in Figure 4.10b and a routine testing was performed, where both sides of the plates were inspected. At first, the frequency dependent crack heating was obtained by sweep excitation in the range of 19 to 23 kHz using a step size of 10 Hz. Every single ultrasound sweep thermography took about 70 s. The maximum excitation level, which is reliably applicable, was individually determined for each plate. If the chosen excitation level is too high, the generator will abortively quit the frequency sweep and turn into an overload state that can only be released by triggering a generator reset.

After the decisive side and the most efficient frequencies were identified for each plate, the generator was manually tuned and set to the highest possible excitation level in order to achieve the maximum instantaneous crack heating due to a mono-frequency pulse of 150 ms duration. The pulse excitation was repeated five times to account for a possible scattering of the heating.

Throughout the investigation a significant spread of the maximum applicable excitation level was noticed. With increasing plate thickness a lower relative power had to be selected in order to keep the generator in operation. The reproducibility in terms of the excitation level during broad band frequency sweeps was quite good when comparing all plates of the same thickness. In case of mono-frequency excitation, the maximum possible generator power varied extremely with frequency. Here, the instantaneously induced power could easily exceed the values used for a sweep by a factor of two or three, especially when testing 6 and 8 mm thick specimens. Figure 4.25 illustrates the scatter regarding all conducted tests of this study. Both sweep and mono-frequency excitation levels decay exponentially with increasing flexural stiffness. Therefore, it is presumed that even much thicker steel components can be excited up to a certain level by the used equipment.

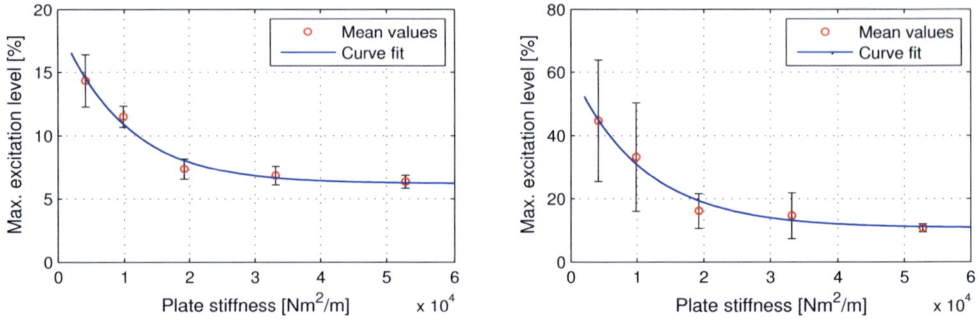

(a) Sweep; $f = 19\ldots23\,\mathrm{kHz}$, $d = 70\,\mathrm{s}$, $\Delta f = 10\,\mathrm{Hz}$

(b) Burst; $f =$ var., $d = 150\,\mathrm{ms}$

Figure 4.25: Thickness dependent maximum power levels achieved during sweep and mono-frequency excitation of plate specimens.

The following Figures 4.26 to 4.30 exemplarily depict the thermographic results obtained for all 10 mm thick plates. Corresponding images for all 6, 8, 12 and 14 mm thick plates can be found in Appendix A.3. Note that those results are sorted by crack depth. For each specimen five subimages illustrate the distributions of the maximum temperature increase due to sweep and mono-frequency excitation, the maximum positive slope achieved during the sweep and the detectability, based on threshold values for max. ΔT. The detectability maps are intended to allow for a quick visual comparison of specimen results. In order to take into account the varying and probably lower emissivity of coated steel members, additional blurring and damping of the thermal response due to thick protective coatings and more complex geometry, conservative detection limits were considered. The hypothetical threshold values were chosen to be 2 K and 5 K for cooled quantum detector cameras ("cryo.") and uncooled bolometer cameras ("bolo."), respectively.

All result images are based on a pixelwise processing, that means each pixel is considered independent of its surrounding. Prior to processing of the derivative $\mathrm{d}T/\mathrm{d}f$, a slight smoothing of all pixel signals with respect to time (or frequency) was conducted to enhance the image contrast. No additional spatial filtering was applied.

Figure 4.26 reveals that a 1 mm deep edge flaw in specimens 10-01-I and 10-01-II cannot be safely detected. Although it was possible to achieve a very low heating of the notch region in both cases (10-01-I with mono-frequency and 10-01-II with sweep excitation), it seems to be more likely that matching a resonant frequency caused a standing wave pattern, which typically leads to such a blurred heating (compare Section 4.3).

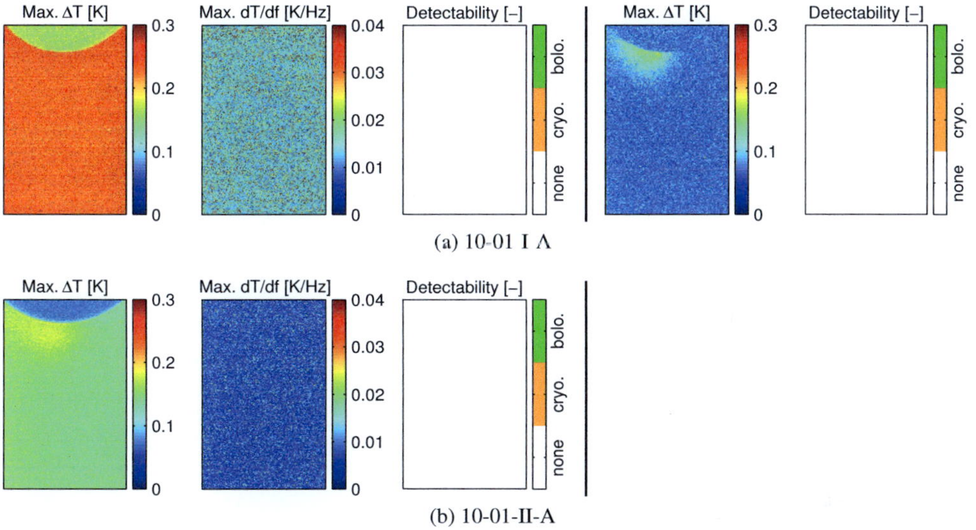

Figure 4.26: Failed crack detection on 10 mm thick HCF plate specimens; crack depth: 1 mm; left: strongest sweep excitation; right: strongest mono-frequency excitation.

The situation changes when the specimens with 3 and 6 mm crack depths are investigated, whereas only specimen 10-06-II features a through-thickness crack. With no doubt, these flaws were clearly detected. However, the heating rate was quite low so that the evaluation of the derivative leads to lower defect contrasts compared to the mapping of ΔT values (Fig. 4.27). Specimen 10-03 perfectly shows the interference between crack heating and a standing wave pattern when excited at a single frequency. In contrast to the sweep excitation, additional heating occurs on the right-hand side of the crack, which cannot be explained by heat originated from the crack path. Due to the changing driving frequency, the ultrasound sweep approach is naturally less prone to such kind of material damping effects. Despite of the relatively small crack depth, specimen 10-06-I shows its thermal hot spot not directly at the fillet edge, as it was found for the plates 10-03 and 10-06-II, but a few millimetres below. This behaviour can be explained since the CMOD of specimen 10-06-I is more than twice than that of the other two considered plates. Hence, a first indication is found that the actual crack opening along the crack path is of great importance regarding crack heating.

Next, three specimens featuring a through-thickness crack with a nominal depth of 12 mm were examined (Fig. 4.28). All defects were clearly imaged, where some very high absolute temperatures could be achieved, especially with the mono-frequency excitation. For appropriate temperature resolution the camera calibration needed to be switched several times (see Section 4.2.4 for the available ranges). The broadest calibration, which needs to be sensitive enough to resolve room temperature but should not reach its saturation point at 150 °C, suffered from a detector non-uniformity that could not be fully compensated. This is the reason for the inhomogeneous detectability map of specimen 10-12-III in case of mono-frequency excitation. The three specimens have the same nominal crack depth and endured quite similar numbers of cycles to crack initiation and final crack depth. Despite of that, the thermal responses reveal a huge scattering of the maximum temperatures achieved and also the hot spot location varies significantly. Again, the varying hot spot position can be explained by different gap sizes between the crack faces. While on plates 10-12-I and 10-12-II a CMOD of 8 μm was measured, specimen 10-12-III features a CMOD of 14 μm, almost twice the value. This leads to a hot spot at about 2/3 of the crack depth in this specific case.

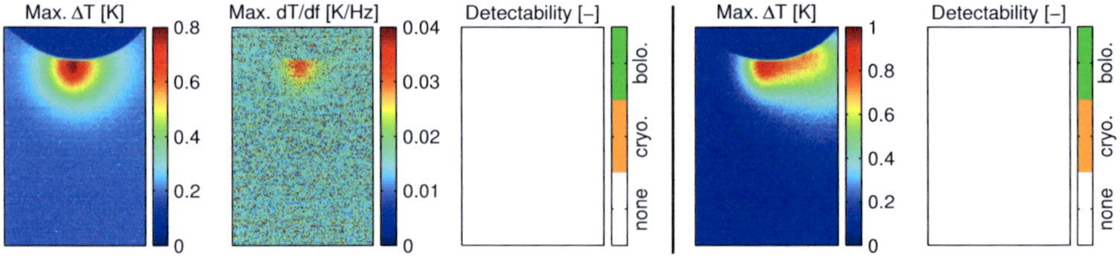

(a) 10-03-B

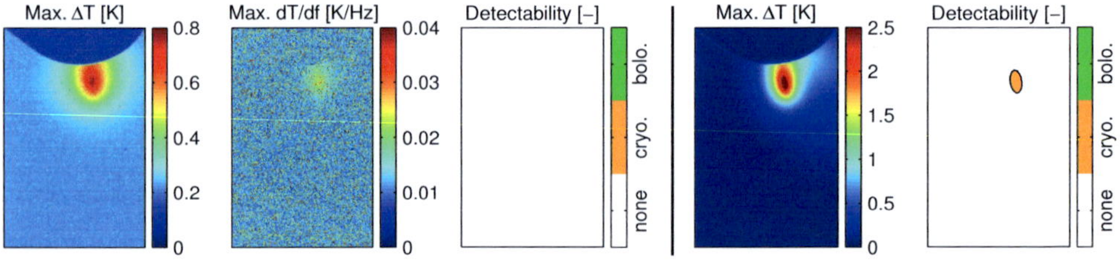

(b) 10-06-I-B

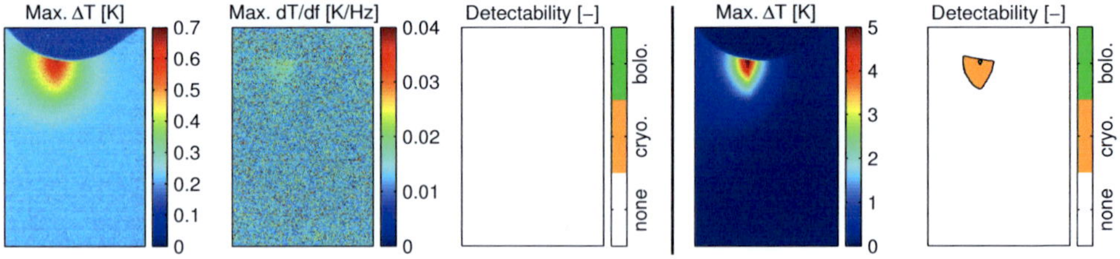

(c) 10-06-II-A

Figure 4.27: Crack detection on 10 mm thick HCF plate specimens; crack depth: 3 to 6 mm; left: strongest sweep excitation; right: strongest mono-frequency excitation.

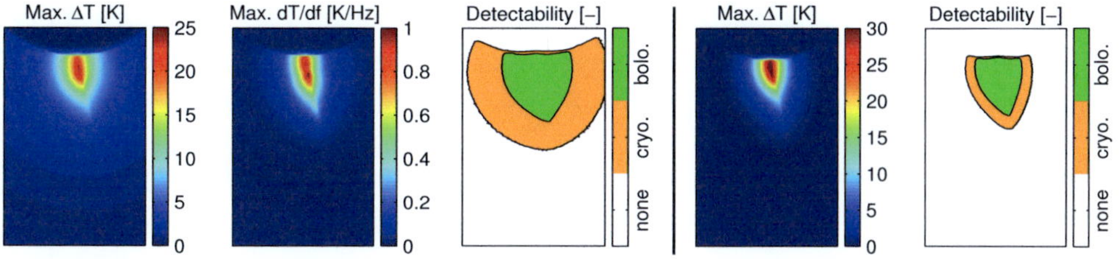

(a) 10-12-I-B

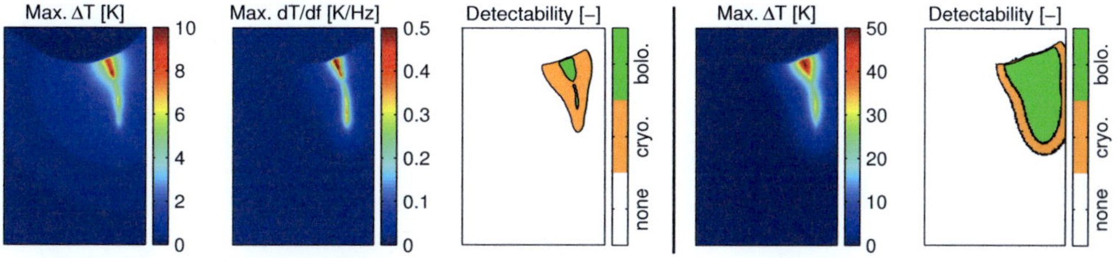

(b) 10-12-II-B

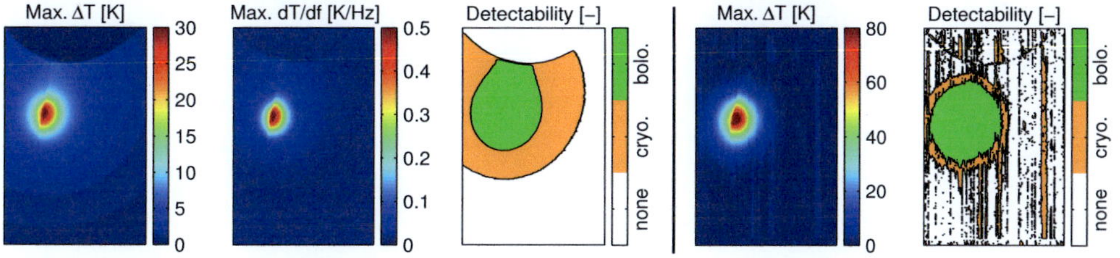

(c) 10-12-III-A

Figure 4.28: Crack detection on 10 mm thick HCF plate specimens; crack depth: 12 mm; left: strongest sweep excitation; right: strongest mono-frequency excitation.

The examination of two plates[2] with a crack depth of 24 mm brought phenomenologically similar results compared to those achieved on 12 mm deep cracks, whereas the thermal hot spot was located more clearly towards the crack tip. Mono-frequency excitation of specimen 10-24-III resulted in an extreme crack heating, which could not be resolved even when using the wide range temperature calibration. The corresponding max. ΔT subimage of Figure 4.29 shows clipping at about 120 K. On the contrary, the specimen 10-40, featuring a 40 mm deep crack and an enormous CMOD of about 780 µm could be heated up only at the crack tip location, reaching a peak temperature increase of just about 2 K. Due to the extremely small size of the hot spot, the detectability of this huge

[2]Specimen 10-24-I was not subjected to the described routine testing. Instead, it was used for a long-term testing procedure described in Section 4.5.

through-thickness crack with a gap between the crack faces of almost 1 mm is similar to or even worse than the 3 mm deep edge flaw previously presented. Note that the camera distance has been enlarged in order to capture the whole crack path when inspecting 40 mm deep cracks.

Finally, the three plate specimens 10-12-IV, 10-12-V and 10-24-IV, which have been excessively loaded in order to generate rough crack faces within less than 10^4 cycles, were tested in the same way. Results are presented in Figure 4.30. Obviously, the thermal response of all plates loaded in the low cycle fatigue regime is very similar to that of the 40 mm HCF crack specimen 10-40. Significant defect heating only occurred near the crack tip, whereas the crack paths are generally more jagged and kinking compared to those of the HCF specimens.

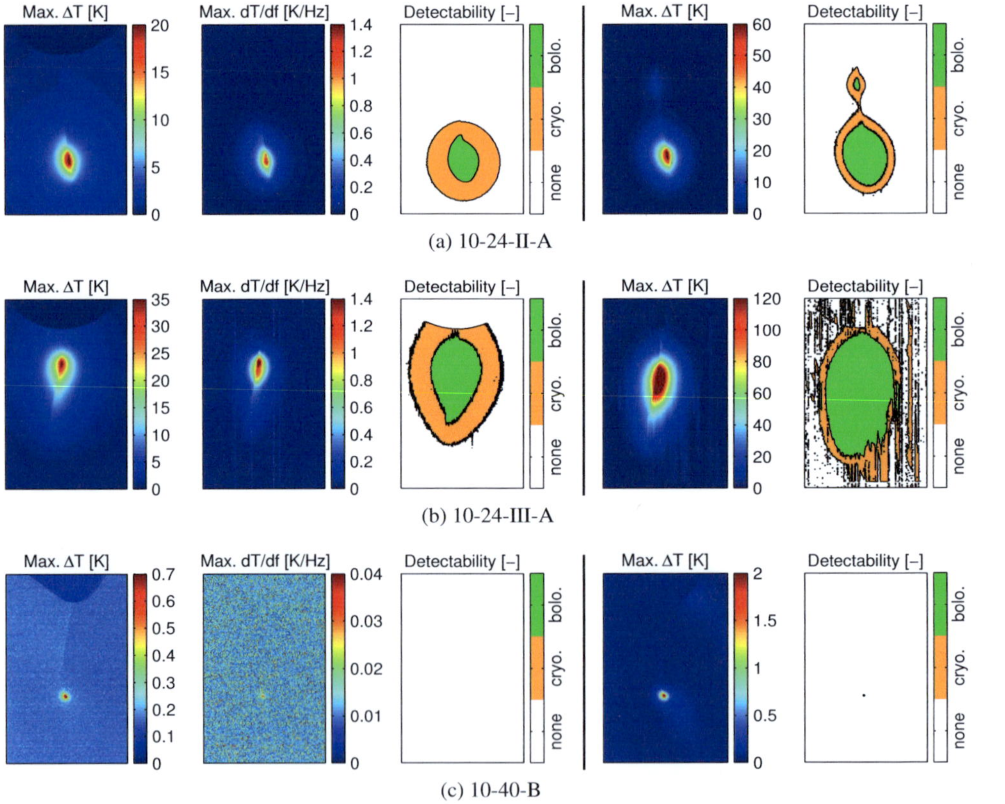

Figure 4.29: Crack detection on 10 mm thick HCF plate specimens; crack depth: 24 to 40 mm; left: strongest sweep excitation; right: strongest mono-frequency excitation.

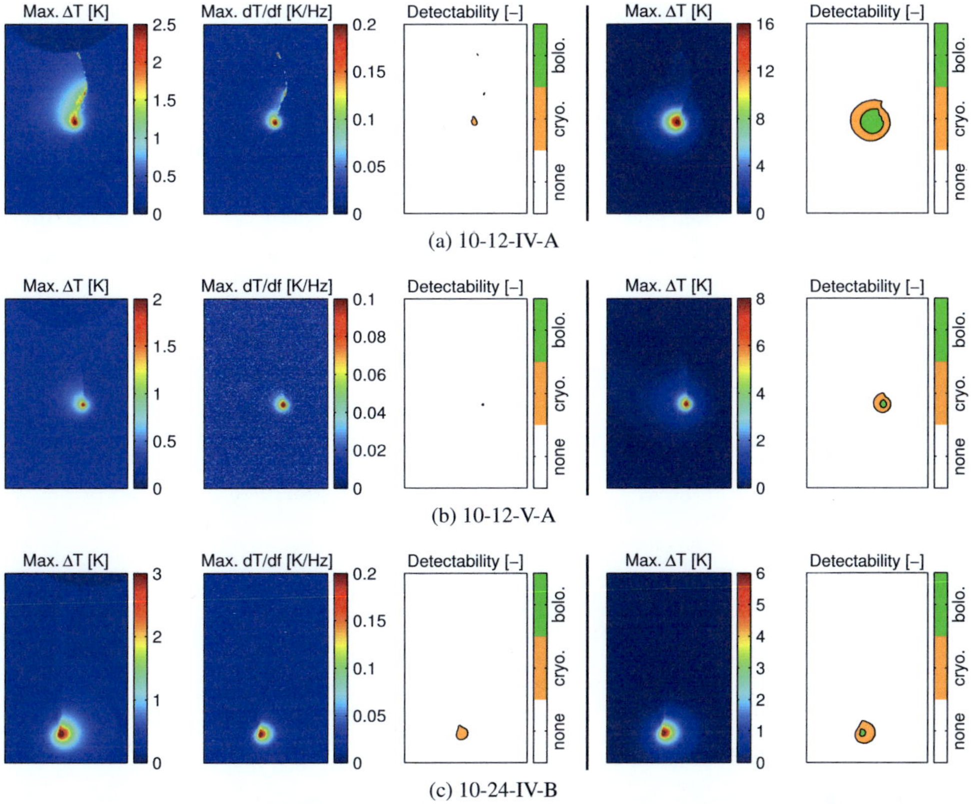

Figure 4.30: Crack detection on 10 mm thick LCF plate specimens; crack depth: 12 to 24 mm; left: strongest sweep excitation; right: strongest mono-frequency excitation.

Up to now it can be concluded that for 10 mm thick plates the medium sized cracks with a depth of 12 and 24 mm are best detected due to extremely high thermal contrasts. Smaller crack depths such as 3 or 6 mm but also the largest crack of 40 mm depth generated much less dissipated heat. If the entire results including those presented in Appendix A.3 are considered, a very similar behaviour can be found for each analysed plate thickness. The smallest flaw size of 1 mm depth was found to be detectable only in case of the 6 mm thick specimen. With increasing plate thickness the detection limit shifts towards larger flaw sizes. Taking into account that standing wave patterns might have interfered with the local crack heating, which leads to rather diffusely warmed up regions, the smallest detectable crack sizes are conservatively estimated and summarised in Table 4.10.

Table 4.10: Smallest detectable crack sizes found by sweep and mono-frequency excitation.

Plate thickness	[mm]	6	8	10	12	14
Limit of detection	[mm]	≤ 1	≤ 3	≤ 3	≤ 3	≤ 6

Figure 4.31 depicts the maximum achieved temperature increases due to mono-frequency excitation depending on crack depth a and plate thickness h. These values represent the absolute maxima of each spatial distribution of max. ΔT. In the majority of cases, only one specimen per combination of a and h was investigated. If more than one plate of the same combination was available, the corresponding results were averaged. Despite of the broad field of parameters and the relatively small number of specimens, a bell-shaped curve of values max. ΔT over crack depth a can be found for each plate thickness.

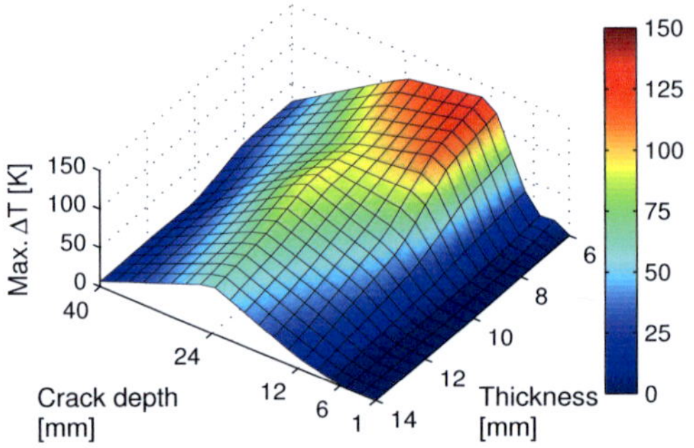

Figure 4.31: Maximum temperature increase due to mono-frequency excitation of all plate specimens tested in the HCF regime.

In general, the thinnest plates led to highest temperature differences of more than 120 K, while in case of the 14 mm thick plates heating was limited to about 60 K. Figure 4.32a illustrates conceivable sets of spline interpolation between the individual results for each plate thickness. From these artificial curves it can be found that the peak values tend to decrease with increasing thickness. Analysis of the measured and interpolated values of max. ΔT over all crack depths revealed that the data approximately suffices a hyperbolic function of the type max. $\Delta T \cdot h = \text{const.}$ Furthermore, the crack size leading to the (interpolated) max. ΔT seems to be shifted towards greater values with increasing plate thickness. Both the tendencies can be considered by fitting a Gaussian curve with argument $(a/h)^{1/3}$ to all the values max. $\Delta T \cdot h$ as shown in Figure 4.32b. The maximum temperature increase achieved by a 150 ms pulse of mono-frequency excitation can then be approximated by the following expression:

$$\text{max. } \Delta T = \frac{r}{h} \cdot \mathrm{e}^{-b \cdot \left((a/h)^{1/3} - m \right)^2} \quad , \qquad a = 1 \ldots 40\,\text{mm} \quad , \qquad h = 6 \ldots 14\,\text{mm} \qquad (4.17)$$

The constants r, b and m have been found by a least squares optimisation to be 924 K mm, 18.3 and 1.24, respectively.

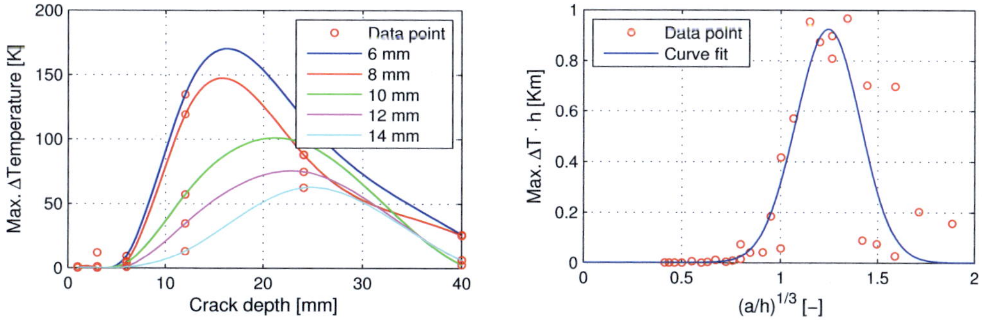

(a) Spline interpolation of maximum ΔT values (b) Gaussian curve fitted to $\Delta T \cdot h$ over $(a/h)^{1/3}$

Figure 4.32: Interpolation of max. ΔT values due to mono-frequency excitation and model fit.

For the sake of completeness, Figure 4.33 depicts the maximum values of the temperature increase ΔT and the first derivative $\mathrm{d}T/\mathrm{d}f$ achieved during ultrasound sweep thermography of all specimen featuring HCF cracks. Both plots are consistent with the results of the mono-frequency experiments but, however, the data surfaces are less smooth and therefore they are not further analysed.

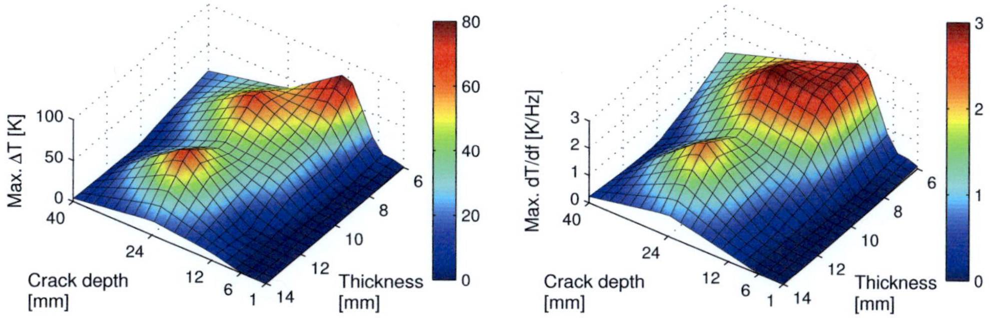

(a) Maximum ΔT (b) Maximum $\mathrm{d}T/\mathrm{d}f$

Figure 4.33: Maximum temperature increase and first derivative achieved by sweep excitation of all plate specimens tested in the HCF regime.

4.4.7 Influence of the prestressing conditions of the crack region

Based on the previous section one can presume that the local gap between the crack faces is crucial for the amount of frictional heat generated. To further analyse the influence of the actual crack opening displacement on the maximum heat dissipation, the test setup was modified. The servo-hydraulic testing machine previously utilised to generate cracks in the LCF regime was used to apply an accurately defined static preload to the plate specimens. This way, the crack mouth opening displacement can be easily controlled and measured by means of a digital microscope camera (DNT DigiMicro 2.0 Scale). In order to specify the actual prestressing of the crack faces in the case of compressive loading, strain gauges have been applied orthogonally to the crack path on that side of the specimen which is not monitored by the infrared camera. A total of six plates with a thickness of 10 mm and crack depths ranging from 1 to 24 mm have been prepared and mounted in the testing machine. The overlapping fork-like grip adapters require the ultrasound transducer to be screwed near the centre of the specimens. Here, coupling position 4 was employed (compare Fig. 4.1). The complete setup is depicted in Figure 4.34a, including a mounted specimen with coupled transducer as well as the infrared camera behind the testing machine. Figure 4.34b illustrates the detailed positioning of the strain gauges. On the one hand the grip adapters limit the available coupling positions due to their dimensions but on the other hand they conveniently prevent the specimens from buckling when being compressed. While the plate is subjected to a constant machine force ranging from -40 to 60 kN an ultrasonic sweep was performed at each static load level. The force was varied in steps of 2 kN and the frequency sweep parameters were the same as described before. For all specimens an excitation level of 8 % was used.

Microscopic images have been taken during the investigation of each plate. Digital image processing and measurement of the CMOD has been done by means of a software provided by the camera manufacturer. Figure 4.35 exemplarily shows some aligned images of plate specimen 10-12-1 covering the chosen load range. Under compression the crack is tightly closed and under tension an increasing force almost proportionally opens up the crack mouth. A selection of thermographic result images of the same plate is depicted in Figure 4.36. The single mappings of max. ΔT are equally spaced regarding the machine force and limited to the range of forces leading to significant crack heating. Considering all available result images reveals that the location of the greatest temperature increase continuously moves from the notch bottom towards the crack tip with stepwise increasing machine force. This trend can be recognised for all tested plates. The corresponding images are located in Appendix A.4.

(a) Testing machine, hydraulic grips, mounting fork adapters, plate specimen and infrared camera

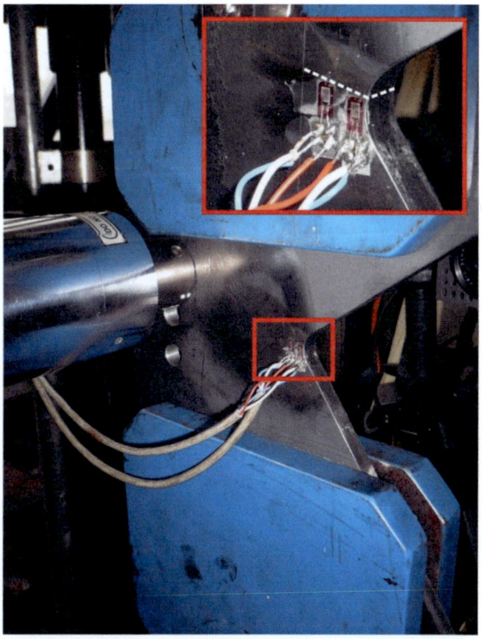

(b) Plate specimen constrained by pin-joints, screw-coupled transducer and wired strain gauges

Figure 4.34: Test setup for vibrothermography and strain gauge measurements under static preload using a servo-hydraulic testing machine MTS 810.25.

For the six investigated specimens the whole set of 51 ultrasound sweep thermography sequences was analysed at three to four discrete points A, B, C and D distributed along the crack path. Focusing on specimen 10-12-I again, plotting max. ΔT at each point against the machine force in Figure 4.38b confirms the following behaviour: Point A, located near the fillet, shows its strongest heating at a low tensile machine force. If either a compressive force or a strong tensile force is applied, the local heating is negligible. Points B and C exhibit very similar properties. Their corresponding curves of max. ΔT over force are just shifted towards higher tensile preloads. The point D, located next to the crack tip, shows its strongest thermal response at high tensile forces. Interestingly, the vicinity of the crack tip is still detectable when points A, B and C do not heat up any longer.

Figure 4.38a depicts the load dependent crack mouth opening displacement (CMOD) and the axial strain next to the fillet radius. From these measurements on specimen 10-12-I it is concluded that a gap of 4 to 6 µm between the crack faces leads to strongest frictional heating. Such a crack opening displacement can be locally achieved along the crack path by means of an increasing tensile force. The strain gauge readings reveal the actual load transmission, which is characterised by a linear relation between pressure on the crack faces and compressive loading. In case of tensile forces, the slope abruptly drops to zero and no further tensile load is transmitted next to the fillet edge.

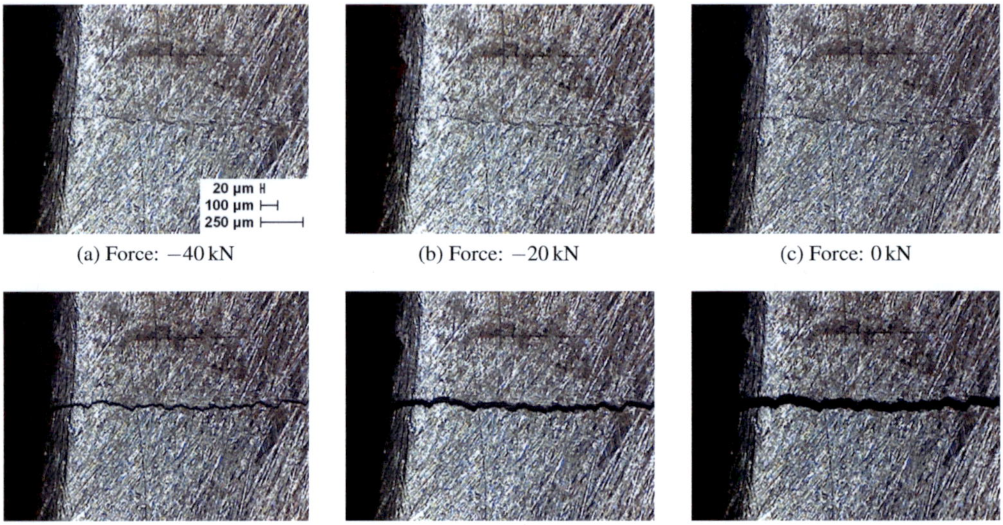

(a) Force: −40 kN (b) Force: −20 kN (c) Force: 0 kN

(d) Force: 20 kN (e) Force: 40 kN (f) Force: 60 kN

Figure 4.35: Crack mouth opening displacement at varying machine forces; specimen 10-12-I-B.

Regarding further results illustrated in Figures 4.37 and 4.38, it can be noted that the machine load required to open a crack clearly rises with decreasing crack depth. Due to the fixed width of the used strain gauges (width of backing film: 3 mm), the measured load transmission approaches that of a sound specimen when the flaw size decreases. Figure 4.37a to 4.38c prove a smooth transition from specimen 10-01-II that fully bears the tensile loading to specimen 10-12-III that reveals diode-like behaviour. The axial strain measured on specimens 10-06-II and 10-12-1 proportionally increases with the load up to a certain point in the tensile regime and then remains at a constant level. This behaviour clearly indicates a residual stress relief, which is more pronounced the smaller the crack depth is. Considering the strain measurement on specimen 10-01-II at 60 kN yields an elastic load transmission factor of about 4.76 MPa/kN. Assuming that the measuring grid of the outer strain gauge has a distance of 1.5 mm (half of the film width) to the fillet edge, this value is in perfect agreement with the static finite element solution in Section 4.1.1. Here, interpolation yields a stress/load factor of 4.84 MPa/kN.

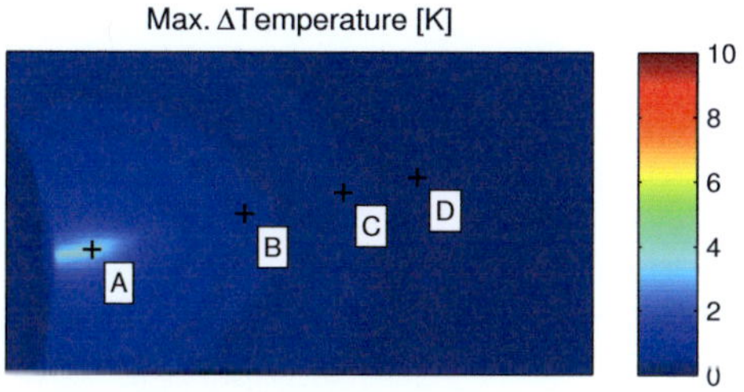

(a) Force: 2 kN

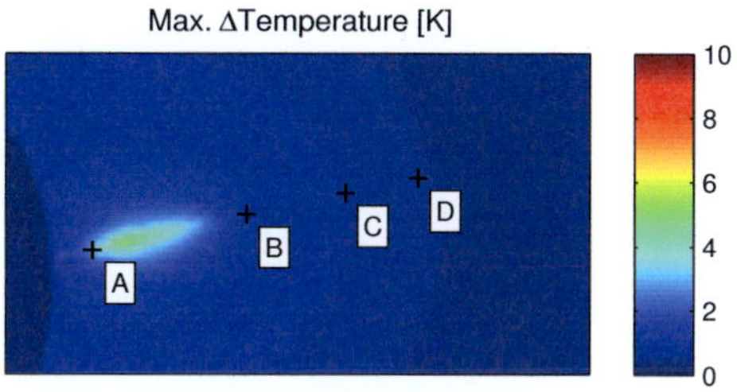

(b) Force: 8 kN

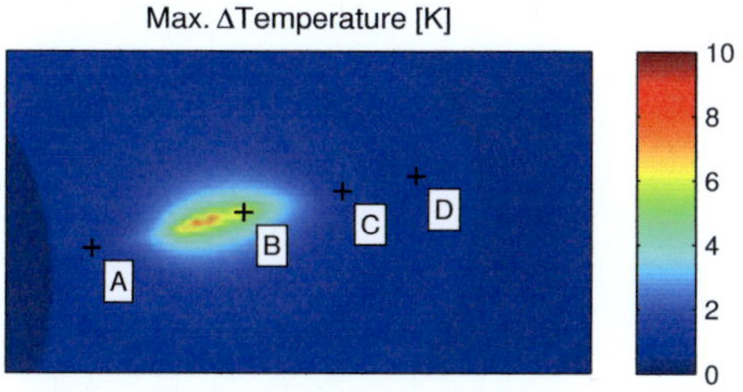

(c) Force: 14 kN

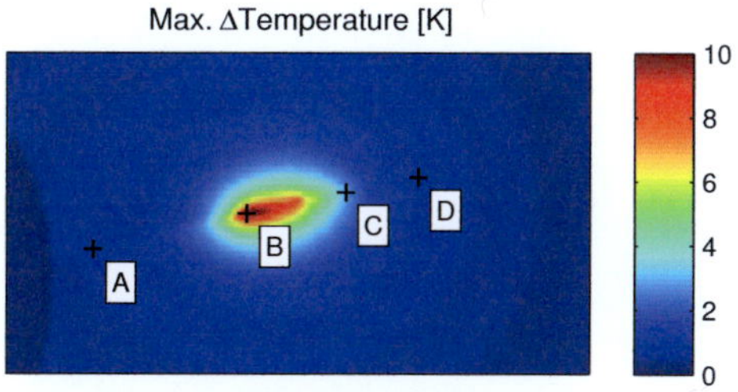

(d) Force: 20 kN

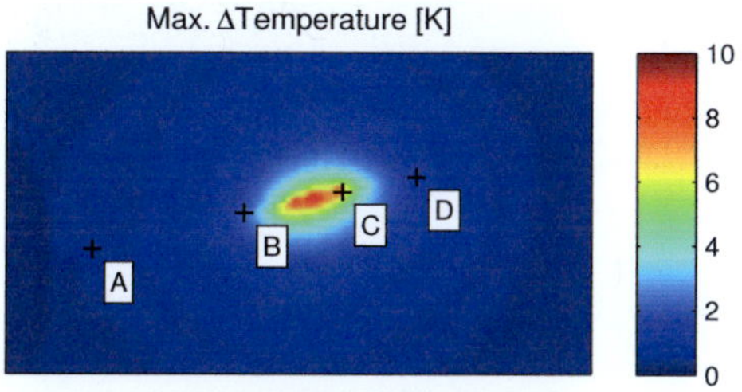

(e) Force: 26 kN

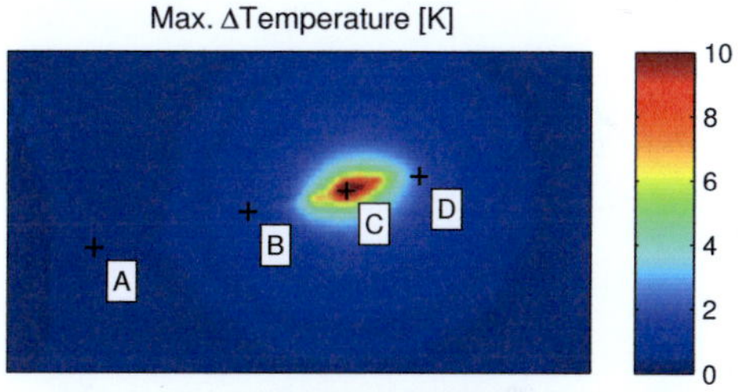

(f) Force: 32 kN

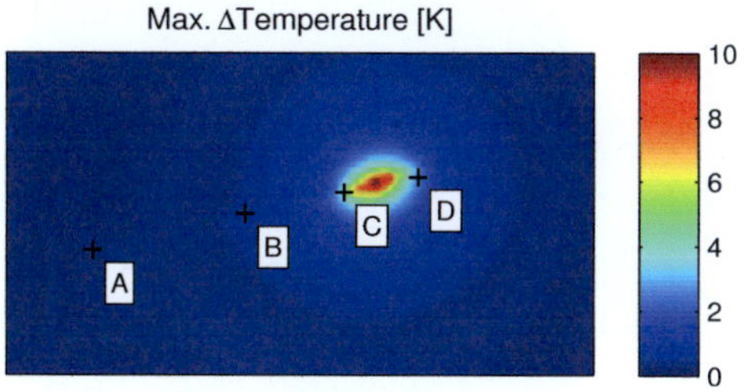

(g) Force: 38 kN

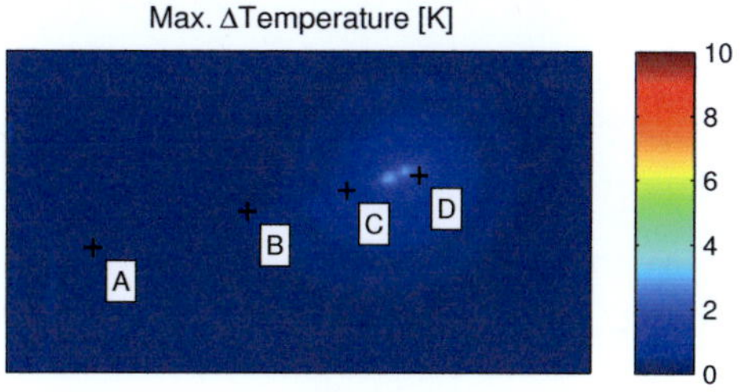

(h) Force: 44 kN

Figure 4.36: Distribution of max. ΔT achieved at different preloading states; specimen 10-12-I-B.

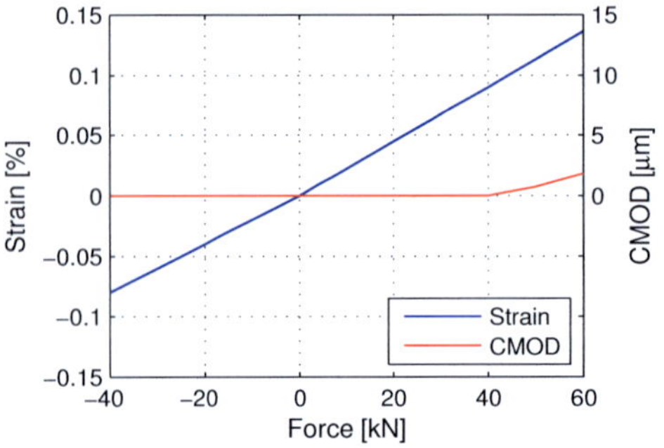

(a) Strain and CMOD; 10-01-II

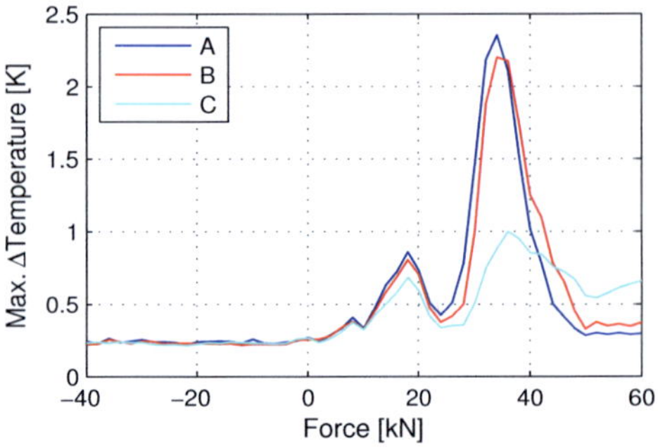

(b) Max. ΔT over force; 10-01-II

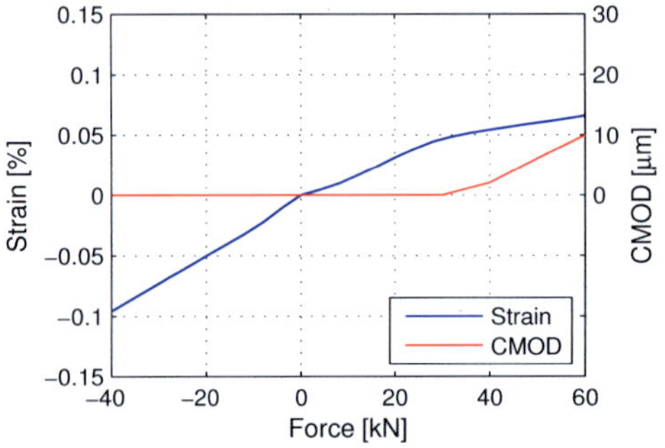

(c) Strain and CMOD; 10-03

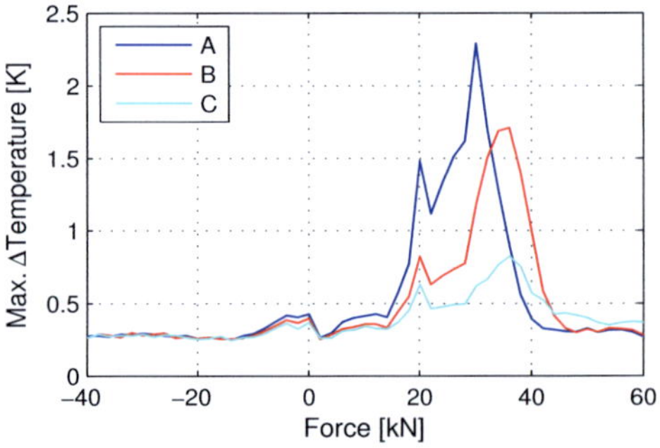

(d) Max. ΔT over force; 10-03

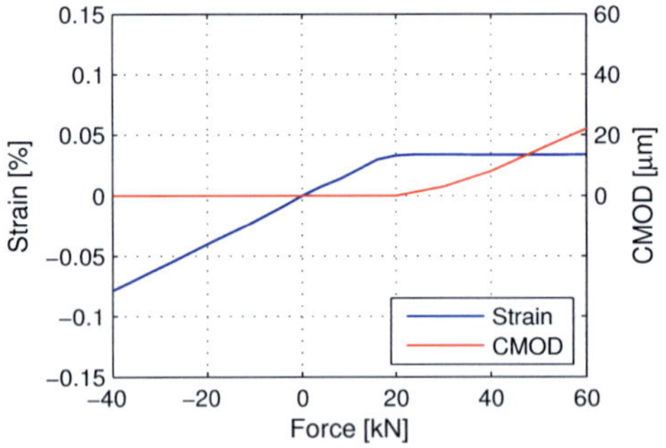

(e) Strain and CMOD; 10-06-II

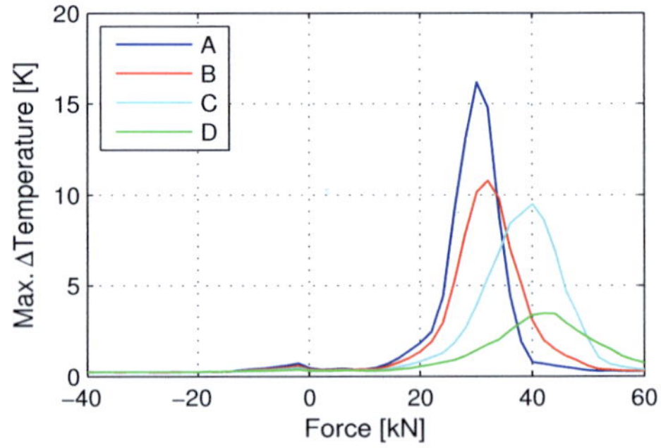

(f) Max. ΔT over force; 10-06-II

Figure 4.37: Axial strain near the fillet edge, crack mouth opening displacement and max. ΔT at points A, B, C and D achieved at varying machine forces; specimens 10-01-II, 10-03 and 10-06-II.

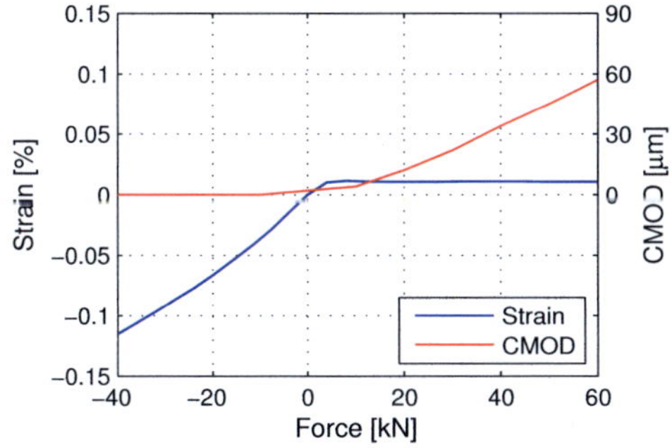

(a) Strain and CMOD; 10-12-I

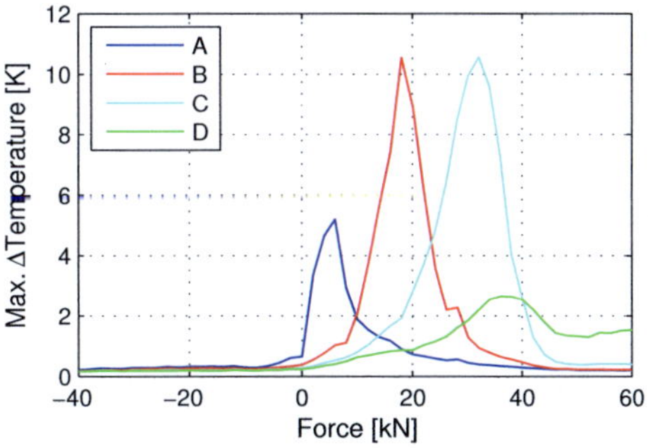

(b) Max. ΔT over force; 10-12-I

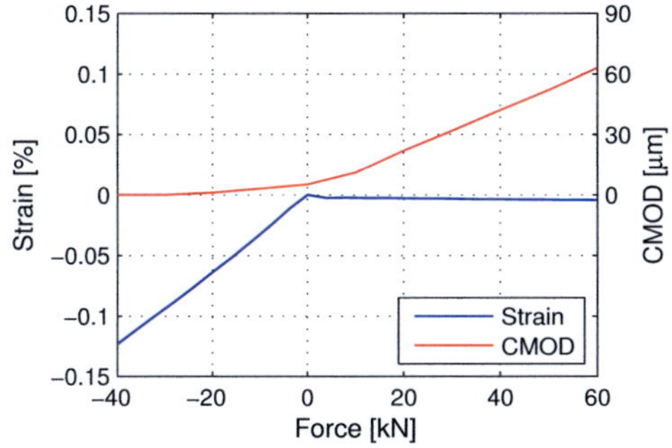

(c) Strain and CMOD; 10-12-III

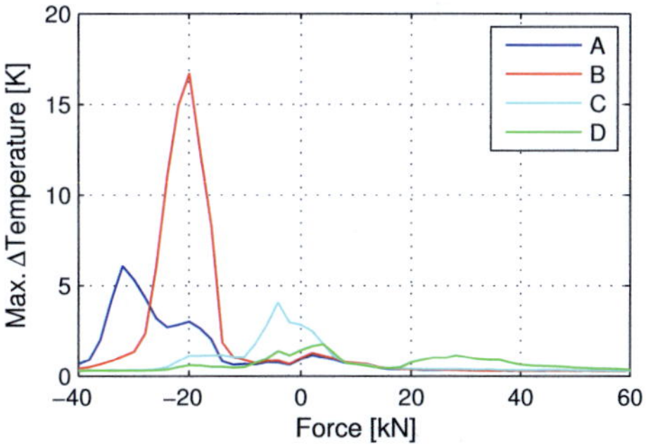

(d) Max. ΔT over force; 10-12-III

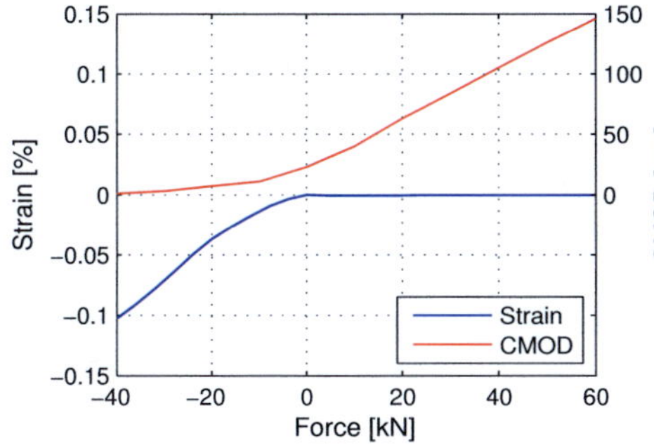

(e) Strain and CMOD; 10-24-II

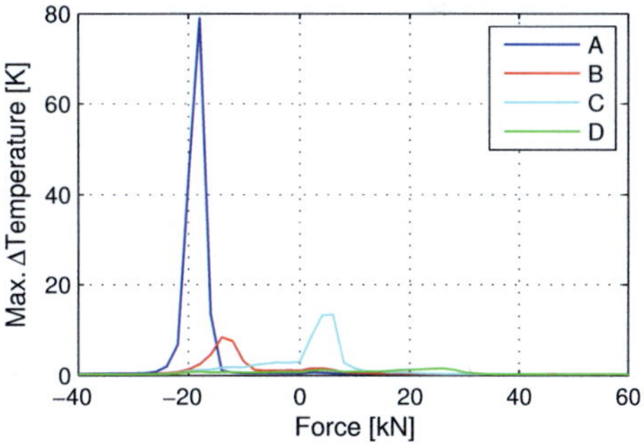

(f) Max. ΔT over force; 10-24-II

Figure 4.38: Axial strain near the fillet edge, crack mouth opening displacement and max. ΔT at points A, B, C and D achieved at varying machine forces; specimens 10-12-I, 10-12-III and 10-24-II.

Corresponding to these results, the region of forces that leads to significant crack heating continuously shifts from high tension to high compression with increasing crack depth. The tendency of detecting the fillet edge at lower tensile or even compressive forces and the crack tip at high tensile forces can be found in all plots of max. ΔT against the applied machine force.

Comparing the results achieved for both 12 mm deep cracked specimens 10-12-I and 10-12-III reveals a major deviation regarding the heating relevant range of forces. Again, this discrepancy is explained by the different CMOD values of 8 and 14 µm and the different residual stresses acting at the initial state. As an outcome, plate 10-12-I requires tensile loading and plate 10-12-III requires compressive loading to enable strong defect heating.

From the present study it is concluded that a gap of 2 to 10 µm between the edges of a crack generally allows for the strongest defect heating. The results clearly indicate that macroscopic normal pressure acting on the crack surfaces may fully suppress relative motion of the crack sides. On the other hand, it is found that a local crack opening of more than about 10 µm prevents the crack faces from contacting themselves. Clearly, this value is depending on the actually achieved vibration amplitudes. In both cases, frictional heating is prohibited and therefore the crack detection is rendered impossible.

These findings are in perfect agreement with the experimental results achieved by Renshaw et al. [RHT08], Lu et al. [LHN$^+$07] as well as Rothenfusser and Homma [RH05]. Renshaw and colleagues investigated the effects of tensile prestressing of a bar with semi-elliptical surface crack on the location of heating [RHT08]. They explained the moving hot spot by means of a crack closure model commonly used in fracture mechanics, which accounts for closure stresses due to the residual plastic wake of the growing crack. Depending on the actual distribution of the closure stress along the crack path, applied tensile stresses partially compensate for the compressive residual stresses and open the crack to some extent. Unfortunately, Renshaw et al. do not give any estimations for the local CMOD. However, their explanation surely applies to the presented experiments.

Consistently, it can be concluded that the initial stress state of the crack region is an important factor regarding the possible crack heating. Figure 4.37b shows that even a 1 mm deep corner flaw in a 10 mm thick plate can be clearly detected, once an appropriate force is acting on it. In the specific case, a maximum temperature increase greater than 2 K was achieved at 35 kN in contrast to a questionable heating of about 0.2 K in the unloaded state. Therefore, Table 4.10, summarising the smallest detectable crack sizes as found from testing without external preload, can only provide a rough but conservative estimation of the detection limits.

4.5 Effects of long-term insonification

In this section the long-term detectability of cracks is investigated. The described experiments aim at the analysis of two effects. These are the variation of the thermal response due to excessively repeated ultrasound excitation and the development of morphological damages to the crack faces.

4.5.1 Thermal response due to repeated excitation

So far, very few investigations regarding the long-term repeatability of crack detection by means of ultrasound excited thermography have been reported. Therefore, plate specimen 10-24-I and several older plates, which were investigated earlier by Mehdianpour [Meh03], have been repeatedly excited using the stationary test setup described in Section 4.2.1. Here, position 1 was employed again for screw-coupling of the transducer. The five additional specimens, which were also loaded in the HCF regime, have exactly the same dimensions but crack depths varying from 8 to 56 mm. Within this work they are arbitrarily labelled as specimen M-I to M-V.

After performing a broadband frequency sweep in the range of 19 to 23 kHz, each specimen was subjected to intense sweep excitations focusing on the most efficient frequency for crack heating. The insonification and the synchronous infrared acquisition have been fully automated by means of a customised LabVIEW programme. The routine included a loop for starting and controlling the stepwise frequency sweep and for start/stop triggering of the camera acquisition. In case of a generator overload, the whole system is no longer operational. To ensure that the device is ready-to-operate, it is preventively reset after each loop pass. A minimum number of 1000 sweep repetitions was desired. The nominal repetition rate of one sweep per minute prevented the transducer from overheating. The first long-term excitation run of specimen 10-24-I failed due to a programming error that led to memory leakage. Because of the increasing memory utilisation the computer system slowed down extremely. As a consequence, the individual sweeps took longer each time and were finally not completely captured by the infrared camera since a recording time limit was enabled. Table 4.11 summarises the frequency ranges, durations, numbers of repetitions and the excitation levels of the performed long-term experiments as well as the actual crack depths and plate thicknesses. All infrared sequences have been processed and the maximum temperature increase was extracted. After repeated testing, the six specimens have been carved into pieces so that the two crack faces are separated.

Table 4.11: Specimen properties and parameters for repeated thermosonic testing.

Specimen	h_{act} [mm]	a_A [mm]	a_B [mm]	Frequency [kHz]	Exc. level [%]	Sweep duration [s]	Repetitions [−]
10-24-I	9.96	24.1	23.9	21.0...22.2	7	$12\ldots>225^2$	400^2
M-I	10.45	55.6	55.6^1	21.7...22.2	10	5	1700
M-II	10.22	28.4	28.4^1	19.9...20.5	7	6	1500
M-III	10.18	28.1	28.1^1	19.9...20.5	10	6	1000
M-IV	10.17	21.4	21.4^1	21.7...22.4	8	7	1200
M-V	10.17	8.5	0	21.9...22.4	16	5	1500

[1] Approximately the same crack depths on both sides
[2] Sweep progressively slowed down due to a programming error

Regarding the maximum thermal response during repeated excitations it is found that the heat generation of all investigated samples significantly decays with increasing number of excitations. Figure 4.39 presents an aligned composition of the plate M-IV front view (photograph and thermogram), the crack face and selected images of max. ΔT. In addition, five pixel positions A to E have been defined along the crack path. Evaluation of max. ΔT at position E next to the fillet manifests an exponential reduction in heating with respect to the number of performed sweeps, as illustrated in Figure 4.40a. The reproducibility concerning the achieved temperature is surprisingly good, whereas the scatter within the first hundred repetitions is obviously increased compared to the last sweep excitations. Considering the thermal response along the crack path, specimen M-IV manifests a uniform temperature decay at pixel positions A-E as shown in Figure 4.40b. The thermal hot spot did not move during the testing procedure.

Similar results have been obtained from specimens M-III and M-V. Interestingly, the development of max. ΔT achieved on specimen M-III shows the same exponential decay first but increases at about 300 sweeps and decreases again after 600 repetitions. Although the edge flaw of specimen M-V generated much lower temperature increases, its qualitative decay of max. ΔT over the number of sweep excitations is in perfect agreement with those observed on plates M-III and M-IV. The smooth decrease of the maximum temperature change can be approximated by an exponential function of the form

$$\text{max. } \Delta T = a \cdot \mathrm{e}^{-b \cdot x} + c \quad , \tag{4.18}$$

where a, b and c are free coefficients and x denotes the number of excitations.

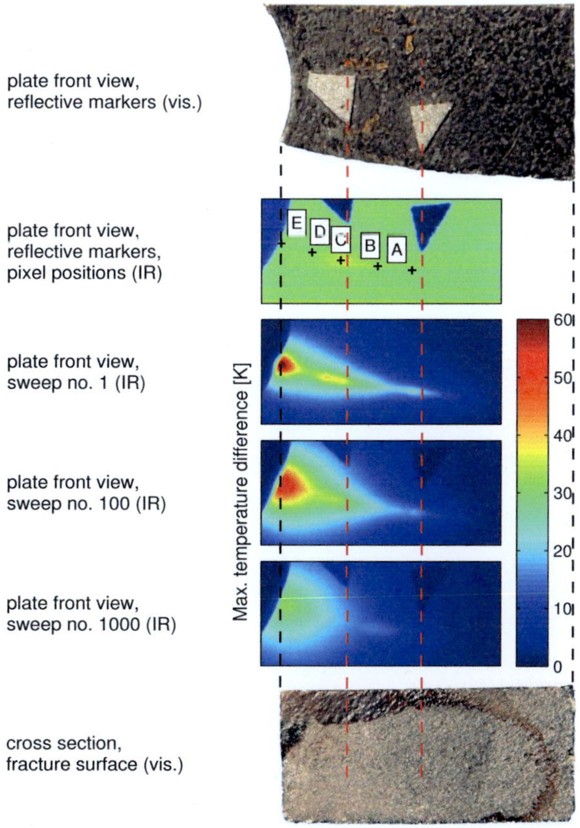

Figure 4.39: Distributions of max. ΔT aligned with the plate front view and fracture surface; specimen M-IV.

Testing of the plates M-III and M-V revealed some outliers as can be seen in Figure 4.41. Since these individual values of max. ΔT are significantly lower (below 50 % of the regularly achieved level) the author believes that these are resulting from a spontaneous generator shut-off due to overloading. Unfortunately, this point cannot be clarified retrospectively because monitoring of the generator output signal that indicates an overload shut-off was not implemented in the software. However, manual analysis of the suspicious infrared sequences revealed a sudden drop of temperature at about 30 % of the sweep duration. In contrast to the other recordings, no further heating could be found after this event. Most likely, this behaviour resulted from an incomplete frequency sweep.

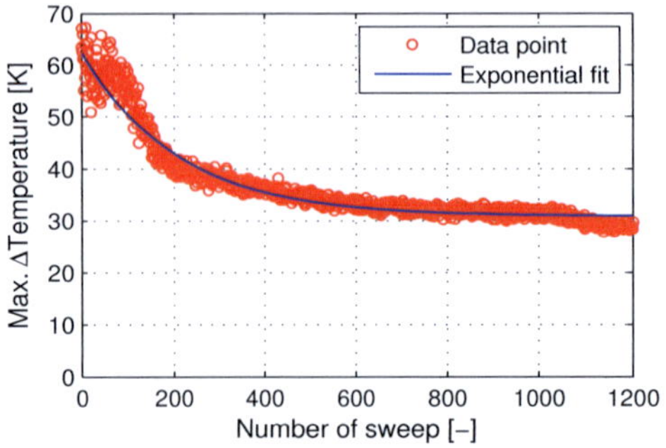

(a) Max. ΔT and exponential curve fit; point E

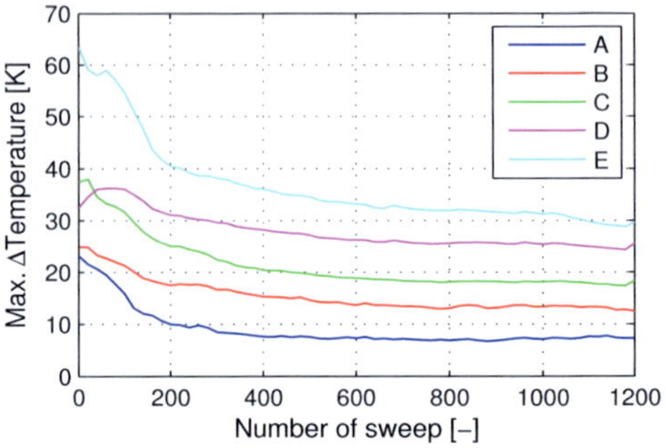

(b) Smoothed values of max. ΔT; points A-E

Figure 4.40: Decrease of max. ΔT with increasing number of sweep runs; specimen M-IV.

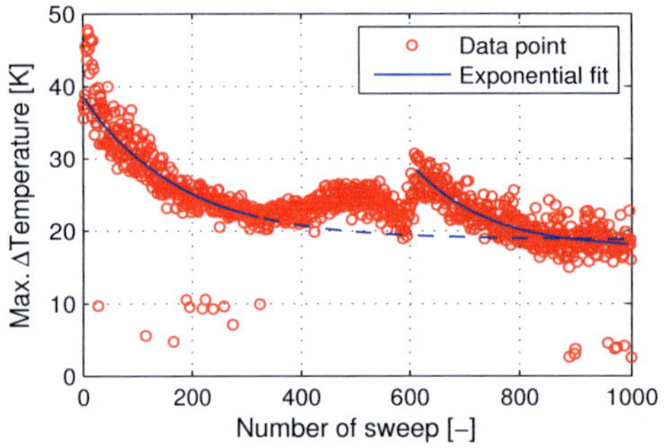

(a) Max. ΔT and exponential curve fit; specimen M-III

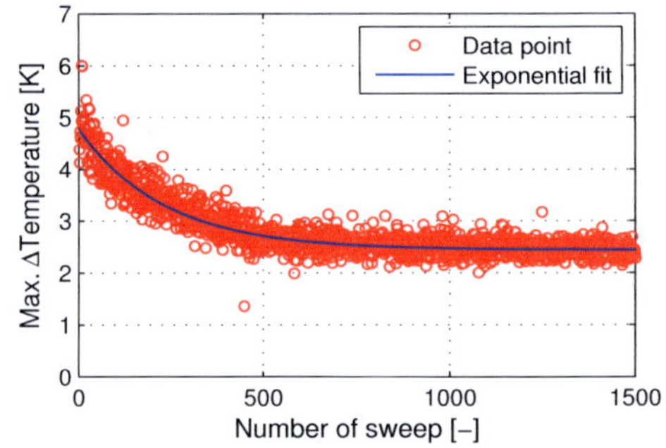

(b) Max. ΔT and exponential curve fit; specimen M-V

Figure 4.41: Decrease of max. ΔT with increasing number of frequency sweeps; specimens M-III and M-V.

A standard least-squares fitting of the experimental data would be strongly affected by the present outliers because squaring the large residuals magnifies their influence. Therefore, a robust least-squares minimisation is applied which incorporates iterative bisquare weighting of the observed max. ΔT values. The weighted sum of squared residuals r_i

$$S = \sum_i w_i \cdot r_i^2 \tag{4.19}$$

is minimised, where the weight factors w_i depend on the residuals itself. At first, all w_i are set to 1.0. If a data point is close to the initially fitted curve, it is rated as high-quality data and gains in weight. In contrast, an extremely large residual indicates data of poor quality and gets a lower weight factor.

The optimisation procedure requires several iterations until the fit converges. The weight function $w(i)$ according to Beaton and Tukey (1974) is defined as [HW77, DO89]:

$$w_i = \begin{cases} \left(1 - u_i^2\right)^2 & \text{for } |u_i| \leq 1 \\ 0 & \text{for } |u_i| > 1 \end{cases}, \qquad u_i = \frac{r_i}{B \cdot \sigma} \tag{4.20}$$

The parameter B is known as a tuning constant, which is equal to 4.685 according to [HW77]. An estimation of the standard deviation of residuals r_i is commonly given based on the median absolute deviation (MAD):

$$\sigma = \frac{\text{MAD}}{0.6745} = \frac{\text{Med}\{|r_i|\}}{0.6745} \tag{4.21}$$

The division by 0.6745 is made so that for normally distributed data σ is a robust estimate of the standard deviation [SCR88]. Further down-weighting of high-leverage points is achieved by an additional normalisation of r_i prior to evaluation of equation 4.20. This refined approach is described by DuMouchel [DO89] and implemented in MATLAB [The08].

Table 4.12 presents the found coefficients of the exponential decay curves.

Table 4.12: Exponential fit coefficients of long-term temperature decrease.

Specimen		a [−]	b [−]	c [−]
M-III	(1st branch)	19.42	-5.82×10^{-3}	18.91
	(2nd branch)	10.61	-7.97×10^{-3}	17.89
M-IV	(Pos. E)	31.37	-4.81×10^{-3}	30.87
M-V		2.32	-4.31×10^{-3}	2.44

As opposed to the previously described behaviour, the thermal hot spots detected during the tests of specimens M-I and M-II changed their position along the crack path. In both cases, the hot spots steadily relocated from the crack tip towards the fillet where the fatigue crack initiated. Figures 4.42a and 4.43a illustrate the moving hot spot of specimens M-I and M-II, respectively. A possible explanation for this behaviour might be found if one considers localised wearing and abrasion of crack face asperities.

It is very likely that the residual crack closure stress, which has its maximum near the crack tip, prevents the crack faces from rubbing/clapping displacements in the vicinity of the crack front. Therefore, the thermal hot spot will not shift towards the crack tip. Instead, it is conceivable that several asperity peaks in the current friction zone are worn down sequentially leading to a certain stress relief. The compressive prestress of the crack faces is partially reduced, and neighbouring asperities are becoming decisive regarding frictional heating. Thus, the thermographically detected hot spot is moving towards the notch base.

The graphs in Figure 4.42b and 4.43b show the smoothed developments of max. ΔT at the five discrete positions along the moving hot spots of specimens M-I and M-II, respectively. Each individual curve still manifests an exponential decay. Clearly, the steady relocation of the heating zone affects the trend curves. Based on the sweep number at which the maximum thermal response at each point A to E is achieved, the hot spot relocation can be estimated.

It is found that positions A to C are passed rather quickly. Then, the rate of propagation significantly lowers. An automated extraction of the hot spot pixel position from each single infrared record confirmed this finding. After a steady progress, the absolute relocation apparently converges asymptotically as shown in Figures 4.42c and 4.43c. The conversion from pixel positions to millimetres is based on the distance measurement of the two reflective markers, which have been applied prior to testing.

The observed behaviour suggests that the asperity region which is initially relevant for heating is worn quickly so that a redistribution of the friction zone occurs and neighbouring roughness peaks mainly contribute to the heat dissipation. Probably, the local gap between the crack faces remains larger than at the initial state so that the frictional heating is degressive, resulting in an exponential decay over the number of excitation passes.

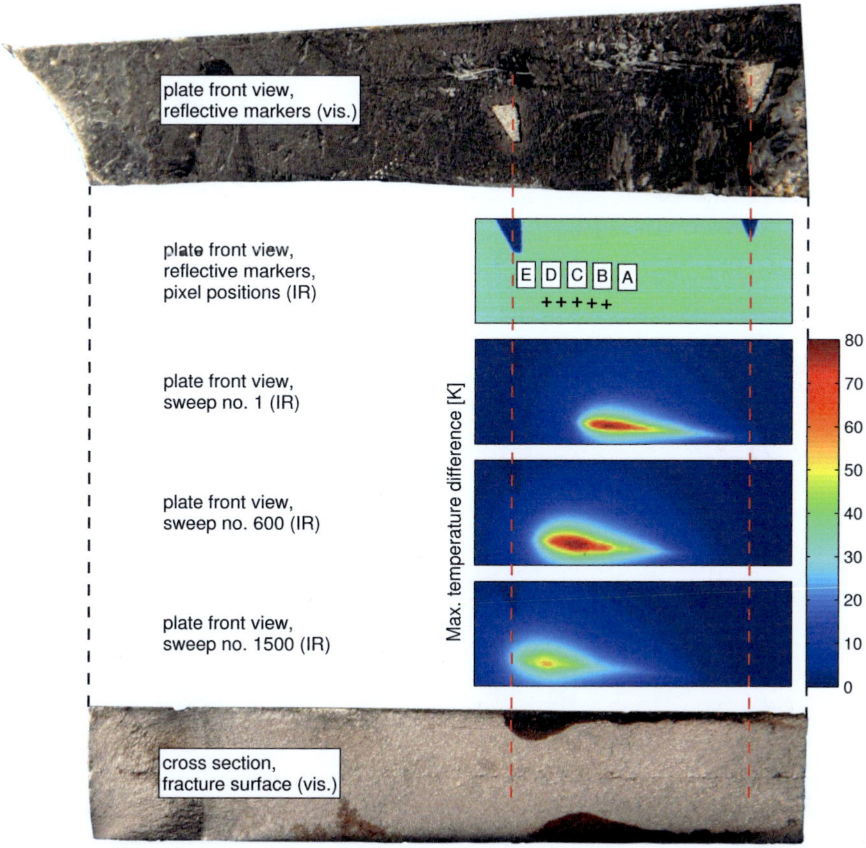

(a) Distributions of max. ΔT aligned with the plate front view and fracture surface

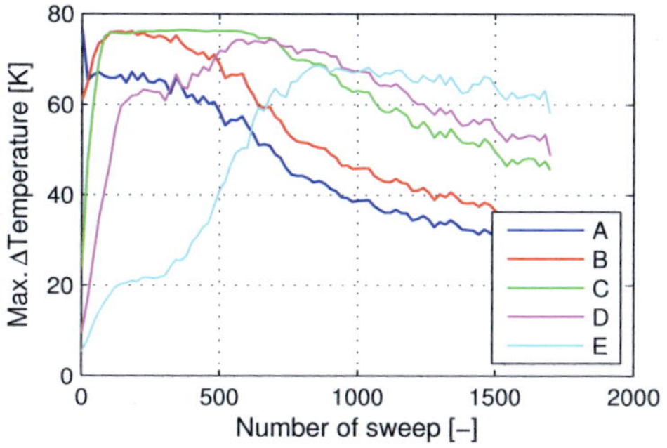

(b) Evolution of max. ΔT; points A-E

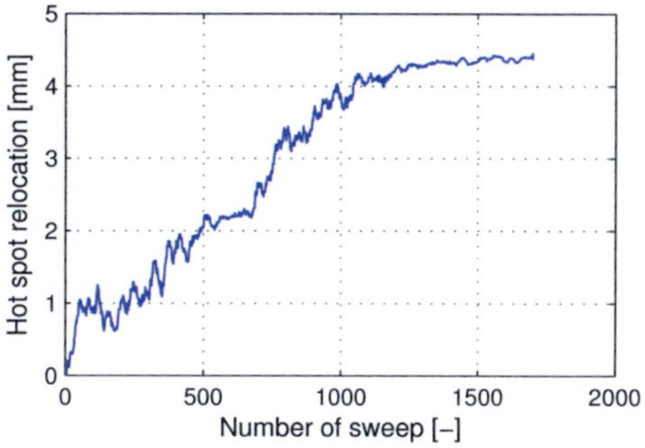

(c) Evolution of hot spot relocation

Figure 4.42: Distributions of max. ΔT, plate front view and fracture surface, degressive thermal response and hot spot relocation; specimen M-I.

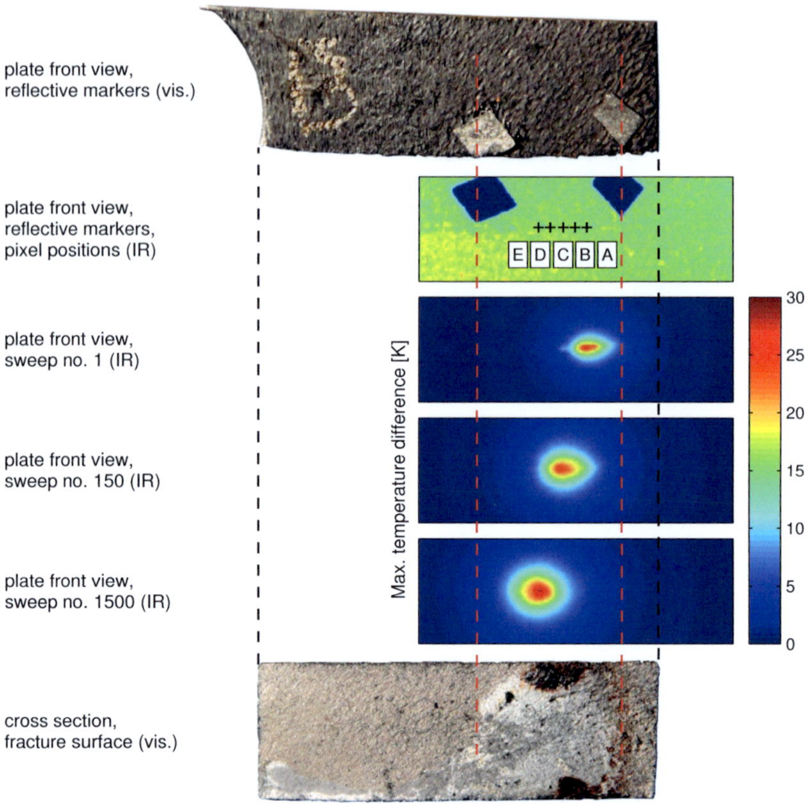

(a) Distributions of max. ΔT aligned with the plate front view and fracture surface

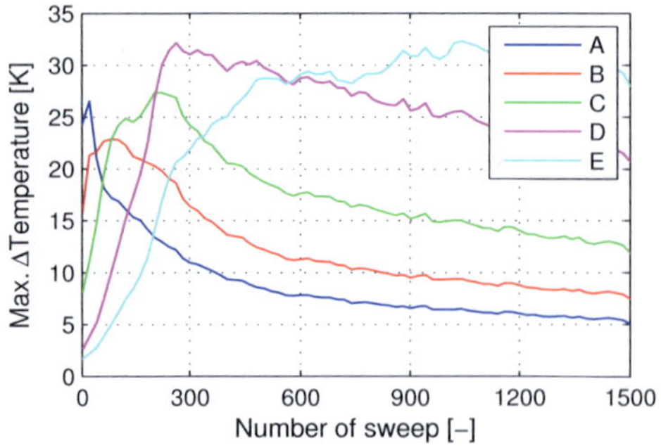

(b) Evolution of max. ΔT; points A-E

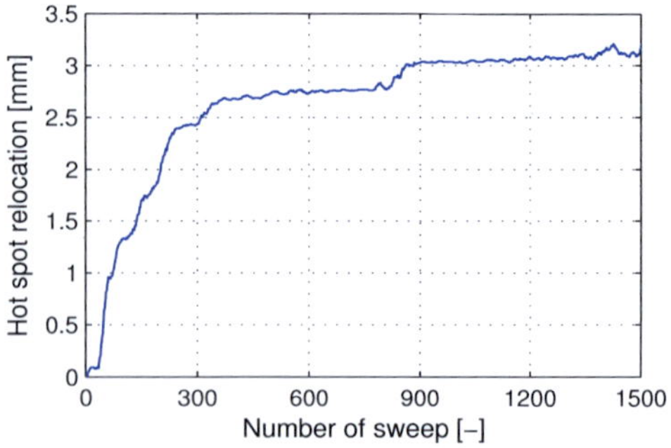

(c) Evolution of hot spot relocation

Figure 4.43: Distributions of max. ΔT, plate front view and fracture surface, degressive thermal response and hot spot relocation; specimen M-II.

4.5.2 Crack initiation and wearing of crack faces

The aligned photographs of the fracture surfaces presented in the previous section reveal that the locations of thermographic crack indication perfectly correspond to those surface regions which are visibly altered in their structure. All investigated specimens manifest some highly reflective metallic regions, which are surrounded by extended red brown areas, as can be seen in Figures 4.39, 4.42a and 4.43a. During inspection of the plate specimens, right after each test series was finished, red brown particles could be observed that accumulated on the plate next to the crack path. These findings give evidence that certain wear mechanisms must have occurred. From a visual inspection of all fracture surfaces it is believed that oxide layers have been formed.

The tribological damages have been investigated by means of confocal laser microscopy. Figure 4.44 exemplarily illustrates regular surface morphologies far from the crack face boundaries and the identified wear regions. The fracture surface manifests a granulous structure with well defined asperities. In contrast, the reflective regions in Figure 4.45, which have been probably subjected to wear mechanisms, reveal much smoother and featureless morphologies. In Figure 4.45b just a few sharply bounded surface irregularities can be found that look like craters. These might be the result of adhesive wear if asperities break off from one surface and adhere to the mating surface.

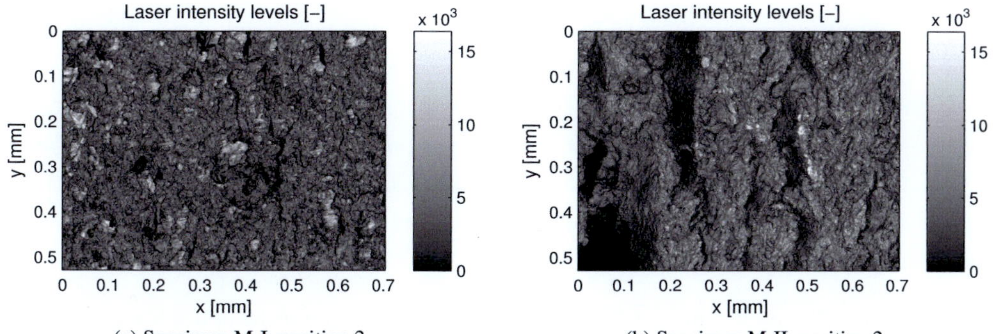

(a) Specimen M-I, position 2 (b) Specimen M-II, position 2

Figure 4.44: Laser intensity levels of regular fracture surfaces (20× magnification).

Renshaw and colleagues [RHTA11] recently presented SEM images of fracture surfaces of brass and titanium (TI 6-4) after repeated insonification. Adhesive wear, oxidation and melting could be identified although bulk temperature increases did not exceed 16 K. According to the statements of Renshaw on oxidation, tightly attached or even consolidated oxide layers are unlikely if the local heating did not approach the material's melting temperature. However, it is well known from other works, e.g. [CH10, SR10, AAK91], that flash temperatures of contacting asperities can reach values greater than 1000 °C. Taking the enormous microstructural heating into account, it is believed that the same mechanisms of wear occurred during the present experiments since the bulk heating was significantly greater compared to that achieved by Renshaw et al.

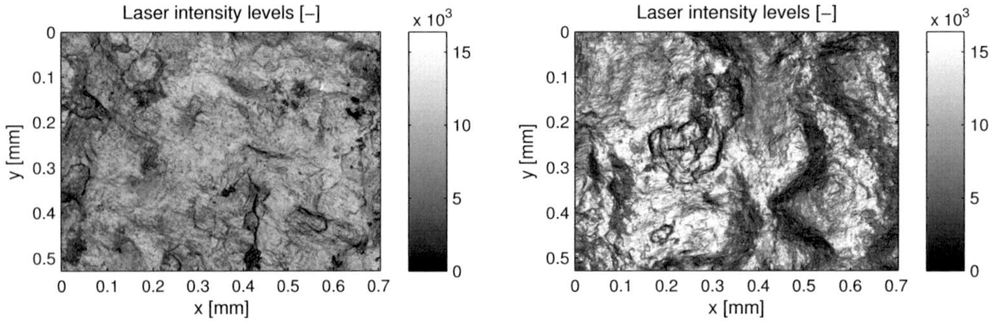

(a) Specimen M-II, position 1

(b) Specimen 10-24-I, position 1

Figure 4.45: Laser intensity levels of metallic/oxidised regions identified on fracture surfaces (20× magnification).

In order to quantify the smoothing of the hot spot regions, the surface roughness was evaluated based on the height profile, which is an outcome of the confocal laser microscopy. Roughness measurements usually involve different steps of raw data processing prior to determining any roughness parameters. According to ISO 4288, the required single measurement length l_r itself depends on roughness parameters R. So, if the measured roughness falls outside the corresponding range of values, l_r is to be adjusted and a new measurement is performed. Finding the valid length l_r is therefore an iterative procedure.

As a first step in data processing, simple geometric functions, e.g. a plane or a sphere, are fitted to the sampled 3D surface profile. Subtraction of the fit function ensures levelling of the profile so that the inclination due to non-ideal mounting of the extracted crack half is compensated for. Noise reduction was achieved by three consecutive filtering passes using a 7×7 median filter first, then a 5×5 Gaussian filter and again a 7×7 median filter. Tilt correction and denoising was automated by means of the software Keyence VK Analyzer. Automated estimations of the average height of the roughness profiles Ra and the maximum height of the profile Rz indicated an appropriate measurement length of 2.5 mm. Since this length is not available at all hot spot regions showing wear, the next lower nominal value $l_r = 0.8\,\mathrm{mm}$ and the corresponding total measurement length $l_n = 5\,l_r$ was chosen according to ISO 4288. Further processing using MATLAB involved the definition of several line profiles of length l_r sampled along the exported map of height values.

In order to obtain the roughness profile, a low-pass and a high-pass filter are applied. A discussion of the weighting function of a phase correct Gaussian filter according to ISO 11562 can be found in [MR09]. The function s that defines the filter characteristic in the time/spatial domain is given by

$$s(x) = \frac{1}{\alpha\,\lambda_{\mathrm{co}}} \cdot \exp\left[-\pi\left(\frac{x}{\alpha\,\lambda_{\mathrm{co}}}\right)^2\right] \tag{4.22}$$

with $\alpha = \sqrt{\ln 2/\pi} = 0.4697$ and x being the position from the origin of the function and λ_{co} the (low-pass) cut-off wavelength. Figure 4.46a shows a graph of the normalised weighting function. Muralikrishnan and Raja derive the amplitude transmission characteristic $S(\lambda)$ by performing the Fourier transform of the continuous function $s(x)$ [MR09]:

$$S_{\mathrm{LP}}(\lambda) = \int_{-\infty}^{\infty} \frac{1}{\alpha\,\lambda_{\mathrm{co}}} \cdot \exp\left[-\pi\left(\frac{x}{\alpha\,\lambda_{\mathrm{co}}}\right)^2\right] \cdot \exp\left(\mathrm{i}\,\lambda\,x\right)\,\mathrm{d}x = \exp\left[-\pi\left(\frac{\alpha\,\lambda_{\mathrm{co}}}{\lambda}\right)^2\right] \tag{4.23}$$

The filter transmission of a sinusoidal waveform is exactly 50 % at the cut-off wavelength. This can be easily proven by setting $\lambda = \lambda_{\mathrm{co}}$. An essential characteristic of the Gaussian filter is the complementary definition of a corresponding high-pass filter, which is given through

$$S_{\mathrm{HP}}(\lambda) = 1 - S_{\mathrm{LP}}(\lambda) = 1 - \exp\left[-\pi\left(\frac{\alpha\,\lambda_{\mathrm{co}}}{\lambda}\right)^2\right] \quad . \tag{4.24}$$

Both S_{LP} and S_{HP} are used for cutting off the very short wavelengths $\lambda < \lambda_{\mathrm{s}}$ and also for separating the roughness profile from the waviness and form deviation with $\lambda > \lambda_{\mathrm{c}}$. Figure 4.46b illustrates the transmission characteristics and the wavelength regions with $\lambda_{\mathrm{c}} = 0.8\,\mathrm{mm}$, $\lambda_{\mathrm{s}} = 2.5\,\mu\mathrm{m}$ (according to ISO 3275) and λ_{f} arbitrarily chosen to be 8 mm. In order to obtain correctly filtered data at the profile boundaries, the input profile needs to be extended. This ensures the filter function s can overlap the profile so that the filter's centre point matches the first and the last data point of the profile, respectively, if shifted along the input stream. Therefore, the roughness profile was based on seven parallel and equidistantly spaced line profiles of length l_r selected from the raw surface height map. After removing a linear trend from each profile they have been strung together and treated as one continuously sampled raw profile. With the low-pass cut-off wavelength $\lambda_{\mathrm{co}} = \lambda_{\mathrm{s}}$

the discretised version of the Gaussian filter curve s defined by equation (4.22) is normalised to $s' = s/\sum s$. The primary profile P which contains all wavelengths greater than λ_s is then found by discrete convolution of the normalised weighting function and the sampled raw profile D. A discussion of the convolution integral and its discrete realisation can be found in [Bri97].

$$P_i = \left(s' * D\right)_i = \sum_{n=-\infty}^{\infty} s'_n \cdot D_{i-n} \tag{4.25}$$

The roughness profile R is finally obtained by subtracting the waviness using $s'(\lambda_c)$ from the primary profile P. At this point the resulting profile is truncated corresponding to the original measurement path of length $l_n = 5\,l_r$, rejecting the extensions by the trailing input data and two convolutions.

$$R_i = P_i - \left(s' * P\right)_i = P_i - \sum_{n=-\infty}^{\infty} s'_n \cdot P_{i-n} \tag{4.26}$$

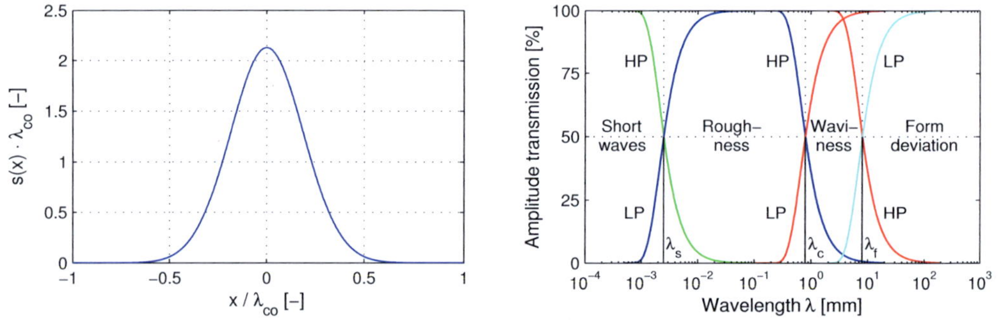

(a) Normalised weighting function of a phase correct Gaussian filter according to ISO 11562

(b) Filter transmission and definition of roughness, waviness and form deviation wavelength bands

Figure 4.46: Filter characteristics and definition of cut-off wavelengths.

Figures 4.47 and 4.48 exemplarily depict the surface morphology, the raw surface profile and the obtained roughness profiles at the regular fracture surface and at the location of visible wear, respectively. From the R profile several measures of roughness can be derived according to EN ISO 4287. The most common parameters Rz and Ra are based on the profile showing peaks and valleys oscillating around the centre line. Rz is simply found by summing up the height of the greatest peak and the depth of the deepest valley within each distance l_r. The mean roughness depth is the average of the five single values. Ra denotes the arithmetic mean of the absolute distances from the centre line. It is well known that the basic roughness measures are not very useful to characterise and classify different types of profiles. More conclusive parameters can be deduced from the material ratio curve Rmr, which is also known as the Abbott-Firestone curve or the bearing area curve (BAC). This curve is constructed based on the total profile heights c and represents the ratio of accumulated sections filled with material and the total measurement length at each profile height c. The centre region of the Abbott curve that contains 40 % of all measured points is found by moving a secant along the curve and computing its gradient. The one with minimal slope is then linearly extrapolated so that its values at Rmr = 0 % and Rmr = 100 % define the core height Rk of the roughness profile. In addition, the peak height Rpk and the valley height Rvk are found by construction of triangles bounded at 0 and 100 % and horizontal lines which intersect the Abbott curve at Mr1 and Mr2. The triangles have the same area as the Rmr curve above and below the core (compare Figs. 4.47d and 4.48d). It should be noted that evaluation of the material ratio curve according to EN ISO 13565 requires a different filtering approach to obtain the roughness profile. Since this applies only to surfaces having stratified functional properties, the roughness profile R was used instead.

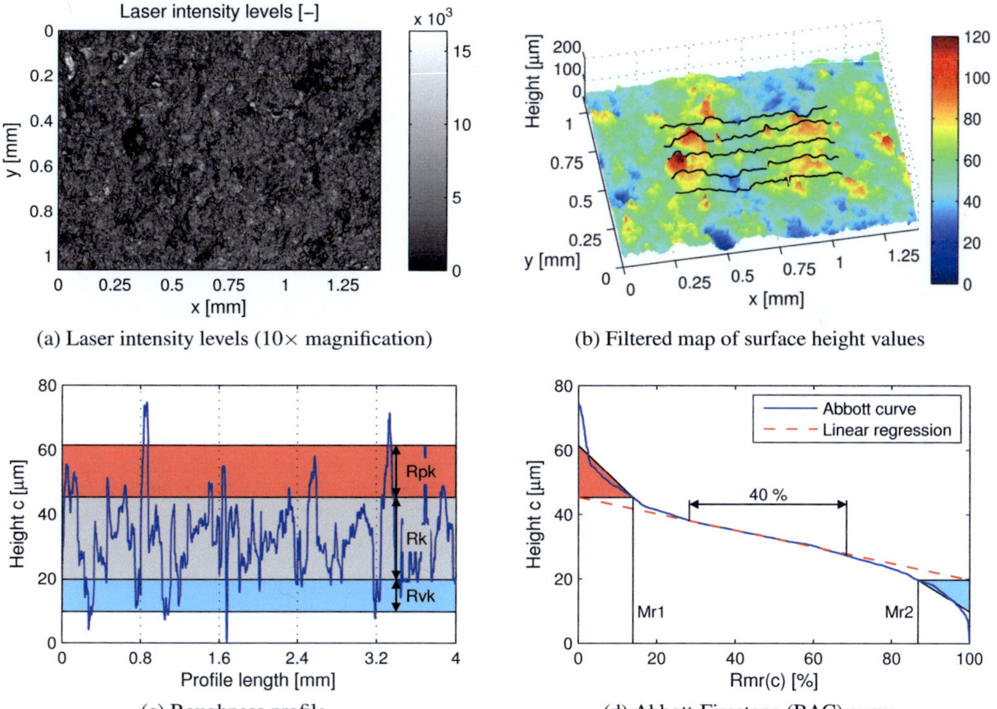

(a) Laser intensity levels (10× magnification)

(b) Filtered map of surface height values

(c) Roughness profile

(d) Abbott-Firestone (BAC) curve

Figure 4.47: Microscopic roughness measurement and evaluation; specimen M-II, position 2.

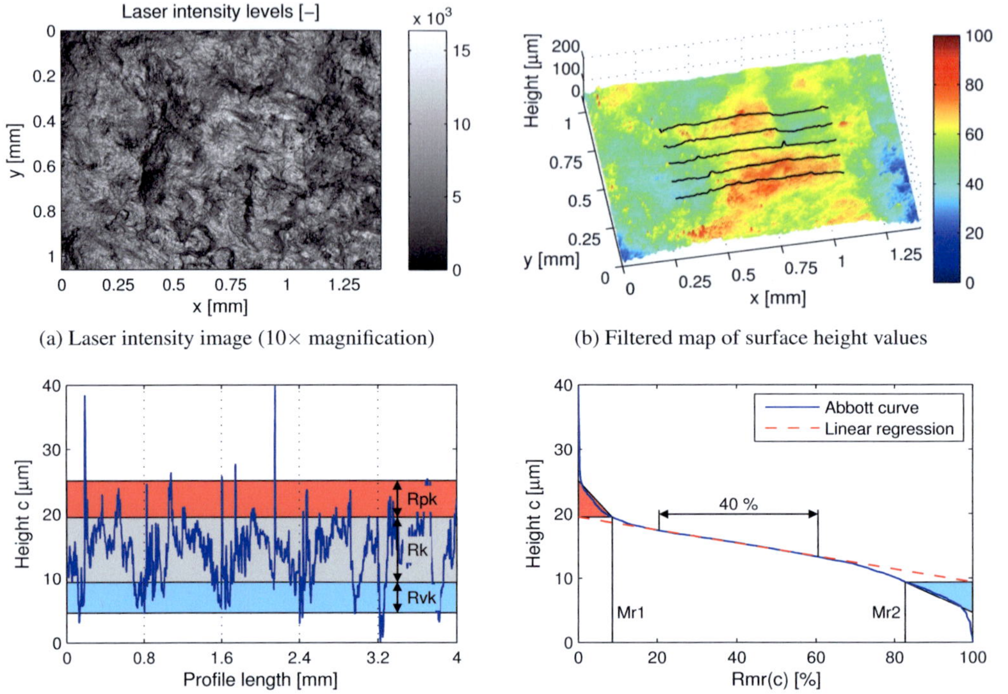

(a) Laser intensity image (10× magnification)

(b) Filtered map of surface height values

(c) Roughness profile

(d) Abbott-Firestone (BAC) curve

Figure 4.48: Microscopic roughness measurement and evaluation; specimen M-II, position 1.

Roughness measurements on specimens M-I, M-II, M-IV and 10-24-I (see also Appendix A.5) clearly proved that the surface regions which could be assigned to the thermal hot spots are significantly smoother compared to a region on the regular fracture surface located at the same crack depth. Furthermore, it was found that the fracture surface region near the crack front is rougher than that next to fillet radius. This was expected since fatigue loading was force controlled and the stress intensity steadily increased with the crack depth. This in turn leads to a steeper cleavage when approaching greater crack depths. Table 4.13 summarises all roughness parameters evaluated within this study. Unfortunately, the present findings cannot be directly compared to the work of Renshaw et al. [RHTA11] since they did not find locally restricted wear regions that can be identified by the naked eye. Instead, the cited investigation on surface cracks in titanium bars revealed that the locations of crack heating correspond to an extended fretting band with a higher roughness than the surrounding regions. Renshaw mentions that adhesive wear might cause a reduction or an increase of interface roughness, whereas melting is likely to reduce asperity peaks.

Table 4.13: Results of fracture surface roughness measurements.

Specimen	Position	Ra [µm]	Rz [µm]	Rk [µm]	Mr1 [%]	Rpk [µm]	Mr2 [%]	Rvk [µm]
10-24-I	1 – metallic/oxidised region	5.1	37.7	15.3	5.2	6.5	87.0	8.1
	2 – regular fracture surface	14.2	71.1	47.5	9.6	11.2	91.8	12.7
	3 – region near notch bottom	4.2	24.0	13.0	10.7	6.5	89.7	5.7
	4 – region near crack front	15.3	80.6	46.4	8.3	6.5	87.3	25.3
M-I	1 – metallic/oxidised region	3.2	24.0	9.5	15.2	5.5	93.2	2.3
	2 – regular fracture surface	3.8	31.0	11.8	11.9	5.9	90.6	4.4
M-II	1 – metallic/oxidised region	3.6	26.9	10.1	8.7	5.7	82.7	4.8
	2 – regular fracture surface	9.1	55.8	25.5	13.9	16.2	86.8	9.9
	3 – region near notch bottom	4.9	26.9	16.9	8.0	4.3	91.5	5.2
	4 – region near crack front	8.1	43.6	27.3	9.8	11.5	93.0	8.1
M-IV	1 – metallic/oxidised region	3.6	17.6	13.2	7.0	2.5	94.2	1.9
	2 – regular fracture surface	5.3	33.5	16.0	12.0	6.5	88.7	7.6

It can be concluded that sliding friction causes wear of crack faces. Though being limited to relatively small hot spots, severe surface damages have been identified. It remains unclear to what extent the different mechanisms like fretting, adhesive wear and phase transformation contribute to the alteration of the surface structure. As indicated by oxidation and reduced roughness parameters at the hot spots, the author believes that extreme microstructural flash temperatures occurred during ultrasonic excitation so that adhesive wear and melting probably have been the dominating effects.

In the course of the present study the existing fatigue cracks did not propagate. In case of bulk damage like crack growth, heating at the crack tip vicinity would have been expected but was not observed either. Instead, plate specimen 10-24-1 features an additional surface flaw that has been generated during repeated insonifications. Visual inspection revealed that it initiated from a strongly attrited region of the original fracture surface. Figure 4.49 provides aligned photographic and thermal results.

Regarding the thermal signature, the secondary crack apparently did not contribute significantly to the overall heating at first sight. The temperature distributions of max. ΔT, however, give evidence for a strongly asymmetric heat generation. The steep gradients, particularly found at sweep number 400, can be clearly attributed to the material discontinuities caused by the primary and the secondary crack. Since the new crack manifests an extremely tilted path, about 20° to the plate surface and pointing towards the metallic area, as seen on the presented fracture surface, it is believed that the secondary crack actually caused the unilateral heating. This would be in agreement with the hot spot relocation from the initiation point to the attrited region. In comparison, specimens M-I and M-II revealed the opposite direction of hot spot movement. Thus, the secondary crack might have even represented the prevailing heat source.

As mentioned before, testing of specimen 10-24-1 suffered from a control error so that the consecutive sweeps progressively increased in duration. Further analysis showed that the first 120 repetitions were not affected and could be performed within 12 s each. Thus, the secondary crack mainly developed within the next 280 excitations, whereas the last sweeps took extremely long and probably exceeded four minutes. Severe damages due to ultrasound excitation can therefore not be excluded in general, even if relatively thick-walled steel plates are examined.

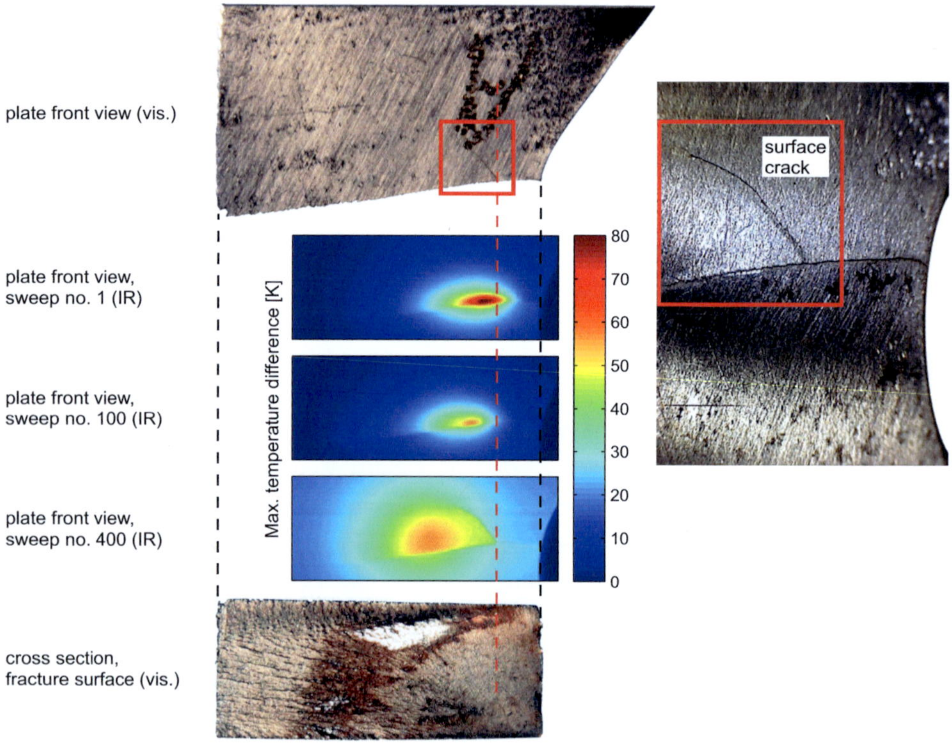

Figure 4.49: Distributions of max. ΔT aligned with the plate front view and fracture surface; indication of a secondary crack; specimen 10-24-I.

4.6 Simulation of in-situ crack detection

In the previous sections isolated plates have been investigated in detail, where the excitation was enabled by means of the screw-coupled transducer. In order to further examine the applicability of ultrasound excited thermography in the field of steel construction, more realistic testing is required. This section focuses on several issues that are most important for in-situ crack detection. These include the dependencies on the coupling location and the structural damping as well as the use of portable excitation methods.

The experiments described in the following were conducted on a 3.2 m long hot-rolled girder (IPE 500, European steel grade S355J2+N) that features fatigue cracks at the fillet radii of two square web openings. Figure 4.50 shows a drawing of the steel beam including thread holes at random positions for direct coupling of the ultrasound transducer. In the late 1990s the fatigue behaviour of this specimen was investigated at the Institute of Steel Structures, Braunschweig University, Germany. Mehdianpour [Meh03] performed three-point bending tests and studied the effects of random load histories. Later, Frenz [Fre10] focused on a more complex setup including load redistributions. Applying a load to the centre of the girder leads to shear deformations of the web openings. Thus, two corners diagonally across from one another exhibit large tensile notch stresses σ_1 while the other ones are compressively stressed at the same time. Although only compressive loads were applied to the present girder, all of the eight corners are cracked due to fatigue testing in the HCF

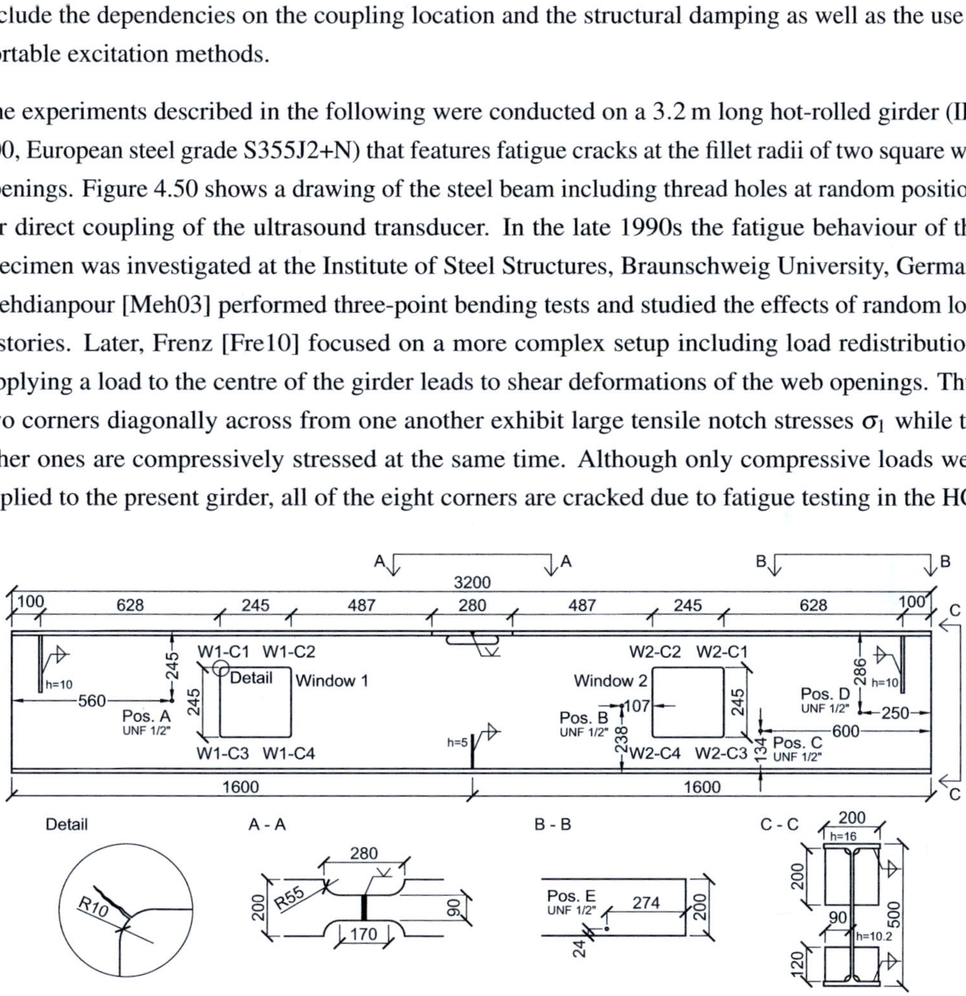

Figure 4.50: Dimensions, cracks and coupling positions of the investigated steel girder IPE 500 made from S355J2+N.

regime. Table 4.14 summarises the denotations, the rather different depths of the cracks and the corresponding primary notch loading when neglecting any creep or ratcheting effects.

Table 4.14: Crack depths measured on the hot-rolled girder and types of notch loading.

Crack	W1-C1	W1-C2	W1-C3	W1-C4	W2-C1	W2-C2	W2-C3	W2-C4
a [mm]	47.7	8.8	10.0	28.6	74.6	6.9	12	36.3
σ_1 [−]	positive	negative	negative	positive	positive	negative	negative	positive

4.6.1 Influence of the coupling position and the structural damping

The investigation of plate specimens in Section 4.4.2 manifests a possible dramatic decay of the achieved defect heating in the order of 75 % if an "inefficient" coupling location is accidentally chosen. To complement this study, the Branson ultrasound transducer was consecutively screwed at the four coupling positions A, B, C and D (compare Fig. 4.50) and all of the eight cracks were thermographically monitored during a 70 s frequency sweep in the range of 19 to 23 kHz.

In a next step, the structural damping of the girder was artificially increased by placing 25 kg bags filled with granular blasting abrasive onto the upper flange and onto the lower flange to establish an additional contact with the web. Figure 4.51 shows the test setup with full damping load applied. Intermediate damping was realised by removing the marked bags. Throughout this study, all frequency sweeps were performed at nearly the same nominal excitation level, namely 9 or 10 % to allow for direct comparison.

Table 4.15 presents the achieved maximum temperature increases max. ΔT in the initial state without extra damping. Two of eight cracks, W1-C4 and W2-C1, which are both located at notches exhibiting tensile stresses under loading, were not detected at all. Defect imaging failed even after thoroughly scanning the whole crack path during repeated tests. With the exception of crack W1-C1, which could not be detected when using coupling position D, the remaining five of eight cracks were reliably detected by means of ultrasound excited thermography. Since the variation of achieved thermal responses reveals relative decreases of about 60 % in case of steadily detected flaws, it is concluded that a single-point excitation might be insufficient to evaluate a metal component. Using the screw-coupled transducer to perform sweep excitations, the relative power level was limited close to the previously chosen level. So, this parameter is probably not capable to fully compensate for inefficient coupling locations.

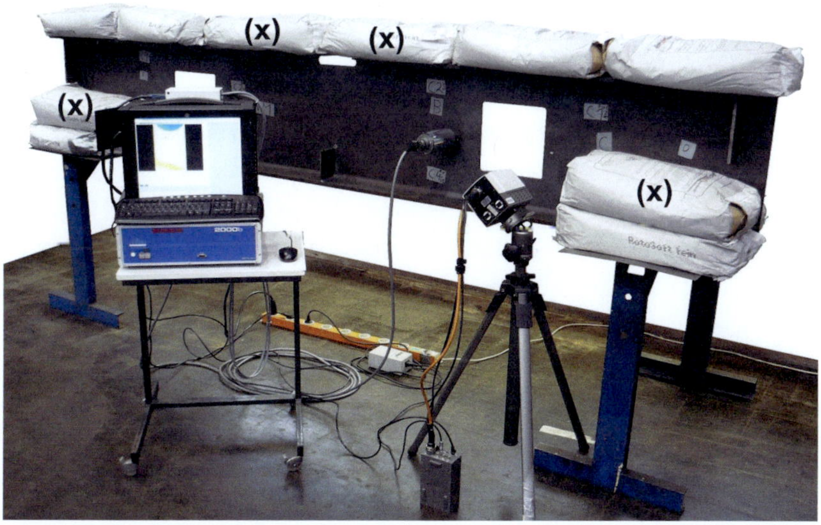

Figure 4.51: Test setup for vibrothermography of a 3.2 m long hot-rolled girder optionally loaded by 8 or 14 bags weighing 25 kg each.

With regard to the previous results, the cracks W1-C4 and W2-C1 were not investigated in the damped setup since a further decrease of the potential crack heating was expected. To quantify the effect of the damping bags, one-point laser vibrometry measurements at one crack side of W1-C3 (number 1 to 5) and at the web (number 6 to 8) have been performed and the free vibration was analysed after excitation at several resonance frequencies. For this purpose the TIB ultrasound system was engaged since it provides a more accurate oscillation regarding the frequency stabilisation and the total harmonic distortion (see Chapter 5 for details). Coupling position C was arbitrarily chosen to be used for excitation.

Table 4.15: Maximum defect heating on the girder without any extra damping [K].

Coupling location	Cracks at window 1				Cracks at window 2			
	W1-C1	W1-C2	W1-C3	W1-C4	W2-C1	W2-C2	W2-C3	W2-C4
A	0.16	0.31	0.76	–[1]	–[1]	1.00	1.65	0.28
B	0.14	0.31	0.55	–[1]	–[1]	0.78	1.68	0.25
C	0.3	0.73	0.89	–[1]	–[1]	0.45	2.61	0.35
D	–[1]	0.33	0.85	–[1]	–[1]	1.21	2.51	0.28

[1]Crack not detected

Several resonance frequencies have been identified by a stepwise sweep and subsequent Fourier transform of the velocity history. Analysis of the amplitude spectra found for the three different damping setups did not reveal any substantial changes to the girder's eigenfrequencies in the considered ultrasound range. Most of them could be exactly retrieved within the DFT frequency resolution of 5 Hz. Those resonance frequencies having the greatest amplitude were excited for a couple of seconds in order to reach a steady state oscillation. After manually triggering the data acquisition using a sampling rate of 500 kS/s, the sine wave input to the TIB system was stopped so that the subsequent vibration decay was reliably recorded within half a second. Figure 4.52a exemplarily depicts the sampled velocity over time. The displacement history in Figure 4.52b, as found by numerical integration, clarifies that the ultrasonic oscillation is superimposed by a low frequency vibration. Applying a 5th-order Butterworth high-pass filter with a cut-off frequency of 10 kHz isolates the high-frequency content, as shown in Figure 4.52c. The decreasing out-of-plane displacement amplitude $\hat{w}$ in the range of 85 to 5 % was chosen to fit an exponential decay curve

$$\hat{w}(t) = C \cdot \mathrm{e}^{-\delta \cdot t} \tag{4.27}$$

with the scale factor C and the decay constant δ. Figure 4.52d presents the curve fit. According to Dresig and Holzweißig [DH05], Lehr's damping ratio ζ is then obtained through

$$\zeta \approx \frac{\zeta}{\sqrt{1-\zeta^2}} = \frac{\delta}{\omega} \quad , \tag{4.28}$$

considering the angular frequency $\omega = 2\,\pi\, f_{\mathrm{exc}}$.

A total of eight frequencies in the standard frequency range have been analysed in the same way. Each of the free oscillation measurements was repeated three times. Regarding the decay constant, it was found that the scatter of individual test results is very small. Relative deviations did not exceed 3 %. The variances of measured displacement amplitudes in the stationary regime were even lower. Figure 4.53 represents the averaged results for the damping ratio and the steady state displacement amplitude.

Comparing the relative damping ratios achieved at different frequencies and setting the fully damped situation to 100 % manifests a consistent increase in damping when bags are attached to and placed on the girder. Full loading results in a value of ζ which is almost twice that of the initial state (Fig. 4.53b). The achieved absolute values of Lehr's damping ratio are at least one order of magnitude lower than common values found in the literature. Dresig and Holzweißig [DH05] give an estimation of 2.5×10^{-3} for the material damping of constructional (low-carbon) steel and mention that ζ will be lower at higher frequencies. Furthermore, the authors present a diagram to illustrate the increase in damping with increasing load amplitude. Petersen [Pet96] estimates the material damping of ferritic steels at low stresses to be about 8×10^{-4}. Discrepancies between published and measured values are clearly owed to the extremely low vibrational amplitudes in the order of 0.1 μm, which were measured on the girder's web.

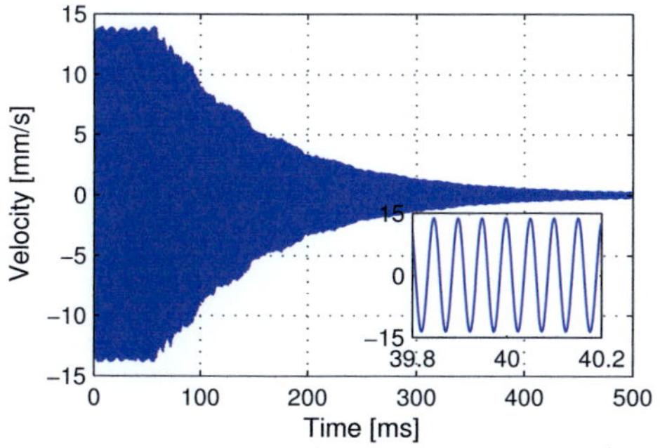

(a) Velocity measured at one side of crack W1-C3 after excitation at 19.56 kHz

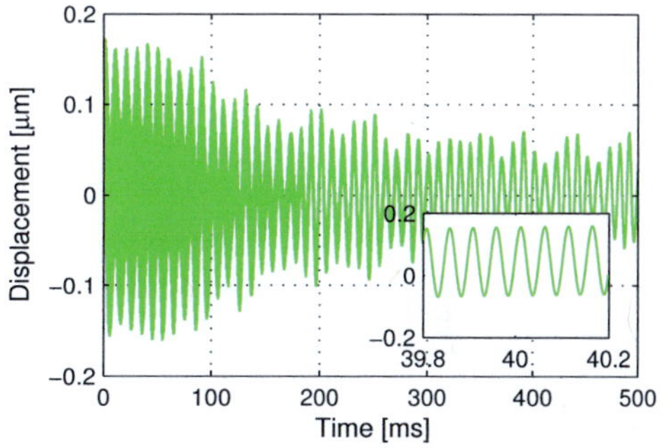

(b) Displacement found from numerical integration of the velocity history

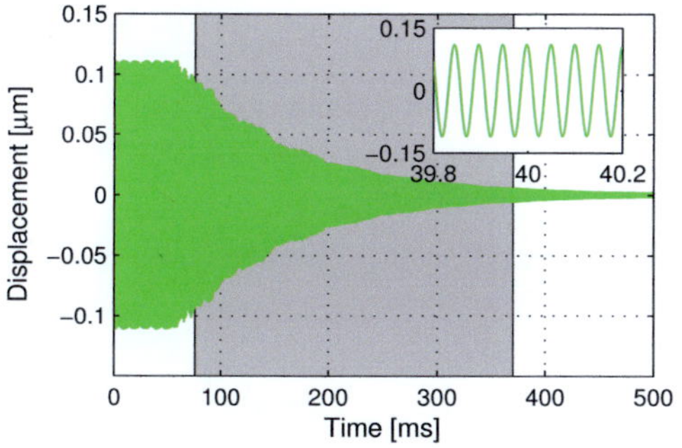

(c) Displacement after 10 kHz high-pass filtering and time interval for decay analysis

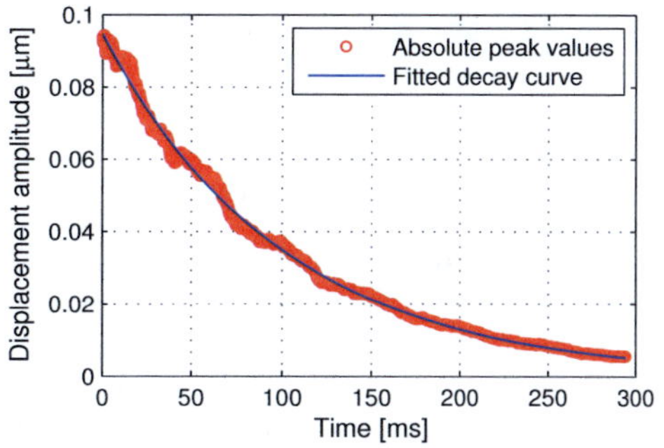

(d) Displacement amplitude decay between 85 and 5 % of the max. level with exponential curve fit

Figure 4.52: Exemplary results and processing of a free oscillation measurement by means of one-point laser vibrometry.

Plotting the displacements expressed as a percentage related to the fully damped situation, Figure 4.53d reveals that individual measurements resulted in smaller amplitudes at lower damping. The reason for that remains unclear. However, the averaged values represented by the straight lines are in agreement with the expected behaviour and indicate that the mean displacement amplitudes are indeed more restricted if the structural damping increases.

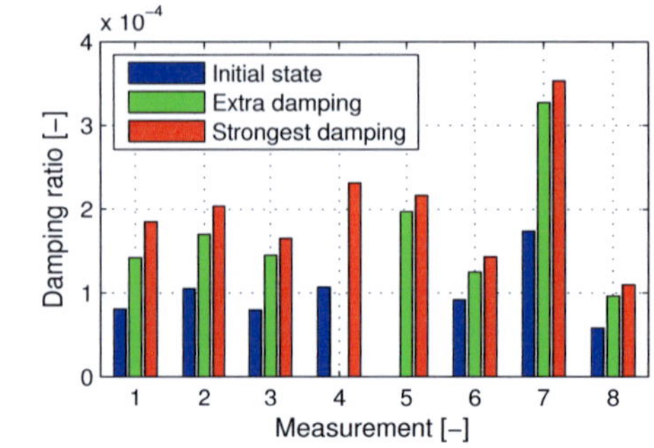

(a) Averaged damping ratios measured at various resonance frequencies

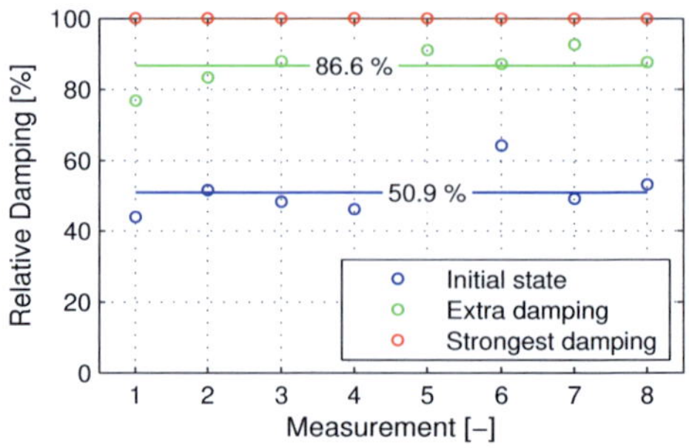

(b) Averaged damping ratios related to values at maximum damping

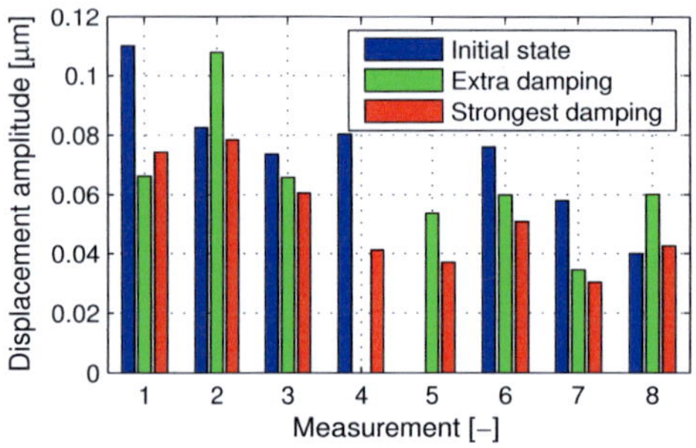

(c) Averaged stationary displacement amplitudes measured at various resonance frequencies

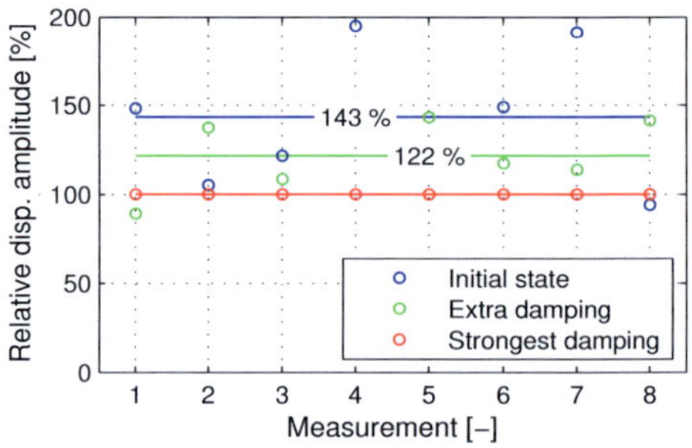

(d) Averaged displacement amplitudes related to values at maximum damping

Figure 4.53: Averaged stationary displacement amplitudes and damping ratios derived from free oscillation measurements.

Focusing on the six cracks which have been detected on the undamped girder, vibrothermographic inspection was continued using the Branson system and two coupling positions B and C, only (compare Fig. 4.50). Positions A and D were no longer accessible due to the attached bags, which should be in areal contact to the web. Despite of the extra damping that was proven effective, no clear tendency could be found regarding the change in crack heating during sweep excitation using the same parameters as before. Contrary to an expected average decrease of max. ΔT values, sometimes relative increases of more than 160 % could be observed under extra damping and strongest damping conditions. Table 4.16 summarises absolute temperature increases and relative values related to the initial state without bags attached to the girder. This study clearly demonstrates the sensitivity of the thermosonic inspection to small changes of the structure.

Table 4.16: Maximum defect heating on the damped girder [K] and relative change [%].

Damping	Coupling location	Cracks at window 1			Cracks at window 2		
		W1-C1	W1-C2	W1-C3	W2-C2	W2-C3	W2-C4
extra	B	–[1]	0.5 +61	1.13 +105	0.95 +22	0.97 −42	–[1]
	C	–[1]	0.72 −1	0.73 −18	1.2 +167	2.21 −15	–[1]
strongest	B	0.17 +21	0.24 −23	0.3 −45	0.44 −44	1.12 −33	0.16 −36
	C	0.23 −23	0.79 +8	0.84 −6	1.18 +162	5.28 +102	0.24 −31

[1]Crack not inspected

As the presented thermographic results are rather inconclusive, the question arises to what extent the weight of the bags may have led to a change of the crack opening displacements. A small alteration of the local prestress could be the reason for varying contacts of the opposing crack face asperities. Although the previously performed low-power excitation using the TIB system did not reveal any changes of the vibration amplitude spectra when bags were placed on the girder, this might be different in case of high-power excitation by means of the Branson system. However, the relatively broad sweep should have compensated a possible shift of the most efficient frequency regarding crack heating. Since the values of max. ΔT achieved from two different coupling locations are diverging when loading the beam, the observed behaviour remains inexplicable to the author. Future investigations on damping should be undertaken by means of a bituminous or light-weight aluminium butyl sheeting glued to the steel structure in order to avoid changing the membrane stress state.

4.6.2 Influence of the coupling method

As it was mentioned earlier, a screw connection between transducer and specimen is not a realistic type a coupling. Hence, the five prototypes of portable excitation devices described in Section 4.2.3 should be compared in terms of the maximum achievable defect heating. For this study the initial setup of the girder without damping bags was employed. Firstly, all coupling devices were placed on the upper flange so that the transducer, booster, sonotrode or the clamps are exactly located at coupling position E (compare Fig. 4.50). Again, ultrasound frequency sweeps have been performed in order to find the maximum heating at crack W2-C3, which was detected best before. Coupling the transducer by means of a screw connection provides reference values for the temperature increase and the applicable generator power.

The small and the large clamp were mounted as tight as possible by manually fastening the bolt with a standard size wrench. When pressing the converter, booster or sonotrode onto the flange via electromagnet and prestressed springs, the static force applied to the tip of the exciter can be estimated based on the measured spring deflections and spring stiffnesses. In case of the converter/booster press, the preload was about 1.8 kN and with the sonotrode press a force of 0.6 kN was applied. Additional tests were conducted using excitation position D on the web. Here, the clamps cannot be employed. Also, the sonotrode press based on two small magnets was found to be insufficiently upheld in case of horizontal installation. Obviously, the weight of the steel transom, together with the complete weld tool results in a bending moment that cannot be carried by the magnets, although the web thickness of 10.2 mm exceeds the minimum plate thickness to enable the nominal magnet forces.

Table 4.17 presents the observed average defect heating based on three sweep repetitions using the reference generator power as well as the maximum value of ΔT achieved with the maximum applicable power level.

Table 4.17: Achieved defect heating and excitation levels using different coupling devices.

Coupling method	Upper flange		Web		Rating
	m. $\overline{\Delta T}$ / P [K] / [%]	m. ΔT / P [K] / [%]	m. $\overline{\Delta T}$ / P [K] / [%]	m. ΔT / P [K] / [%]	
Screwed converter	0.68 / 7	0.83 / 7	5.07 / 9	5.61 / 9	good (ref.)
Small clamp	1.07 / 7	1.75 / 8	–[1]	–[1]	very good
Large clamp	0.24 / 6	0.26 / 6	–[1]	–[1]	poor
Pressed converter	$<$ 0.1 / 7	1.64 / 30	0.38 / 9	4.29 / 40	very good
Pressed booster	$<$ 0.1 / 7	0.15 / 15	$<$ 0.1 / 9	0.25 / 15	poor
Pressed sonotrode	0.64 / 7	0.93 / 7	–[1]	–[1]	good

[1]Testing not possible

Despite the limited number of test results, some cautious conclusions might be drawn which definitely need a revision within the context of future investigations focussing on portable excitation methods. Regarding the coupling position E located on the flange, the small clamp is found to provide efficient excitation that leads to stronger heating than the reference test at the same nominal generator power. Based on the variational coefficient of the achieved heating values, even a smaller scatter could be observed. In contrast to that, coupling with the large clamp barely enabled crack detection. It seems that due to the higher weight and the larger coupling area the clamp itself was excited but very little energy was induced into the girder.

Both magnet based devices pressing the converter or the booster onto to the flange completely failed regarding defect heating when applying the reference generator power of 7 %. However, the converter press allowed to chose a relative excitation power $P > 40\,\%$, which finally resulted in twice the crack heating max. ΔT observed in the reference test. Though similar in construction, the booster press did not perform better at its highest possible excitation level. The author assumes that this outcome is mainly caused by the booster that is highly tuned to have a longitudinal eigenmode at 20 kHz, which is the native working frequency of the original Branson system. This might have counterproductive effects regarding the vibration amplitudes during a broadband frequency sweep. However, exciting the girder by means of the sonotrode press containing the complete welding instrument led to crack heating similar to the reference test but revealed a larger scatter. Compared to the test run using the booster only, the maximum possible excitation level is further decreased, probably because of the sonotrode, which is also built to match the 20 kHz resonance. On the other hand, the catenary sonotrode causes a significant magnification of the vibration amplitude (factor between 1.33 and 1.78). This might compensate the limited generator power to some extent. Concerning the direct excitation of the web, the booster press fails again to generate a significant heating, even at maximum power. In contrast, the converter press again allows for crack detection at a high 40 % relative power with a heating of 76 % compared to the reference test with the screw-coupled transducer.

Based on these results the position of the exciter was varied to simulate a realistic testing situation and to determine the influence of the coupling position once again. The small clamp was relocated 40 times while its position was altered along the left side of the upper flange (compare Fig. 4.50). Similarly, the converter press, which was also proven efficient before, was moved 30 times along the left side of the web. Again, crack C2-W3 was thermographically monitored while applying frequency sweeps with 7 % and 25 % relative power, respectively. The achieved values of max. ΔT are illustrated in the form of histograms in Figure 4.54. Corresponding minimum, maximum and mean values are given in Table 4.18 together with the standard deviation (STD) and the coefficient of variation (CV).

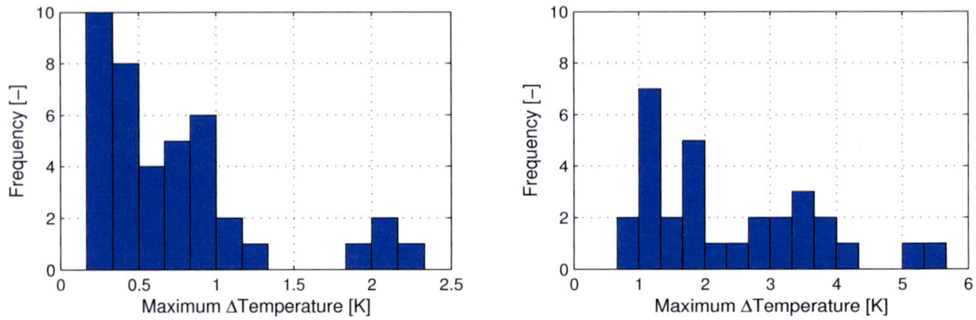

(a) Histogram of max. ΔT values achieved by means of the small converter clamp at the upper flange, bin size = 1/6 K

(b) Histogram of max. ΔT values achieved by means of the converter press attached to the web area, bin size = 1/3 K

Figure 4.54: Histograms of the maximum temperature increases achieved on the steel girder.

Table 4.18: Evaluation of max. ΔT achieved at crack C2-W3 during repeated frequency sweeps.

Type of coupling and location	Number [-]	P [%]	Min. [K]	Max. [K]	Mean [K]	STD [K]	CV [-]
Small clamp at the upper flange	40	7	0.21	2.29	0.71	0.54	0.76
Converter pressed onto the web	30	25	0.81	5.42	2.38	1.30	0.55

Obviously, the presented histograms do not allow any assumptions regarding a certain statistical distribution. For this purpose much more tests would be required. Compared to the limited number of test results the histogram bin sizes are intentionally chosen very small in order to emphasise that in both cases, web and flange excitation, two or three consecutive classes are left empty. Clearly, this study cannot refute this observation being a random effect but once again the structure seems to act as an influence field that offers a slight probability to achieve greatest crack heating and a much greater chance to observe a low defect heating. Nevertheless, there was not a single coupling position that did not lead to a measurable defect heating. Similar to the work of Hassan et al. [HHW$^+$07], ratios between maximum and minimum heating of about 11 were found when varying the excitation position.

From the present study it can be concluded that ultrasound excited thermography definitely has the potential to be further developed towards a fail-safe NDT method. However, two main issues emerged that need to be addressed in the future to finally judge on the practical reliability of the method and to specify a reasonable probability of detection.

Since the coupling position has a major impact on the achievable defect heating, a compensation by advanced excitation systems is required. The proposed coupling of the exciter by means of an electromagnet was proven viable in case of ferritic steel components. A more sophisticated excitation unit should allow for a quick relocation of the transducer tip in a certain range of operation without fully detaching and re-coupling the whole device.

As the previous investigations have shown, strongly gaping cracks could not be detected with the employed equipment. This situation will not change unless greater vibration amplitudes can be realised. Regarding the mobile application on heavy steel structure, this requirement surely represents a challenge. Genest and colleagues [GMFM09] recently investigated the use of an electromagnetic/piezoelectric dual shaker system and detected single edge cracks in aluminium sheets at extremely low frequencies of about 200 Hz. Presuming that a low-frequency excitation can facilitate large amplitude but long wavelength vibrations of the structure, a superposition with high-power ultrasound might improve the thermographic crack detection if the gap between crack faces can be closed periodically.

5 Experimental analysis – Laser vibrometry

This chapter introduces laser vibrometry, a non-contacting velocity and displacement measurement technology, which is used to investigate the actual vibration state of cracked specimens when subjected to ultrasound excitation. After a detailed characterisation of the available transducers, single-point and two-point differential measurements are conducted that aim at the establishment and understanding of the relation between crack face displacements and the thermal response of the defect region. Finally, three-dimensional vibration mode shapes are analysed and compared to theoretical predictions of frictional heating.

5.1 Test equipment and setup for laser vibrometry

This section describes the equipment used for vibration measurements on plate specimens. For this purpose a differential laser vibrometer is introduced together with a two-axis positioning system, which allows raster scanning of the plate. Three-dimensional measurements become feasible since the employed retroreflective film allows to measure surface points even through an angle of incidence that strongly deviates from 90°.

5.1.1 Two-axis positioning system with two-axis rotation stage

The basic test stand used for thermosonic inspection of plate specimens, described in Chapter 4, was complemented by a two-axis linear stage to allow for reproducible positioning of the vibrometer laser sensor. Later on, as described in Section 5.4, the system is used to perform automated raster scans along a pre-defined grid on the plate specimen. Both linear axes are of the same type PI miCos LS-110. The vertical stage is mounted through an assembly bracket on the horizontal axis slide so that a total travel range of $25 \times 25\,\text{cm}^2$ is available. A base plate which is supported by three fine-thread screws allows for relocating and levelling the whole system as shown in Figure 5.1a. The stepper motors of both axes are controlled via a PCI controller PI miCos SMC Corvus and proprietary software. According to the manufacturer, the system's bi-directional repeatability is better than $\pm 1\,\mu\text{m}$ [PI 08].

Since the assessment of a three-dimensional vibration mode shape requires angular laser measurements tilted about two different axes, a two-axis rotation stage was designed and assembled based on a Newport RSP-2T (horizontal) and an RSP-1T (vertical) rotary table. As illustrated in Figure 5.1b, the laser sensor is centred in both axes and can be manually rotated and inclined at least $\pm 45°$ with an accuracy of about $0.5°$.

(a) Two-axis positioning system carrying the laser vibrometer sensor mounted in a two-axis rotation stage

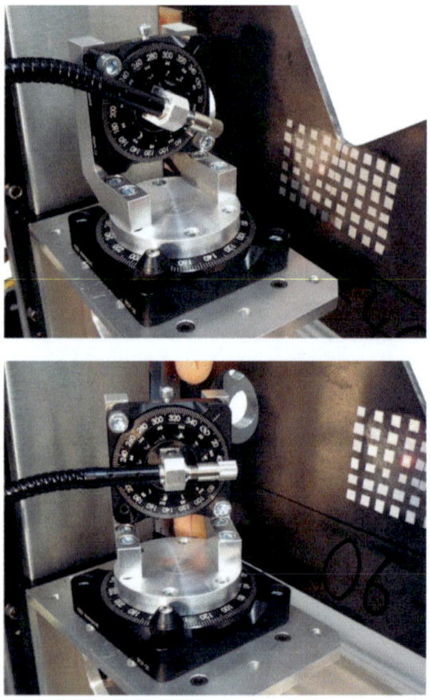

(b) Rotation stage inclined (top) and rotated (bottom) by 30°

Figure 5.1: Extended test stand used for semi-automatic, three-dimensional one-point (scanning) laser vibrometry of plate specimens.

5.1.2 Laser vibrometer

Throughout this work acquisition of local and areal vibration data has been conducted by means of a fibre-optic laser vibrometer Polytec OFV-5000/552, which allows for two-point differential measurements and simple one-point measurements. For single-point sensing a reflector end piece is screwed on one of the fibre ends as can be seen in Figure 5.1a. The controller and the laser head with the fibre-optic cables are shown in Figure 5.2.

Figure 5.2: Polytec OFV-5000 vibrometer controller, differential fibre-optic interferometer OFV-552, OFV-151 reflector for single-point measurements and two magnetic stands with articulated arm for manual positioning of the laser sensors.

Equipped with the digital velocity decoder VD-09, the system provides a wide measurement range of frequencies from 0 Hz (DC) to 2.5 MHz and a variety of high-pass and low-pass filter settings. The maximum velocity that can be resolved is 10 m/s [Pol09], where eight measurement ranges are available covering analogue voltage outputs from 5 to 1000 mm/(s V).

5.1.3 Retroreflective film

In order to achieve low-noise vibrometric velocity signals, a piece of reflective film is commonly attached to the structure and used as a measurement point. In contrast to the ideal mirror characteristics, where the angle of incident equals the angle of reflection, the employed film type 3M Scotchlite contains a retroreflective layer that sends an incoming laser beam back to its origin. Technically, this property is achieved through embedded crystal balls that allow for multiple reflections inside each individual sphere. Microscopic images shown in Figure 5.3 reveal the dense packing of glass balls having an average diameter of 25 to 40 µm.

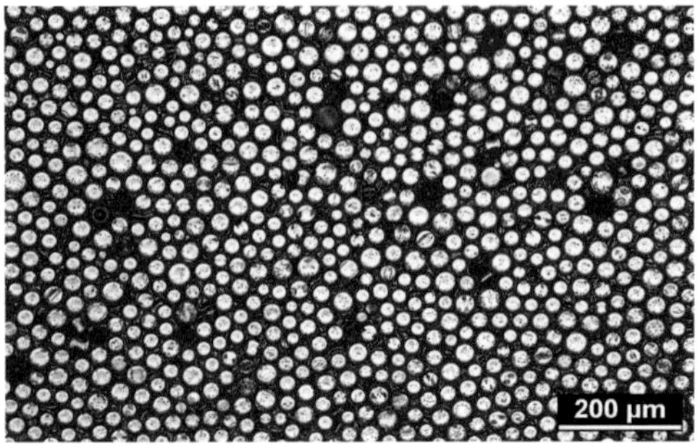

(a) Microscopic image (50× magnification)

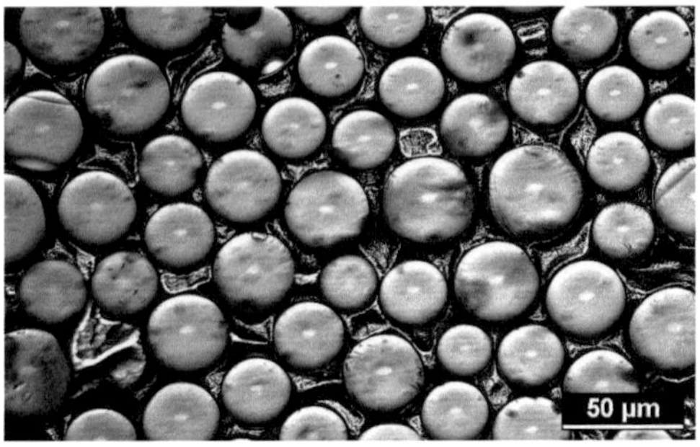

(b) Microscopic image (200× magnification)

Figure 5.3: Microstructure of 3M retroreflective film used for laser vibrometric measurements.

If pure out-of-plane vibrations are of interest, the sensor is pointed orthogonally to the plate surface. In order to acquire three-dimensional vibration mode shapes, the two in-plane components need to be derived from measurements with inclined orientation of the laser towards the plate. A preliminary check of the measurement accuracy considering different angles of incidence is performed since reliable inclined sensing becomes a critical aspect when composing 3D vibration data.

At first, the piston of a servo-hydraulic testing machine was stationary operated in displacement control to provide a precise harmonic movement at a frequency of 3 Hz. The chosen displacement amplitude of 0.2 mm should be resolved by one-point laser vibrometric measurements performed at five angles of incidence up to 45° with respect to the piston's longitudinal axis. All measurements were repeated five times for each adjusted angle α to account for a possible scatter.

The velocity amplitude in laser direction $\hat{v}$ was obtained by means of a Fourier transform with employed Flat Top window based on 8192 samples acquired at a rate of 512 S/s. Figure 5.4a illustrates the averaged amplitudes after normalisation with respect to the orthogonal measurement of $\hat{v}_0$ with $\alpha = 0$ and the relative accuracy regarding the relation $\hat{v}/\hat{v}_0 = \cos\alpha$. It is found that the deviation from the theoretical velocity amplitude is less than 1 %.

The measurement approach was validated in the same fashion for the case of ultrasound oscillations. For this purpose, a piece of retroreflective film was attached to the tip of the Branson transducer. While running the generator at 20 kHz with a relative power of 20 %, again 8192 samples have been acquired from the same set of angles but with a sampling rate of $2^{19} = 524\,288\,\mathrm{S/s}$. Apparently, as found from Figure 5.4b, the accuracy of the measurements is slightly lower compared to the previous results. However, the relative deviations from the theoretically predicted behaviour do not exceed 6 % in any case and the mean values have a relative error below 2 %, which is a very good outcome. The use of retroreflective film for vibrometry measurements is therefore proven valid up to an inclination of 45° with respect to the normal direction.

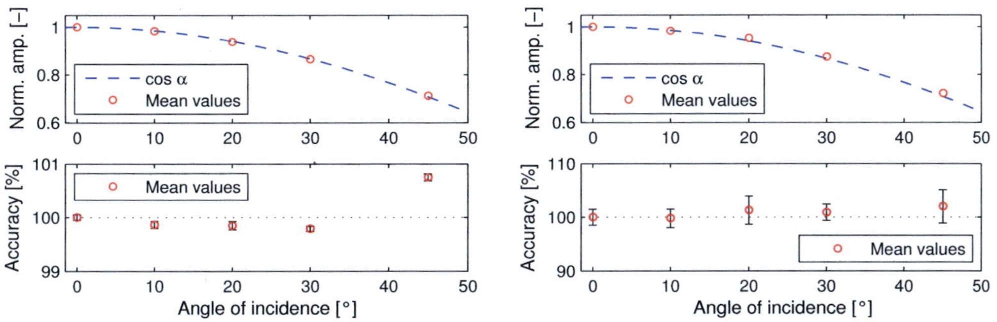

(a) Piston of a testing machine at 3 Hz (b) Branson ultrasound transducer at 20 kHz

Figure 5.4: Normalised velocity amplitudes, accuracy and standard deviations found from vibrometry measurements at varying angles of incidence.

5.2 Characterisation of the ultrasonic transducers

So far, the ultrasound transducers are thought of as perfect oscillators which allow for variation of the working frequency and the oscillation amplitude independently and without any restrictions. In this section the response of the Branson and the TIB transducer is examined in detail. For the correct interpretation of all further experiments it is essential to clarify the following issues:

- The rise and fall time or pulse steepness of an ultrasound burst
- Eigenfrequencies and the frequency response within the main working range
- The actual frequency dependent on the nominal frequency
- The physical oscillation amplitude dependent on the relative power level
- The extent of non-linear oscillations, e.g. the generation of higher harmonics

5.2.1 Start/stop behaviour

The response time of an electromechanical oscillator is a very basic property, which can be crucial for successful vibrothermographic testing, especially if short burst excitations are intended such as, for example, in ultrasound burst-phase thermography.

Both ultrasound systems, the high-power Branson weld generator and the TIB amplifier, were investigated in the same way. The ultrasonic converters have been securely mounted to the aluminium frame shown in Figure 5.1a via two rounded clamping parts. Analysis of the rise and fall times of the transducer oscillation was realised using rectangular tone-bursts at 20 kHz as benchmark signals and single-point measurements of the axial converter tip velocity.

A customised LabVIEW programme has been utilised for controlling the ultrasound systems and for acquisition of the laser vibrometer output signal. The mono-frequency excitations with burst durations of about 120 ms have been sampled at a high rate of 1 MS/s.

Figure 5.5a exemplarily depicts the typical envelope shapes obtained from the Branson converter at 50 % relative power and from the TIB transducer at full power. Obviously, the start responses differ from each other. In case of the Branson system, the time which is required to achieve the desired steady-state vibration amplitude can be understood as the rise time in a strict sense. In contrast, the TIB system shows some overshooting and takes some time to attenuate until a constant amplitude is reached.

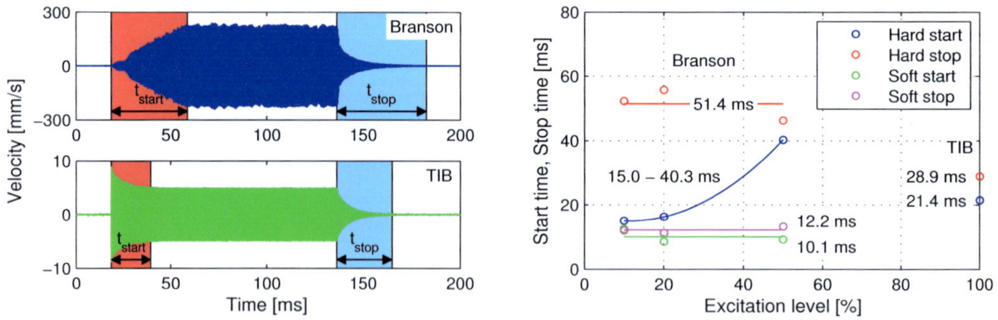

(a) Ultrasound burst at $P = 50\,\%$ (Branson) and $P = 100\,\%$ (TIB); $f = 20\,\mathrm{kHz}$

(b) Power-dependent rise/fall times

Figure 5.5: Burst shape and rise/fall times of the Branson and the TIB transducer.

Comparison of the oscillation contours reveals that the high-power Branson converter suffers from a certain instability regarding its velocity amplitude, whereas the TIB system manifests a perfectly stable vibration amplitude during excitation. As soon as the control signal stops, an exponential decay of the amplitude level can be estimated in both cases. It should be noted that the maximum velocities differ by at least one order of magnitude.

Since the Branson generator has been customised by a modified firmware as described in Section 4.2.2, the system allows to be switched on in two different ways. At first, the desired nominal power level can be adjusted via an analogue voltage prior to sending the digital signal which starts the ultrasound output abruptly. This is referred to as the "hard" start and stop, respectively. A second possibility is to hold the generator in its operating mode (digital signal on) but to adjust the relative power to 0 %. A sudden rise of the analogue DC voltage, which is proportional to the vibration level, leads to a "soft" start. Correspondingly, resetting the amplitude control signal back to 0 V but keeping the generator turned on is referred to as the "soft" stop. In contrast to the Branson generator, the TIB amplifier automatically switches off, whenever the sine input level approaches 0 V, so there are no different start and stop methods available.

Experiments with the Branson equipment including a variation of the power level from 10 to 50 % revealed that the rise and fall times are significantly longer when switched on and off "hard". Figure 5.5b summarises the average start and stop times achieved with both systems. The hard start rise time of the Branson generator increases with the power level from about 15 to 40 ms. The smooth increase of the vibration amplitude is intended by the manufacturer to prevent excessive stressing of the weld tool [Bra04]. However, such a behaviour might be unfavourable for burst-phase thermography if the nominal burst duration is constant but the ultrasound intensity is to be varied. In this case, the relation between maximum vibration level and thermal response is probably influenced just because of the different rise times. In order to eliminate this effect as well as possible, the "soft" start/stop control method was used throughout this work. Here, start and stop times of the freely vibrating Branson transducer are about 12 ms and do not noticeably depend on the relative power. This way, typical burst excitations longer than 100 ms are technically reasonable. Rise and fall times of the TIB amplifier are found to be in the same order of magnitude but about 2 or 3 times longer compared to the Branson system.

5.2.2 Amplitude frequency response

Essential characteristics of an excitation device, such as the used ultrasound systems, are commonly described by its complex frequency response, which contains information on the output magnitude and the phasing related to an input signal over a certain frequency range. In order to prevent the exciter from being permanently operated at its eigenfrequencies and to correctly assign causes of frequency dependent phenomena, such as the crack heating, the amplitude response of both ultrasound systems was analysed in a simplified manner. In the course of thermosonic testing phase lags of the transducer with respect to the input signal are not of interest and cannot be properly determined anyway in case of the Branson system. Therefore, stepped frequency sweeps at constant nominal power are carried out and the velocity of the freely vibrating transducer tip is measured vibrometrically in axial and in transverse direction.

At first, the maximum bandwidth, limited by the Branson generator, 15 to 25 kHz is considered. Clearly, the focus of this investigation is on the stationary oscillations that should set in after a certain number of cycles. Thus, transient oscillations which will only occur immediately after changing the nominal frequency are not to be analysed. To ensure this, the velocity data acquisition at each frequency step is delayed by a constant time interval Δt_1 (compare Fig. 3.30b). Obviously, the total duration of a single frequency step Δt has to exceed the sum of Δt_1 and the time Δt_2 needed to acquire N samples at a sampling rate f_s. An additional safety margin Δt_3 is introduced to allow for buffering and writing data to the hard disk.

For this study the sweep parameters $\Delta f = 10\,\text{Hz}$, $\Delta t_1 = 20\,\text{ms}$, $\Delta t_2 = 120\,\text{ms}$, $\Delta t_3 = 40\,\text{ms}$ were chosen. The delay Δt_1 corresponds to a sufficient number of 300 to 500 cycles to damp out any transient response when the nominal frequency range of $f_{\text{nom}} = 15\ldots25\,\text{kHz}$ is swept. For each of the 1001 frequency set points the transducer velocity was measured using $N = 2^{15}$ S and $f_s = 2^{18}$ S/s, resulting in a DFT frequency resolution of $f_s/N = 8\,\text{Hz}$.

Several test runs showed that a relative power of the Branson system greater than 6 to 7 % leads to a generator shut-off when sweeping the complete spectrum. Thus, the Branson generator was operated at $P = 5\,\%$ in contrast to the TIB system, which could be run at full level control. The actual velocity amplitude was then found by applying the discrete Fourier transform to the vibrometric data obtained for each frequency step. To enhance the amplitude accuracy, the Flat Top window was used. Since the frequency precision of both ultrasound systems is not known, yet, the amplitude frequency response is searched for a peak within the range of $f_{\text{nom}} \pm 500\,\text{Hz}$. All tests have been repeated three times. The maximum transversal movement of the transducer tip was found by incrementally rotating the transducer three times by 30°. Figure 5.6 illustrates the achieved amplitude frequency responses for both systems, where the value at 20 kHz is referred to as the respective reference amplitude.

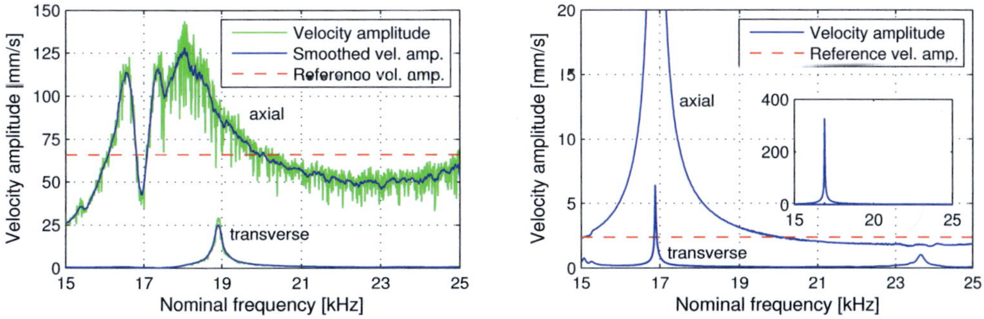

(a) Branson transducer; $P = 5\,\%$ (b) TIB transducer; $P = 100\,\%$

Figure 5.6: Velocity amplitude spectrum of the Branson and the TIB transducer.

The response in longitudinal direction achieved for the Branson system reveals strong fluctuations, especially in the first half of the spectrum. Contrary to what might have been expected, no distinct amplitude peaks could be found. Instead, three mild peaks are observed, where the whole response history exhibits a pronounced scatter. The band from 19 to 25 kHz offers the most steady response with the least overall deviation from the reference amplitude. Because of that and also due to the fact that in transverse direction only one peak exists just below 19 kHz, this frequency range was considered for the experiments on vibrothermography described in Chapter 4. A compromise between time exposure/amount of data for a single sweep and the probability of exciting "efficient" frequencies was found by further limiting the primary working range to $f_{\text{nom}} = 19\ldots23\,\text{kHz}$.

Analysis of the TIB system reveals a clear and decisive amplitude response showing a very sharp peak at about 16.9 kHz. With the peak values of 330 mm/s in longitudinal direction and 6.5 mm/s in transverse direction the transducer can be considered to oscillate almost perfectly unidirectional. However, besides of the resonance peak its velocity amplitude is lower than 3 mm/s. This corresponds to a displacement amplitude of only 0.024 µm at 20 kHz. Compared to the Branson transducer, which was just operated at $P = 5\,\%$, this value is 25 times smaller and probably the reason for not detecting any cracks using the TIB system.

5.2.3 Frequency accuracy

Up to this point it is not proven that the modified Branson system is capable of exactly tuning in the extended frequency band 15 to 25 kHz or at least the primarily used range of 19 to 23 kHz. Moreover, the accuracy of reproducing a desired frequency and a possible instability regarding high-power operation represent issues that need to be clarified for both ultrasound systems prior to the analysis of plate vibrations.

First off all, the frequency accuracy is assessed considering the complete spectrum. Using the same settings as before, the tip velocity of the Branson and TIB transducer in longitudinal direction was measured during a broadband, stepped ultrasound sweep at 5 and 100 % relative power, respectively. Ideally, the transducer's actual oscillating frequency exactly matches the controlled nominal frequency.

Figure 5.7 illustrates the actual frequencies obtained by Fourier transform of the velocity data and the deviation from the nominal frequency. It is found that both ultrasound systems can reproduce the test band, whereas the Branson transducer manifests a linear drift in its actual frequency towards lower values when increasing the nominal frequency. Maximum deviations up to 140 Hz are indicated in Figure 5.7b. Once again, it should be noted that the Branson generator is controlled by DC voltage signals, which do not allow to prescribe a certain waveform. In contrast to that, Figures 5.7c and 5.7d reveal the perfect frequency response of the TIB system. Obviously, the DFT frequency resolution of 8 Hz prevents a more precise determination of the frequency accuracy.

To understand whether the Branson exciter, though showing a drift, provides a stable frequency response independent of the desired excitation level, further test runs are conducted focusing on the working range of 19 to 23 kHz. To compare the overall frequency deviations at varying relative power, the coefficients obtained from linear regression analyses are used. The extent of linear correlation between actual and nominal frequency is indicated by the empirical correlation coefficient r and the coefficient of determination r^2.

Being x_i the n discrete nominal frequencies and y_i the corresponding actual frequencies found by means of a DFT, the variances s_x^2 and s_y^2 and the covariance s_{xy} are given by

$$s_x^2 = \frac{1}{n-1} \cdot \sum_{i=1}^{n} (x_i - \overline{x})^2 = \frac{1}{n-1} \left(\sum_{i=1}^{n} x_i^2 - n\overline{x}^2 \right) \tag{5.1}$$

$$s_y^2 = \frac{1}{n-1} \cdot \sum_{i=1}^{n} (y_i - \overline{y})^2 = \frac{1}{n-1} \left(\sum_{i=1}^{n} y_i^2 - n\overline{y}^2 \right) \quad (5.2)$$

$$s_{xy} = \frac{1}{n-1} \cdot \sum_{i=1}^{n} (x_i - \overline{x}) \cdot (y_i - \overline{y}) = \frac{1}{n-1} \left(\sum_{i=1}^{n} x_i y_i - n\overline{x}\overline{y} \right) \quad (5.3)$$

with the mean values $\overline{x}$ and $\overline{y}$. The empirical correlation coefficient r then reads ([Pap01]):

$$r = \frac{s_{xy}}{s_x \cdot s_y} = \frac{\sum_{i=1}^{n} x_i y_i - n\overline{x}\overline{y}}{\sqrt{\left(\sum_{i=1}^{n} x_i^2 - n\overline{x}^2 \right) \left(\sum_{i=1}^{n} y_i^2 - n\overline{y}^2 \right)}} \quad (5.4)$$

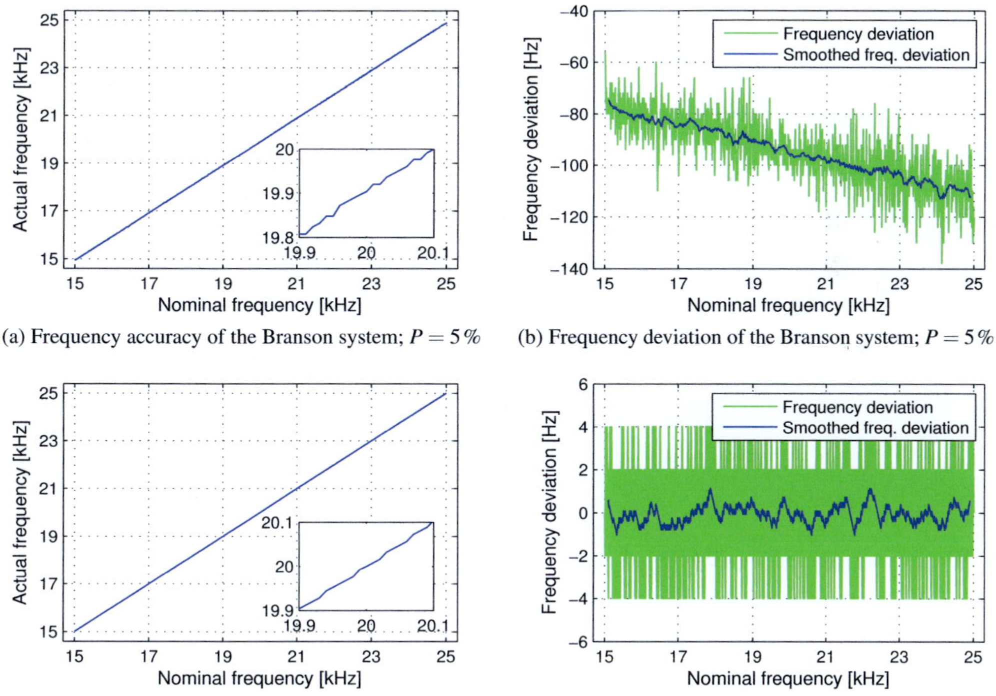

(a) Frequency accuracy of the Branson system; $P = 5\,\%$

(b) Frequency deviation of the Branson system; $P = 5\,\%$

(c) Frequency accuracy of the TIB system; $P = 100\,\%$

(d) Frequency deviation of the TIB system; $P = 100\,\%$

Figure 5.7: Frequency accuracy of the Branson and TIB ultrasound system.

This coefficient can be interpreted as a measure of how good a linear equation characterises the relation between actual and nominal frequency. An absolute value $|r| = 1$ states that the actual frequency is perfectly correlated with the nominal frequency and thus the data points are lying exactly on one line (compare Fig. 5.7d). In the case of $r = 0$ no functional relation between x_i and y_i can be found. Commonly, the coefficient of determination r^2 is used as an indicator that characterises the "overall quality" of the linear functional relationship.

The analysis of both transducers reveals an almost perfectly linear dependency. The calculated coefficient of determination r^2 differs from 1.0 by a value smaller than 1×10^{-4}. Indeed, this proves the strong functional relation but does not indicate whether the residual discrepancies between actual and nominal frequency are due to a "gain" or "offset" error. This point is clarified by an assessment of the regression coefficients. For this purpose the generator's original working frequency of 20 kHz is subtracted from the x and y values so that the data points are centred at $x = 0$. A linear regression is then applied to the shifted data points x_i and y_i in order to calculate the actual slope a and intercept b of the partial regression line

$$y = a \cdot x + b \quad . \tag{5.5}$$

The regression coefficient and slope a is proportional to the empirical correlation coefficient r and can be expressed as ([Pap01]):

$$a = \frac{n \cdot \sum_{i=1}^{n} x_i y_i - \left(\sum_{i=1}^{n} x_i\right)\left(\sum_{i=1}^{n} y_i\right)}{n \cdot \sum_{i=1}^{n} x_i^2 - \left(\sum_{i=1}^{n} x_i\right)^2} = \frac{\sum_{i=1}^{n} x_i y_i - n\overline{x}\,\overline{y}}{\sum_{i=1}^{n} x_i^2 - n\overline{x}^2} = \frac{(n-1)s_{xy}}{(n-1)s_x^2} = r \cdot \frac{s_y}{s_x} \tag{5.6}$$

The interception b corresponds to the offset discrepancy of the actual frequency and is given by:

$$b = \frac{\left(\sum_{i=1}^{n} x_i^2\right)\left(\sum_{i=1}^{n} y_i\right) - \left(\sum_{i=1}^{n} x_i\right)\left(\sum_{i=1}^{n} x_i y_i\right)}{n \cdot \sum_{i=1}^{n} x_i^2 - \left(\sum_{i=1}^{n} x_i\right)^2} = \overline{y} - a\overline{x} \tag{5.7}$$

To ensure that the frequency accuracy is not affected by high excitation levels, the analysis was repeated based on several sweeps within the limited frequency band of 19 to 23 kHz but with increasing relative power. Table 5.1 presents the coefficients in the form of $1 - r^2$, $1 - a$ and b. The results for the TIB system are also given for comparative purposes. Here, the true frequency accuracy is even better than the given values due to the limited spectral resolution of the DFT. Regarding the frequency precision of the Branson transducer, the achieved regression coefficients do not indicate any dependency on the system power. However, with increasing excitation level a growing tendency of plateau-like steps in the actual frequency history during sweeps could be observed. Thus, the overall error measure $1-r^2$ significantly increases when exceeding $P = 20\,\%$. The reason for this behaviour remains unclear.

Table 5.1: Frequency accuracy of the Branson and the TIB transducer indicated by regression coefficients.

Branson	P	[%]	5	10	15	20	30	40	50	60
	$1-r^2$	$[10^{-4}]$	0.22	0.22	0.20	0.26	0.49	0.85	1.30	2.66
	$1-a$	$[10^{-3}]$	2.7	3.2	3.4	3.1	3.1	2.7	2.2	1.5
	b	[Hz]	-95.7	-95.6	-95.0	-95.3	-95.7	-96.1	-96.5	-97.6
TIB	P	[%]	30	40	50	60	70	80	90	100
	$1-r^2$	$[10^{-4}]$	0.04[1]	0.04[1]	0.04[1]	0.04[1]	0.04[1]	0.04[1]	0.04[1]	0.04[1]
	$1-a$	$[10^{-3}]$	0.04	-0.13	-0.04	0.01	0.00	0.08	0.23	0.13
	b	[Hz]	-0.04	-0.07	-0.04	-0.11	0.06	-0.08	0.25	0.19

[1]Accuracy limited due to the chosen DFT frequency resolution

Knowing the coefficients a and b allows to compensate for the global frequency drift and offset via a correction function that is taken into account when adjusting the desired frequency. Though, experiments have shown that especially the frequency offset slightly differed over time. This is most probably due to changes of the initial converter temperature. Since the actual frequency is indeed quite volatile when focusing on a small band, it was chosen not to employ a global correction function but to perform a preceding frequency sweep each time mono-frequency testing is involved, e.g. for vibrometric measurements of mode shapes at resonant excitation.

5.2.4 Input/output linearity

One of the essential transducer properties that has been assumed in Chapter 4 to hold true is the linearity of the output vibration level with respect to the input signal. Focusing on the limited working range, individual velocity measurements have been conducted at 19, 20, 21, 22 and 23 kHz, where the nominal excitation level was incremented after each test. The steady state amplitudes achieved for both systems are illustrated in Figure 5.8.

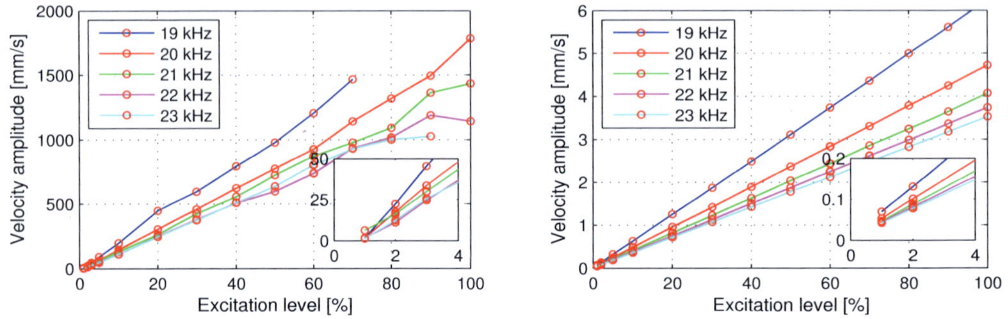

(a) Branson transducer

(b) TIB transducer

Figure 5.8: Input/output linearity of the unloaded Branson and TIB transducer.

In case of the Branson system, a generator shut-off was experienced at different excitation levels when operating at 19 and 23 kHz. However, the overall linearity up to the relative power of 60 % is fairly good. A closer look at the very low levels reveals a small offset error. Hence, the threshold behaviour found during vibrothermography with increasing excitation level is at least partly caused by the non-linear system response.

Corresponding analysis of the TIB linear amplifier resulted in absolutely stable input/output ratios for each of the test frequencies. No offset error could be found. A comparison of the maximum achieved velocity amplitudes at 20 kHz reveals that the Branson welder surpasses the TIB system by a factor of 400.

5.2.5 Harmonic distortion

The last point of the transducer characterisation is a distortion analysis. It is well known that acoustic non-linearities, e.g. a clapping or rubbing interface in a solid body generates higher harmonics of an input sound wave [BMR78, Sol98, SKB02]. Such a harmonic analysis, which can also be used to prove the existence of an active mechanical non-linearity, will be complicated if the excitation itself has a rich spectrum and contains higher harmonics with respect to a base frequency.

It is useful to introduce an additional parameter which acts as an integral measure of the distortions of an apparent mono-frequency (sine) signal. The total distortion factor TDF and the total harmonic distortion THD are most commonly used in the field of electrical engineering to quantify the extent of non-linear signal distortions.

In order to express the strength of a non-linear disturbance caused by a single higher harmonic oscillation, the harmonic distortion HD_i can be defined as the ratio of the RMS value U_i of the ith harmonic (2nd, 3th, etc.) and the RMS signal level at the fundamental frequency. Instead, relating U_i to the root-sum-square value of all harmonics including the fundamental frequency leads to the distortion factor DF_i.

$$\mathrm{HD}_i = \frac{U_i}{U_1} \quad , \qquad i = 2,3,\ldots,\infty \tag{5.8}$$

$$\mathrm{DF}_i = \frac{U_i}{\sqrt{\sum\limits_{n=1}^{\infty} U_n^2}} = \frac{U_i}{\sqrt{U_1^2 + U_2^2 + U_3^2 + \ldots + U_\infty^2}} \quad , \qquad i = 2,3,\ldots,\infty \tag{5.9}$$

Based on these definitions the total harmonic distortion THD is the ratio of the root-sum-square value of all the higher harmonics and the RMS level at the base frequency. Expressed in percentage, this integral value is a well-known specification used as a quality indicator in the field of audio components like amplifiers. Although THD is a non-standardised parameter, it is recognised by the International Electrotechnical Commission (IEC) in the form of:

$$\mathrm{THD} = \frac{\sqrt{\sum\limits_{n=2}^{\infty} U_n^2}}{U_1} = \frac{\sqrt{U_2^2 + U_3^2 + U_4^2 + \ldots + U_\infty^2}}{U_1} \tag{5.10}$$

According to the definition of DF_i, the total distortion factor TDF relates the intensity of all higher harmonics to the root-sum-square value of all harmonics including the fundamental frequency. Böge and Plaßmann [BP07] give the relation between TDF and THD:

$$\mathrm{TDF} = \sqrt{\frac{\sum\limits_{n=2}^{\infty} U_n^2}{\sum\limits_{n=1}^{\infty} U_n^2}} = \sqrt{\frac{U_2^2 + U_3^2 + U_4^2 + \ldots + U_\infty^2}{U_1^2 + U_2^2 + U_3^2 + \ldots + U_\infty^2}} = \frac{\mathrm{THD}}{\sqrt{1 + \mathrm{THD}^2}} \tag{5.11}$$

Both concepts are originally developed for RMS values U. In this work the formulas are directly used in terms of amplitudes. In practice, the detection of harmonic distortions is limited by the sampling rate of the data acquisition, the quantisation error and the noise level of the analysed signal. Most often the higher order terms are greatly reduced in amplitude since high-frequency contents are naturally damped out strongly.

In order to assess higher harmonics in the transducer oscillation output, the velocity sampling rate and the number of samples was set to $f_s = 2^{19}$ S/s and $N = 2^{16}$ S, respectively. According to the Nyquist-Shannon sampling theorem, these settings theoretically allow for the detection of the 11th harmonic, when considering a maximum base frequency of 23 kHz. The distortion analysis is done by performing frequency sweeps with increasing excitation levels. Power and frequency dependent distributions of the velocity amplitudes according to the 1st, 2nd, 3rd, 4th, 5th and 6th harmonic are illustrated in Appendix A.6.

Based on these results Figure 5.9 presents the THD for both ultrasound systems. It is found that the Branson transducer vibrations are strongly superimposed by higher harmonic oscillations. For several frequencies the total harmonic distortion was determined to be 25 % at low power and up to 80 % at high-power excitation. In contrast, the TIB transducer was proven to vibrate almost free of any distortion. Here, the average THD is below 0.5 %.

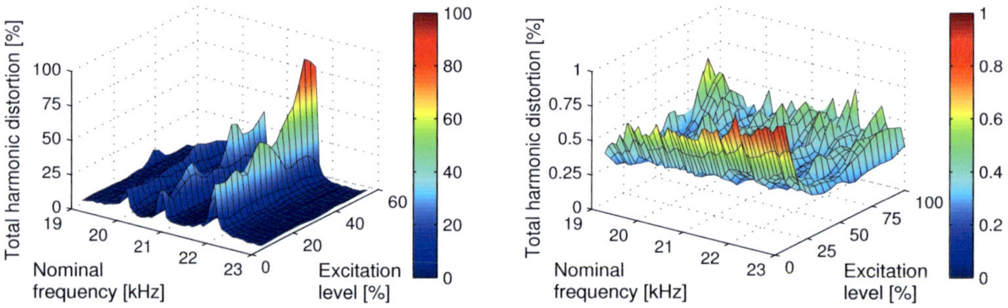

(a) Branson system

(b) TIB system

Figure 5.9: Total harmonic distortion of the Branson and TIB ultrasound system based on the evaluation of the fundamental tone and the first five overtones.

5.3 Single-point and two-point differential laser vibrometry

This section focuses on vibrometric measurements of the crack sides combined with thermal imaging during ultrasound excitation of plate specimens. Since vibrothermographic routine testing of all available plate specimens in Chapter 4 proved that plastic straining at the crack tip is not the main reason for heat generation, the following investigations are primarily on assessing relative displacements and relating these to frictional heating.

5.3.1 Correlation of crack face displacements with defect heating

First of all, the question arises to what extent relative movements of the crack sides occur and whether they are correlated to the defect heating. Homma [Hom07] states that high infrared intensities can only be achieved at resonant frequencies but does not present an overall correlation of the IR signal and the vibration amplitude histories synchronously achieved during a frequency sweep.

In order to clarify this point, plate specimen 06-12, which showed intense heat dissipation when tested in Section 4.4.6 (see Fig. 4.31 and Appendix A.3 for detailed results), has been equipped with retroreflective stickers. To obtain the tangential relative movement according to crack mode II (in-plane shear deformation) and crack mode III (out-of-plane shear deformation) the film was attached close to the crack path on the front side and the fillet of the plate. Using the basic test stand described in Chapter 4, the infrared camera was moved away from the notch centre and then rotated slightly to give way to the laser sensors. These were hold in place by two small magnetic stands with articulated arm (see Fig. 5.2). Figure 5.10 illustrates the laser orientations in case of two-point measurements.

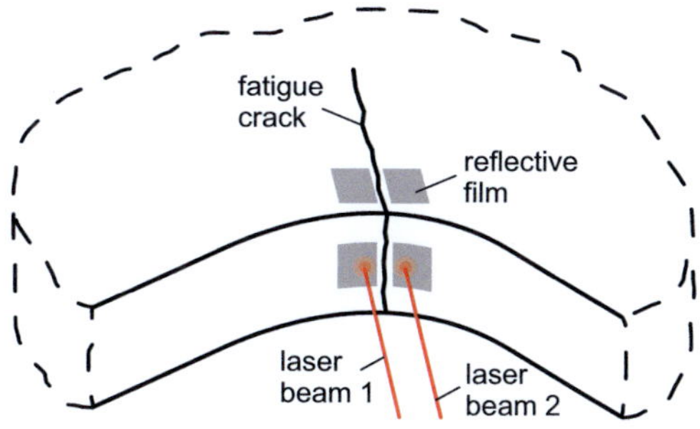

(a) Mode II measurement

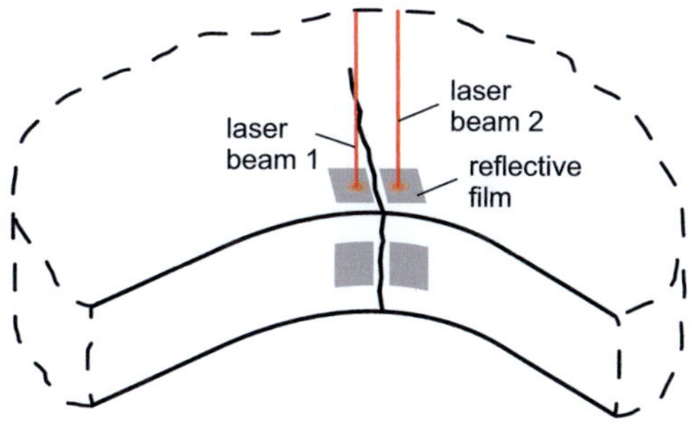

(b) Mode III measurement

Figure 5.10: Two-point differential laser vibrometry according to the crack opening modes II and III.

The plate has been subjected to a sweep excitation for several times, where single-point and two-point velocity measurements have been carried out in the mode II and III configuration. Again, the standard frequency range of 19 to 23 kHz was used with increments of 10 Hz. During repositioning of the laser sensor the plate itself and the infrared camera were not moved. The fillet face was not inspected thermographically.

A custom LabVIEW programme was employed to trigger the infrared camera and to record the vibrometer voltage output. At each of the 401 frequency steps 2^{15} samples were taken at a high sampling rate of 2^{19} S/s, which theoretically allows to quantify overtones up to the 11th harmonic velocity amplitude. Vibrometric data acquisition was delayed by 20 ms (Δt_1, compare Figure 3.30b) to ensure that a steady state vibration is recorded but not the transient oscillation. The Flat Top window was applied to the data stream to achieve an improved DFT amplitude accuracy.

Figure 5.11a and 5.11b show the hot spot temperature increase ΔT, averaged over a 3×3 px neighbourhood, and the differential velocity amplitudes according to mode II and III. After removing a slight linear trend from the heating history, a certain correlation of the data can be visually found. The normalised cross-correlation allows to quantify the overall similarity of both signals. A basic requirement is to have the same time step width or the same number of samples N in the two data streams of equal duration. Therefore, an interpolation of the temperature evolution onto the velocity data grid (downsampling) is performed.

Unbiased estimates of the sample cross-correlation function at lag numbers r are then defined as

$$R_{xy}(r) = \frac{1}{N-r} \sum_{i=1}^{N-r} x_i y_{i+r} \quad , \qquad r = 0, 1, 2, \ldots, m \quad , \quad m < N \tag{5.12}$$

with x_i being the velocity amplitude values and y_i the corresponding temperature samples after subtracting the mean values ($\bar{x} = \bar{y} = 0$) [BP00]. The autocorrelations R_{xx} and R_{yy} given through

$$R_{xx}(r) = \frac{1}{N-r} \sum_{i=1}^{N-r} x_i x_{i+r} \tag{5.13}$$

$$R_{yy}(r) = \frac{1}{N-r} \sum_{i=1}^{N-r} y_i y_{i+r} \tag{5.14}$$

yield the (biased) variances s_x^2 and s_y^2 for the lag $r = 0$. Finally, the sample cross-correlation coefficient function reads

$$\rho_{xy}(r) = \frac{R_{xy}(r)}{\sqrt{R_{xx}(0)} \cdot \sqrt{R_{yy}(0)}} = \frac{R_{xy}(r)}{s_x \cdot s_y} \tag{5.15}$$

and takes values in the range $-1 \leq \rho_{xy} \leq 1$. Two perfectly correlated signals are indicated by a peak value $\rho = 1$, no matter if they are shifted by a time delay (e.g. due to thermal diffusion).

Figure 5.11c and 5.11d depict the correlation coefficient functions found for the two-point mode II and III measurements regarding positive and negative lag numbers (see [The08] for the used algorithm implemented in MATLAB). Besides of the normalised cross-correlations of ΔT and the differential velocity amplitudes at the base excitation frequency, the next five overtones (velocity amplitude II to VI) and the total harmonic distortion THD are also considered.

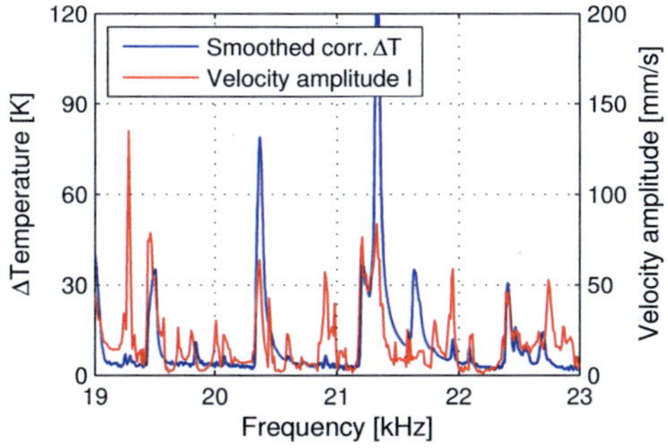

(a) Corrected ΔT evolution and mode II differential velocity amplitude

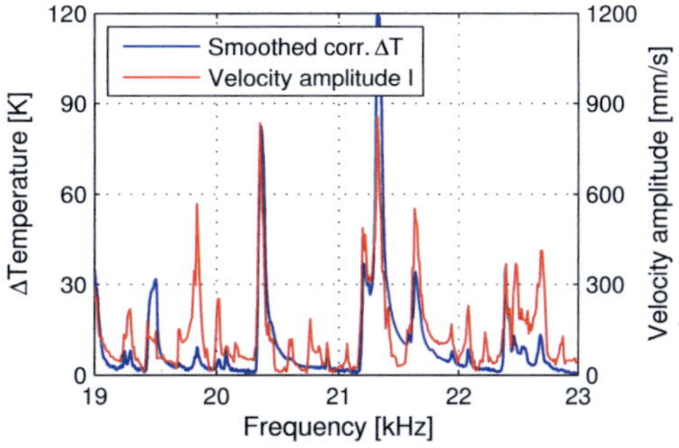

(b) Corrected ΔT evolution and mode III differential velocity amplitude

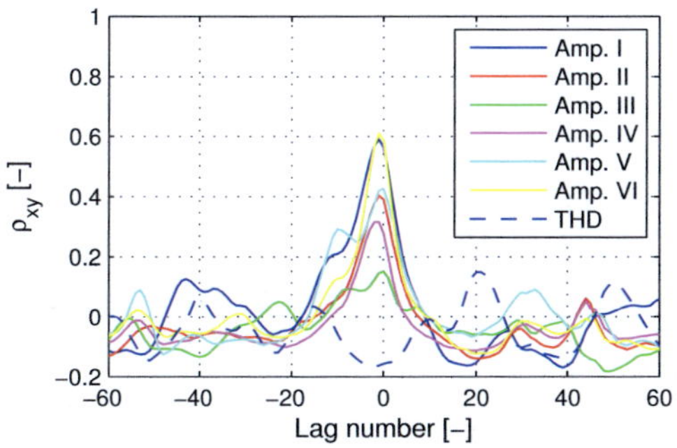

(c) Normalised cross-correlation of mode II velocity amplitudes I to VI, THD and corrected ΔT

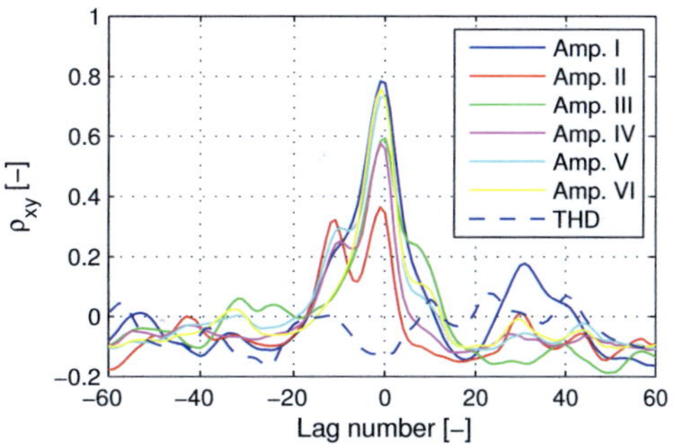

(d) Normalised cross-correlation of mode III velocity amplitudes I to VI, THD and corrected ΔT

Figure 5.11: Corrected ΔT evolutions and differential velocity amplitudes according to mode II and III (a, b) and correlation coefficient functions of harmonic amplitudes and defect heating; specimen 06-12-A; $P = 13\,\%$.

It is concluded that increased crack mode displacements are clearly correlated to the defect heating, where the mode III relative displacement is more significant in terms of magnitude and the peak correlation coefficient. The small shift of the peak value towards negative lag numbers ($r = -1$) indicates the delay in crack heating, which is caused by heat diffusion from the true hot spot to the plate surface. One lag unit corresponds to 0.17 s, based on the frequency steps and the total sweep duration. Obviously, a strong thermal response occurs very likely if resonance frequencies are matched during the sweep excitation. Regarding the heating and velocity amplitude histories, it is found that most of the velocity peaks can be clearly assigned to a distinct temperature increase. However, the "efficiency" of crack heating at different eigenfrequencies varies extremely.

The analysis of higher harmonics revealed that the intensities of overtones are also correlated to the defect heating. In case of the mode II measurements, the 6th harmonic shows even better correlation than the amplitude at the base frequency. A clear differentiation between the occurrence of even and odd harmonics, as predicted by Solodov and Pfleiderer [Pfl06, SKB10], is not possible. Contrary to what might have been be expected, the integral measure of the harmonic distortion does not reveal any correlation at all. Therefore, at the moment of crack heating, the movement of the crack edges in tangential direction is not significantly distorted by frictional effects. More likely, the rising amplitudes of higher harmonics are just related to the strength of the fundamental oscillation. This was indicated by computing the cross-correlation of the first and all higher harmonic amplitudes resulting in peak values of ρ_{xy} between 0.5 and 0.65 (mode III) and between 0.26 and 0.55 (mode II) at zero lag.

The same analysis was carried out for the mode II and III single-point measurements of both crack sides but generally resulted in lower correlation coefficients compared to the differential measurements. Thus, relative displacements are clearly proven most useful to be further investigated towards a predictive indicator of frictional heating.

5.3.2 Influence of the initial temperature

Prior to further investigation of the relationship between relative displacements and crack heating the "softening" behaviour observed in Chapter 4 is analysed again by means of additional vibrometric measurements. During consecutive sweep thermography runs on specimen 06-06 using stepwise increasing relative power a shift of the most efficient ultrasound frequency for defect detection towards lower frequencies was found.

Similar to the previous section, vibrometric data acquisition and processing was done for each discrete frequency step in order to compare the evolutions of the temperature increase and the vibration amplitude. Corresponding to Figures 4.24a and 4.24b that show the hot spot heating depending on the nominal frequency, Figures 5.12a and 5.12b depict the mode III single-point velocity amplitude of one crack side and the maximum achieved amplitude for each sweep, respectively. Both representations are in perfect agreement. The peak frequencies leading to the maximum heating and vibration amplitude coincide when plotted against the initial plate temperature, which was extracted from the first image of each infrared sequence (Fig. 5.12c). Ignoring the first data points found at low excitation levels, the linear approximations have a slope of -17.64 Hz/K and -17.55 Hz/K, respectively. Accordingly, the linear regression shown in Figure 5.12d yields a high coefficient of determination $r^2 = 0.95$.

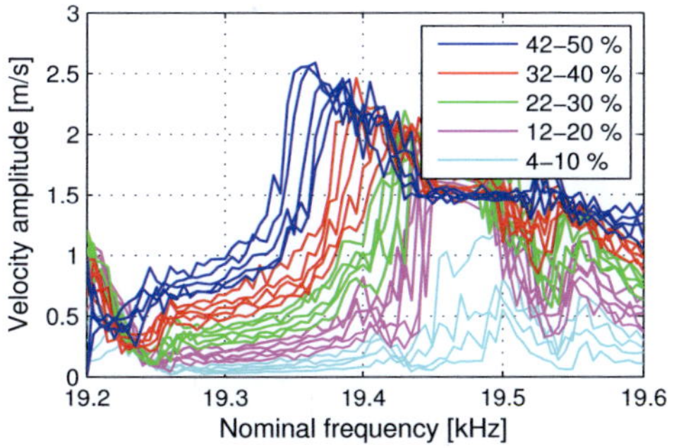

(a) Histories of single-point velocity amplitude originated from sweep excitations with varying power

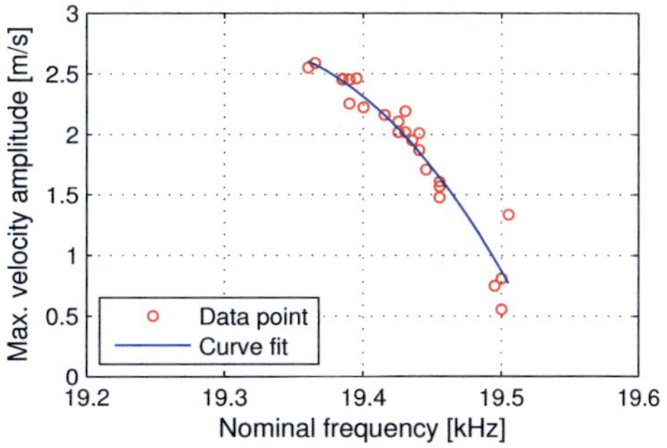

(b) Achieved maximum velocity amplitudes

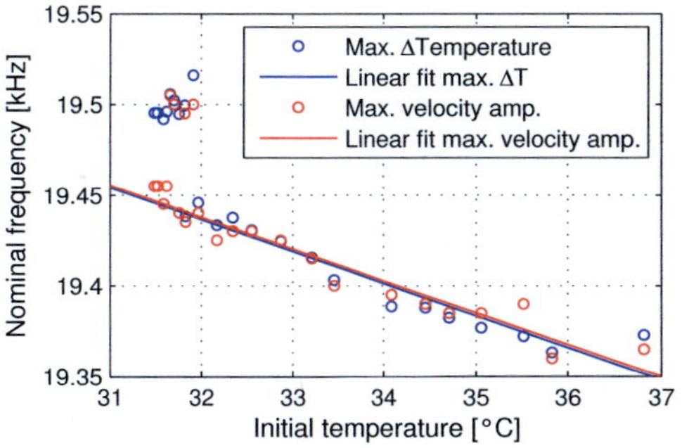

(c) Decreasing velocity amplitude and ΔT peak frequencies with rising initial temperature

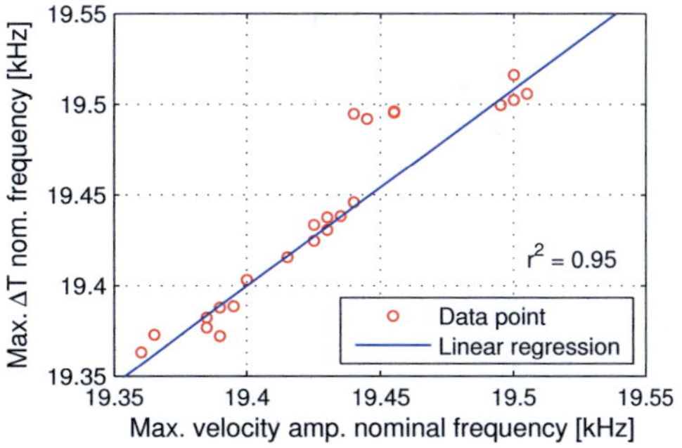

(d) Regression of velocity amplitude and ΔT peak frequencies

Figure 5.12: Shift of the vibrational response to sweep excitation towards lower frequencies and comparison with the crack heating; specimen 06-06.

The experimental results do not leave any doubt that excitation of a resonant frequency is required in order to achieve a significant crack heating. Obviously, the rising initial temperature T_0 causes the observed decrease of the natural frequency. Attempting to rule out any other possible effects leading to a frequency shift, the gradients of the fitted curves in Figure 5.12c are compared with that predicted by the classical plate theory. For this purpose, temperature dependent material properties of low-carbon steel S355J2+N, given by Peil and Wichers [PW04, PW05], are taken into account. Recalling equation 3.22, several resonance frequencies $f_{m,n}$ of a rectangular plate being simply supported along the edges have been calculated. With its initial dimensions of $660 \times 220 \times 6\,\mathrm{mm}^3$ at 20 °C eigenfrequencies below and above the current working range are found by choosing the out-of-plane vibrational mode via the integer numbers of half sine waves to be $n = 5,6$ and $m = 3n$. Table 5.2 summarises the relevant material properties at 20 °C and 50 °C, the plate stiffnesses B and the corresponding natural frequencies $f_{15,5}$ and $f_{18,6}$. Since the coefficient of thermal expansion does not vary much at low temperatures, a constant value of $11 \times 10^{-6}\,\mathrm{K}^{-1}$ is assumed. The predicted slopes $\Delta f/\Delta T_0$ are in the range of -5.6 to -8.1 Hz/K. Considering the nominal test frequency of 19.45 kHz, interpolation results in a value of -7.3 Hz/K, which is only about 42 % of the approximated slope that was found experimentally. Hence, it seems very likely that at least one additional effect is involved, which contributes to the steady decrease of the natural frequency.

The characterisation of the uncoupled Branson converter in Section 5.2.3 did not indicate any drift of the converter's actual frequency due to increasing generator power. A rather flat amplitude spectrum and a reasonable frequency response was observed when testing the complete and the limited working range, respectively. The situation might change if the transducer is attached to a plate specimen so that discrete system resonances and strongly magnified velocity amplitudes are excited. Hence, the actual frequency that prevails in each step of the sweep excitation is extracted by means of the Fourier transform. Figure 5.13a presents the discrepancy between measured and nominal frequency at high-power excitation. Focusing on the used narrow band, a significant deviation from the ideal behaviour (slope $a = 1$, offset $b = 0$) can be found, which is more pronounced at higher

Table 5.2: Temperature dependent properties and eigenfrequencies of a rectangular plate.

T_0 [°C]	E^1 [GPa]	ρ^1 [kg/m^3]	ν^1 [−]	B [N m^2/m]	$f_{15,5}$ [kHz]	$\Delta f_{15,5}/\Delta T_0$ [Hz/K]	$f_{18,6}$ [kHz]	$\Delta f_{18,6}/\Delta T_0$ [Hz/K]
20	202.513	7837	0.277	3948.2	14.869	-5.6	21.412	-8.1
50	196.663	7828	0.287	3861.5	14.702		21.170	

[1]Averaged material properties according to Peil and Wichers [PW04, PW05]

excitation levels and fades at low levels. Obviously, the "true" natural frequency of the vibrating system at about 19.38 kHz is effectively excited within the nominal frequency range of 19.45 to 19.5 kHz. The previously observed jump of about 50 Hz is therefore fully explained by the tendency of a mechanical system to vibrate at its natural frequencies, even if these are not exactly matched by the exciter. This behaviour is in agreement with all thermosonic experiments, since a band of about ± 20 to ± 30 Hz around each natural frequency was found to be effective for crack detection.

Concerning the temperature dependency discussed before, it can be found from Figure 5.13b that the actual frequency shift of -24.1 Hz/K is even greater compared to the slopes determined based on nominal frequencies. Therefore, it seems very likely that due to the long-term use of the coupled transducer the inner piezo stack warmed up and consequently detunes the converter's vibration characteristic. This might strongly affect the overall system response. Note that the shift of the plate's natural frequency was estimated based on the classical plate theory, neglecting any external influences. Unfortunately, the piezo temperature could not be monitored directly since there is no access to the high-voltage parts inside the converter chassis.

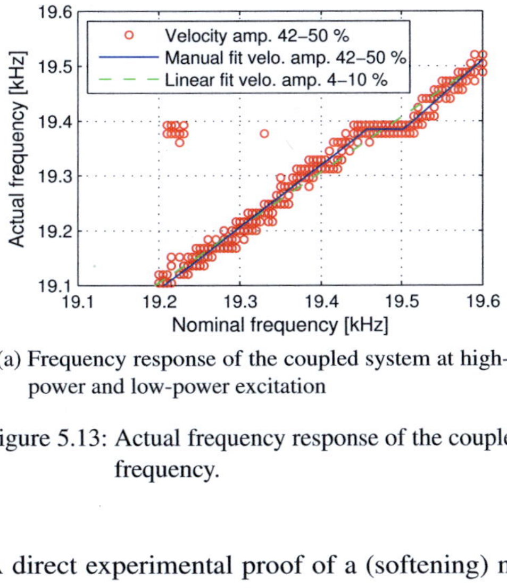

(a) Frequency response of the coupled system at high-power and low-power excitation

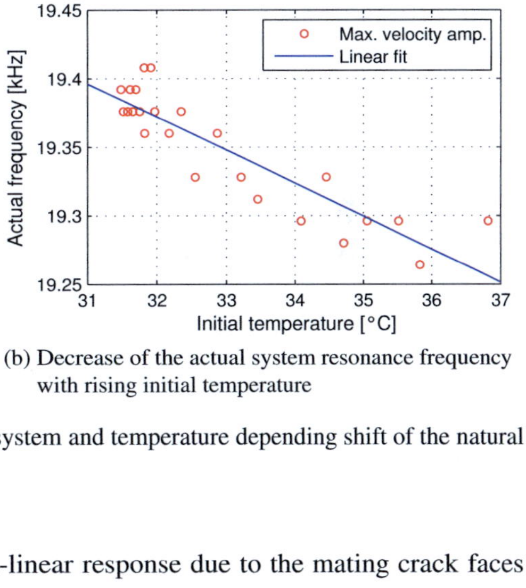

(b) Decrease of the actual system resonance frequency with rising initial temperature

Figure 5.13: Actual frequency response of the coupled system and temperature depending shift of the natural frequency.

A direct experimental proof of a (softening) non-linear response due to the mating crack faces seems to be very difficult since all other interfering effects mentioned before need to be ruled out or quantified exactly. Such an analysis is beyond the scope of this work and requires a great effort regarding the absolute temperature monitoring of the whole plate and interior of the transducer, the use of highly accurate and high-power excitation devices, vibration measurements and reproducible initial conditions.

5.3.3 Relationship between velocity amplitude and frictional heating

Clearly, the frictional heating is one of the most essential mechanisms to be clarified in order to develop a consistent understanding of thermosonic testing. The presented relations between relative excitation power and achieved crack heating in Chapter 4 can only give a rough estimation of the underlying frictional mechanism since the actual vibrational response was not analysed.

Relating the thermographic und vibrometric data achieved from 24 sweep excitations of specimen 06-06 (compare sections 4.4.5 and 5.3.2) demonstrates that the interpretation of ΔT-velocity diagrams can be rather complicated. Figure 5.14a presents all data pairs of the (resampled) crack heating and the single-point velocity amplitude. Since the time histories of crack heating and vibration intensity are strongly correlated, one would expect a clear functional relation when plotting the temperature increase against velocity amplitude. However, due to the transient thermal response, which involves a certain time delay especially during cooling phases, the scatter plot manifests a strong spreading. Consequently, those irrelevant data pairs have been removed. Figure 5.14b shows only the maximum values of ΔT and the maximum velocity amplitude achieved during each sweep. However, it remains unclear which type of functional relation is most appropriate. Candidates are a power law, a power law that turns into a linear function, a shifted (bi)linear relation or other combinations. Clearly, a broader data basis is required for further investigations.

In order to assess the prevailing relation between vibrational displacement or velocity and the resulting defect heating, specimens 10-03, 10-12-I, 10-12-III and 10-24-III have been subjected again to the ultrasound excitation, where the laser vibrometer was employed to acquire velocity

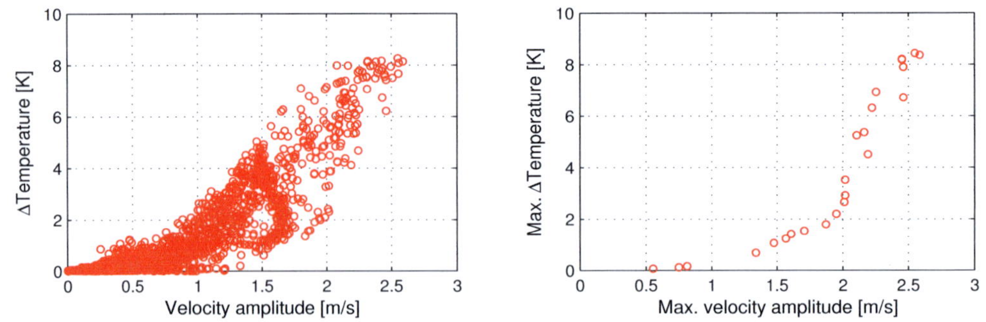

(a) Resampled heating and one-point velocity amplitude data pairs achieved during 24 sweep runs

(b) Maximum values of temperature increase and velocity amplitude achieved during each sweep run

Figure 5.14: Thermal response of specimen 06-06 depending on the mode III velocity amplitude.

data from both sides of the crack. The experiments are conducted in the same fashion as described in Section 4.4.4 using (mono-frequency) tone-burst excitations, where the ultrasound frequency was updated occasionally. Here, single-point and two-point measurements were carried out only in mode III configuration since the previous study (see Section 5.3.1) revealed a better correlation between out-of-plane displacements and defect heating than between in-plane motion and heating. For each excitation level the vibrometric measurements were triggered after a 20 ms delay in order to record stationary oscillations. The burst duration of 150 ms allows for an acquisition of 2^{16} samples at a rate of 2^{19} S/s. All tests were repeated five times to enhance reliability.

First of all, the linearity of the vibration amplitude with respect to the excitation level is questioned once again since in Section 5.2.4 only the uncoupled transducer was analysed. However, plotting the velocity amplitudes against relative excitation power in Figures 5.15a to 5.15d confirms a system inherent threshold at about 1 to 2 %, below which no structural vibration arises. In all four cases a linear regression of the data points reveals a reasonable linearity between velocity amplitude and relative power, once the threshold is exceeded.

When relating the achieved maximum crack heating to the measured steady state velocity amplitude, a variety of functional relationships have been found that can be interpreted based on the model calculations presented in Chapter 3. Due to the small crack size of specimen 10-03, no attempt has been undertaken to measure the differential velocity between both sides of the crack. Instead, single-point measurements of one crack side have been conducted. The thermographic and vibrometric results are presented in Figure 5.16a. Curve fitting of the data points using a robust algorithm reveals that a power law of the form $y = a \cdot x^{2.65}$, which was previously found to represent the pure micro-slip friction behaviour, indeed results in a good approximation of the measurements. For comparative purposes the best quadratic ($a \cdot x^2$) and cubic ($a \cdot x^3$) fit functions are also plotted.

Regarding the larger crack depths of the plates 10-12-I, 10-12-III and 10-24-III, two-point velocity measurements were conducted to directly obtain the differential velocity of both crack sides, which corresponds to the interfacial mode III shear deformation. The retroreflective film stickers have been attached right next to the crack and the laser sensors were pointed as close as possible to the crack path to minimise a possible interference due to superimposed plate bending displacements.

The results obtained from plate 10-12-I are depicted in Figure 5.16b and manifest a similar relation as found on specimen 10-03. Again, the power law with exponent 2.65 fits the available data points very well. However, it seems that the ΔT values achieved at high velocities do not follow the increasing slope of the fit curve. This possibly indicates a transition from micro-slip to Coulomb friction. Hence, an alternative (but not optimised) relation with a linear branch leaving the curve at about 300 mm/s is also displayed in the diagram. Considering the ultrasound frequency of 21.71 kHz, this corresponds to a displacement amplitude of 2.2 μm.

Although having the same nominal crack depth, a totally different situation can be found when testing plate specimen 10-12-III. The achieved relationship of hot spot temperature increase and differential velocity amplitude is shown in Figure 5.16c and clearly indicates an initial gap between the contacting crack faces. A mode III relative displacement amplitude of more than 6.7 μm is required to achieve a significant heating. Increasing the generator power beyond this point resulted in a strong crack heating, similar to specimen 10-12-I. In contrast to that plate, the slope of the ΔT-velocity curve progressively increases so that a shifted power law curve fit with exponent 3.5 leads to a much better approximation of the measured data. This functional relation is in agreement with the model calculation in Chapter 3, considering tangential Coulomb friction with linearly increasing normal contact modulation and an initial gap between the contacting bodies.

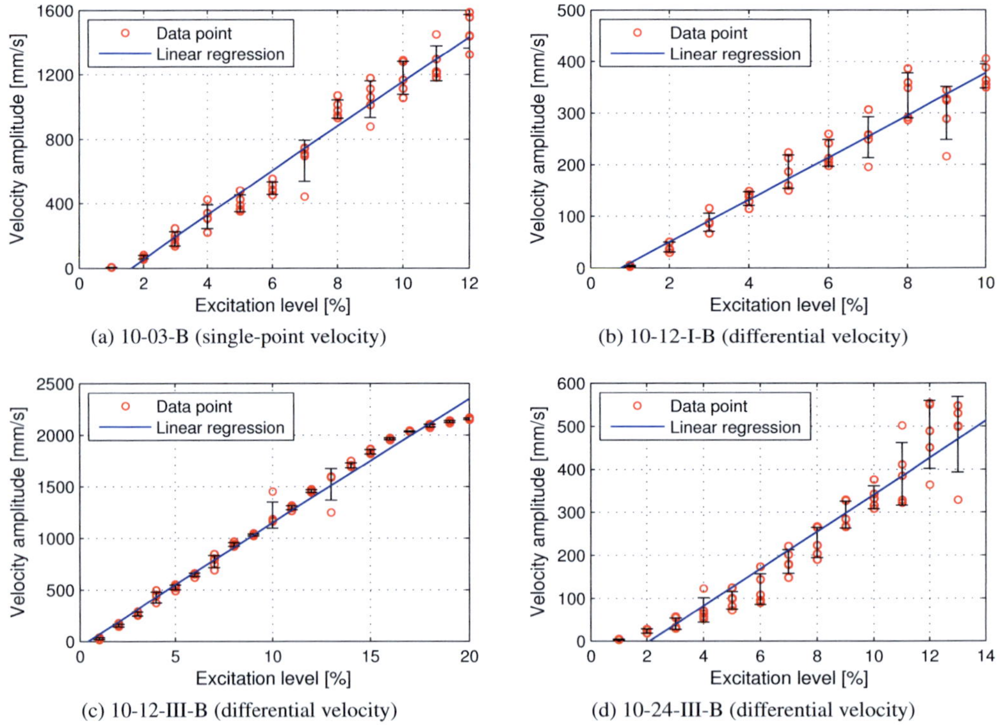

(a) 10-03-B (single-point velocity)

(b) 10-12-I-B (differential velocity)

(c) 10-12-III-B (differential velocity)

(d) 10-24-III-B (differential velocity)

Figure 5.15: Linearity between relative power level and specimen velocity amplitude.

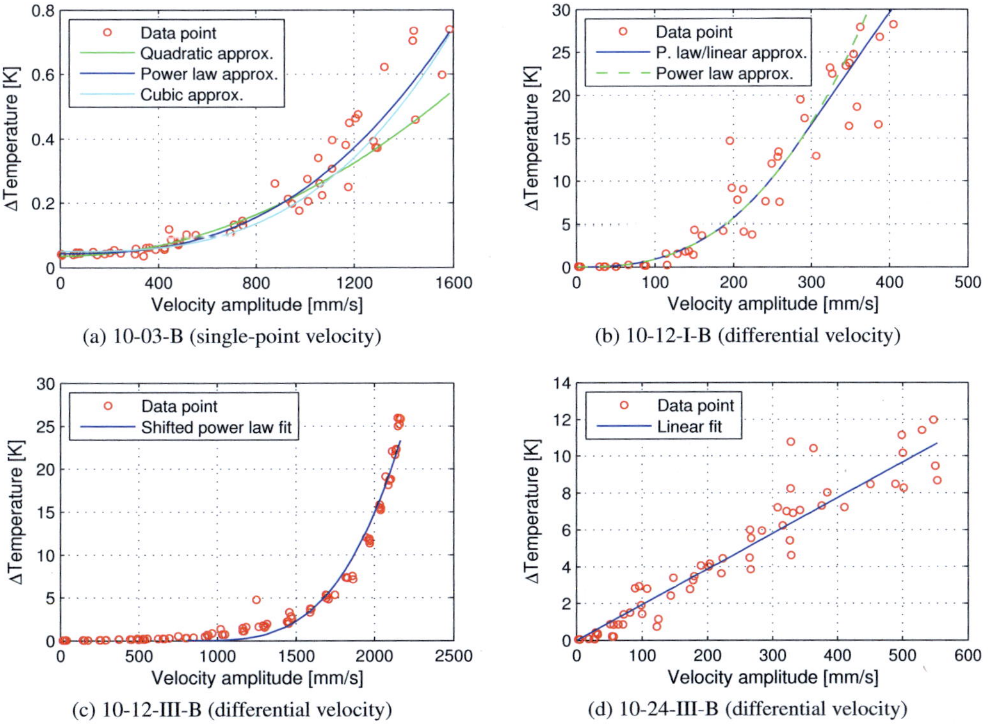

(a) 10-03-B (single-point velocity)

(b) 10-12-I-B (differential velocity)

(c) 10-12-III-B (differential velocity)

(d) 10-24-III-B (differential velocity)

Figure 5.16: Thermal response to mono-frequency excitation depending on the velocity amplitude.

The investigation of specimen 10-24-III manifests another different behaviour concerning the thermal response. Against expectations, an almost perfectly linear relation between hot spot heating and velocity amplitude is found for the 24 mm deep crack as illustrated in Figure 5.16d. Based on the theoretical modelling, this outcome allows two different interpretations: Pure Coulomb friction without any variation of the normal contact pressure would lead to a linear relation between power dissipation (or heating) and relative displacement (or velocity) amplitude. On the other hand, a high compressive stress corresponds to a positive offset of the contact surface approach. As can be seen from Figure 3.11c the ΔT-velocity relation again turns linear with increasing approach or contact pressure. Since the thermal hot spot which has been detected in this study is much closer to the crack tip than that shown in Figure 4.29b (at about 3/4 of the crack depth), the author believes that a significant residual compressive stress is likely to cause the achieved linear relation. Regarding the model calculation and the results in Figure 3.11c, a prestress of 3.5 MPa would be sufficient so that the ΔT-velocity relation appears linear up to the achieved displacement amplitude of 4 μm.

The presented results clearly demonstrate that determination of a specific unifying frictional law that would explain all of the four different relations achieved is hardly possible on the basis of single-point and two-point differential vibrometry. However, a first attempt has been made to interpret the vibrothermographic results considering Coulomb friction, the micro-slip effect and a normal contact modulation. Further diagnosis requires synchronous measurements of the relative displacements according to crack mode I, II and III.

5.3.4 Phase sensitive measurement of crack mode II and III displacements

In the previous sections a reasonable correlation between mode II/III vibration intensity and defect heating was found during sweep excitation. However, it remained unclear whether each single peak observed in the ΔT history can be unambiguously assigned to an increased shear deformation at the crack interface or if a pure bending mode of the plate that involves mode I relative displacements heats up the crack as well. In order to identify the type of crack face displacement, phase sensitive velocity measurements of both sides of the crack are performed.

The fibre-optic laser vibrometer allows for single-point measurements and two-point differential measurements. Thanks to these options there are essentially two different approaches to determine the phase relation between the oscillations of both crack edges:

- Synchronised acquisition of each single-point laser vibrometry signal and a reference signal of the ultrasound generator
- Acquisition of each single-point laser vibrometry signal and acquisition of the two-point differential vibrometry output signal

Synchronised measurement of the generator reference signal and a single-point velocity can be considered as the direct approach. Since the Branson generator does not provide a high-frequency reference, a custom-made toroidal coil was installed by threading it on the high-voltage conductor inside the device. The electromagnetic field around the conductor induces an AC current in the electric coil. The voltage drop across an ohmic resistor connected in parallel indicates the phasing of the excitation. Vibrometry of one point then results in the velocity amplitude and the phase shift with respect to the ultrasound excitation. This approach has the advantage that it can be repeated for any number of additional points on the structure. Since there is a temporal reference signal available, the phase angle of each oscillating point is directly evaluated. Thus, complete vibration mode shapes of a surface can be obtained. The method requires a very stable excitation frequency.

The second option for achieving amplitude values and the phasing of two points requires three consecutive measurement passes: Single-point vibrometry of both target points and a two-point differential vibrometry. The absolute phase lag between both points is unambiguously determined by the three amplitude values. This approach is not suitable for scanning a whole surface. As an advantage, it can be used if there is no reference signal available. The method requires a good excitation reproducibility in terms of vibration level.

Within a first investigation of the crack face displacements only one point on each side of the crack should be measured according to crack mode II and III. Since the ultrasound frequency strongly influences the thermal response, the sweep excitation is performed using the same parameters (frequency band, step width and duration) as described in Chapter 4. After installation of the toroidal coil the ohmic resistance was tuned and fixed to provide a suitable compromise between sensitivity and safety against voltage overload (max. voltage of analogue input: 10 V) during ultrasound excitation. However, many test runs have been performed and it was found that the output signal of the coil was heavily distorted at very few discrete frequencies. The reason for this phenomenon could not be identified. Because of that the second measurement approach for determining the phase relation was chosen.

Under the assumption that both crack faces 1 and 2 perform sinusoidal oscillations of the type

$$y_1 = \hat{y}_1 \sin(\omega t + \varphi_1) \quad , \qquad y_2 = \hat{y}_2 \sin(\omega t + \varphi_2) \tag{5.16}$$

the measured two-point differential oscillation is given by the superposition

$$y_{\mathrm{diff}} = y_1 - y_2 = \hat{y}_{\mathrm{diff}} \sin(\omega t + \varphi_{\mathrm{diff}}) \quad . \tag{5.17}$$

Since all laser vibrometric measurements are independent of each other, only the amplitude values $\hat{y}_1$, $\hat{y}_2$ and $\hat{y}_{\mathrm{diff}}$ are obtained directly. The phase relation which is determined by φ_1, φ_2 and φ_{diff} remains unknown. However, these values are not explicitly required in order to assess the temporal lag between the crack edge oscillations according to crack modes II and III. Instead, only the absolute phase shift $|\varphi_2 - \varphi_1|$ between both crack sides is of importance. The vector representation as shown in Figure 5.17 clarifies that the absolute phase shift between the oscillations with amplitudes $\hat{y}_1$ and $\hat{y}_2$ is determined when the differential amplitude is also known.

The superposition is physically possible if $\hat{y}_{\mathrm{diff}}$ lies within the interval

$$|\hat{y}_1 - \hat{y}_2| \leq \hat{y}_{\mathrm{diff}} \leq \hat{y}_1 + \hat{y}_2 \quad . \tag{5.18}$$

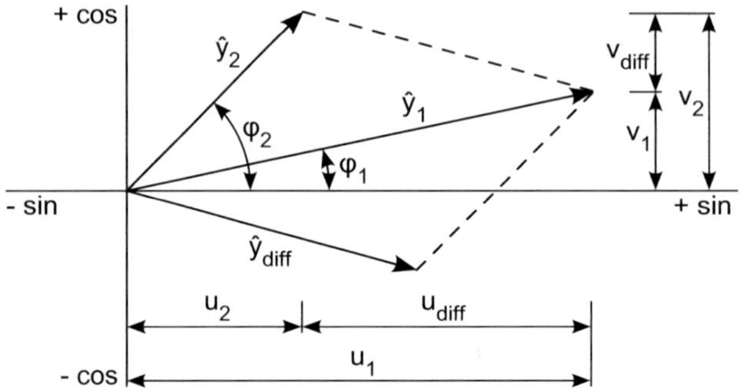

Figure 5.17: Superposition of two oscillations in the vector diagram.

If the oscillations y_1 and y_2 are perfectly in-phase ($\varphi_1 = \varphi_2$), $\hat{y}_{\mathrm{diff}}$ is equal to $|\hat{y}_1 - \hat{y}_2|$ and for the out-of-phase situation ($\varphi_1 = \varphi_2 \pm \pi$) $\hat{y}_{\mathrm{diff}}$ is equal to $\hat{y}_1 + \hat{y}_2$.

Making use of the components

$$\begin{aligned} u_1 &= \hat{y}_1 \cos\varphi_1 \quad , \qquad & u_2 &= \hat{y}_2 \cos\varphi_2 \quad , \\ v_1 &= \hat{y}_1 \sin\varphi_1 \quad , \qquad & v_2 &= \hat{y}_2 \sin\varphi_2 \quad , \end{aligned} \tag{5.19}$$

the differential amplitude $\hat{y}_{\mathrm{diff}}$ can be expressed as:

$$\begin{aligned} \hat{y}_{\mathrm{diff}}^2 &= u_{\mathrm{diff}}^2 + v_{\mathrm{diff}}^2 = (u_1 - u_2)^2 + (v_2 - v_1)^2 \\ &= (\hat{y}_1 \cos\varphi_1 - \hat{y}_2 \cos\varphi_2)^2 + (\hat{y}_2 \sin\varphi_2 - \hat{y}_1 \sin\varphi_1)^2 \\ &= \hat{y}_1^2 \left(\cos^2\varphi_1 + \sin^2\varphi_1\right) + \hat{y}_2^2 \left(\cos^2\varphi_2 + \sin^2\varphi_2\right) \\ &\quad - 2\,\hat{y}_1\,\hat{y}_2 \left(\cos\varphi_1 \cos\varphi_2 + \sin\varphi_1 \sin\varphi_2\right) \end{aligned} \tag{5.20}$$

Plugging the trigonometric identities (see [BSMM97])

$$\cos^2\varphi + \sin^2\varphi = 1 \quad , \qquad \cos\varphi_1 \cos\varphi_2 + \sin\varphi_1 \sin\varphi_2 = \cos|\varphi_2 - \varphi_1| \tag{5.21}$$

into equation (5.20) finally leads to the absolute phase shift between oscillations 1 and 2:

$$|\varphi_2 - \varphi_1| = \arccos\left(\frac{\hat{y}_1^2 + \hat{y}_2^2 - \hat{y}_{\mathrm{diff}}^2}{2\,\hat{y}_1\,\hat{y}_2}\right) \tag{5.22}$$

The proposed measurement procedure was applied to several plate specimens, where mode II and III displacements of the crack edges were acquired. Analysis of the three velocity histories (y_1, y_2, y_{diff}) and the thermal response achieved during consecutive sweep excitations revealed that most of the peak values $\mathrm{d}T/\mathrm{d}f$ are clearly correlated with increased velocity amplitudes. In case of plate resonances, a standing wave pattern occurs so that two points on the specimen move either perfectly in-phase or out-of-phase, unless one of the measurement points represents a nodal point of the mode shape. Evaluation of the conditional equation (5.22) indeed proved the absolute phase lag to be always close to 0° or 180°, which means that the single-point velocity amplitudes are subtracted or added, respectively, to yield the two-point differential amplitude.

Figures 5.18 and 5.19 exemplarily depict the histories of the positive values $\mathrm{d}T/\mathrm{d}f$ and the velocity amplitudes according to mode II and III, found for specimen 06-12 during the six individual sweep runs. The most significant peak values have been analysed and highlighted according to the following scheme. Corresponding results of plates 10-12-I, 10-12-III and 10-24-III can be found in Appendix A.7.

At first, only the frequencies which allow for thermographic crack detection are considered:

- Frequencies that lead to a large differential amplitude are shaded blue if both crack edges move in-phase.
- Frequencies resulting in a large differential amplitude due to an out-of-phase movement of the crack sides are highlighted green.
- Frequencies which are obviously heating up the crack due to relative displacements according to the respective other crack mode are marked yellow.

Secondly, frequencies are regarded at which any of the three velocity amplitudes shows a peak but yet there is no considerable defect heating:

- Frequencies that lead to large one-point velocity amplitudes but very low differential amplitudes are shaded light grey.
- Frequencies that manifest a large differential amplitude are coloured in dark grey.

As a general finding it should be noted once again that the maximum velocity amplitudes according to crack mode III exceed the mode II velocities by a factor of five. Moreover, almost all of the heating/displacement relevant frequencies identified by the mode II measurements could also be assigned to a corresponding velocity peak value according to mode III. The frequency of 19.44 kHz represents the only exception. Here, the mode III differential amplitude is very low but the mode II differential amplitude clearly shows a peak during the sweep excitation. In summary, four frequencies were found in the mode II analysis that lead to a relatively high differential amplitude but do not significantly heat up the defect. In contrast, all frequencies that reveal increased mode III differential amplitudes match the heating relevant frequencies within the investigated band. However, a direct proportionality between velocity amplitude and $\mathrm{d}T/\mathrm{d}f$ values cannot be found. The thermal response strongly varies from one frequency to another although the same vibration intensity, regarding mode III, is reached.

In conclusion, it is questionable if mode II displacements contribute to the crack heating at all, especially in case of a relatively thin plate specimen such as 06-12. Based on this study, it seems that most of the mode II peak displacements do not exceed the threshold level which is required to cause of resolvable temperature increase. Since all conspicuous frequencies indicate the presence of mixed modes instead of pure in-plane or out-of-plane shear modes, the question arises to what extent the phasing between mode II and III displacements and the crack mode I deformation influences the resulting defect heating.

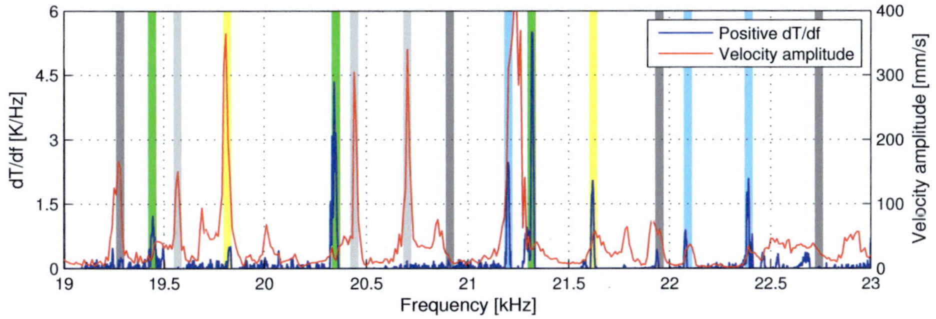

(a) Histories of $\mathrm{d}T/\mathrm{d}f$ and mode II velocity amplitude of crack side 1

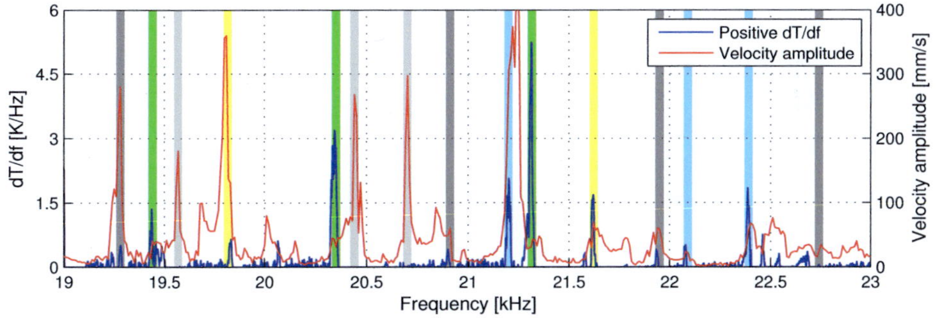

(b) Histories of $\mathrm{d}T/\mathrm{d}f$ and mode II velocity amplitude of crack side 2

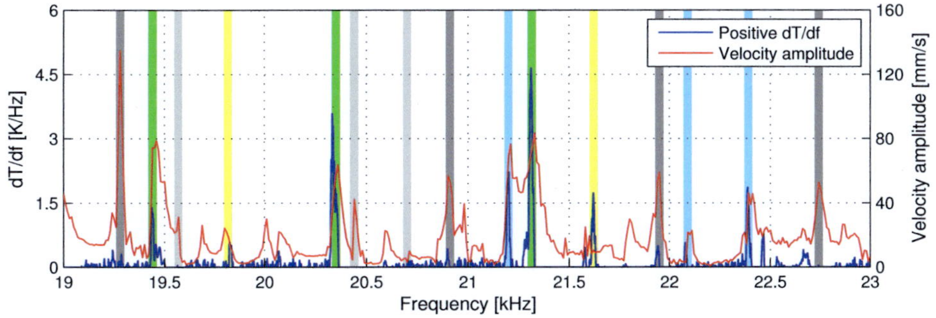

(c) Histories of $\mathrm{d}T/\mathrm{d}f$ and mode II differential velocity amplitude of crack sides 1 and 2

Figure 5.18: Histories of $\mathrm{d}T/\mathrm{d}f$ and mode II velocity amplitudes; specimen 06-12-A; blue: large diff. amplitude and heating, crack sides in-phase; green: large diff. amplitude and heating, crack sides out-of-phase; yellow: mode III is determinative of the diff. amplitude and heating; light grey: low diff. amplitude, no heating; dark grey: large diff. amplitude, no heating.

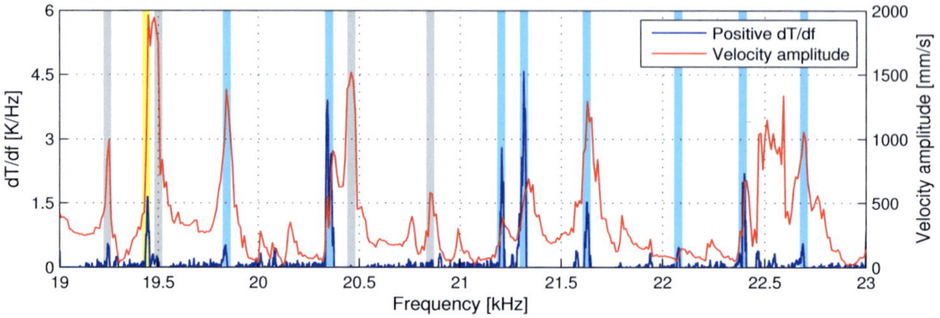

(a) Histories of $\mathrm{d}T/\mathrm{d}f$ and mode III velocity amplitude of crack side 1

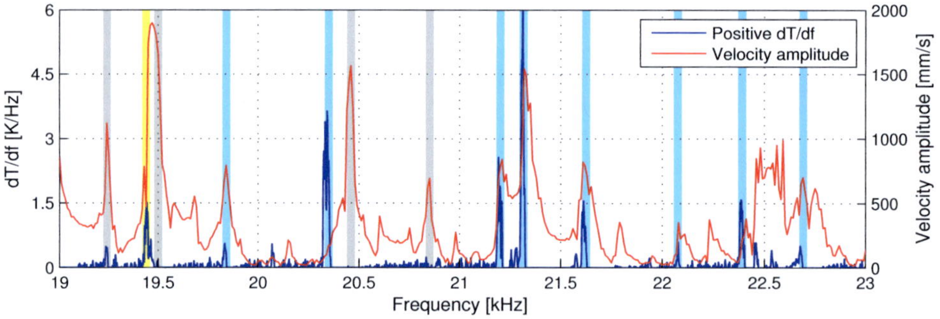

(b) Histories of $\mathrm{d}T/\mathrm{d}f$ and mode III velocity amplitude of crack side 2

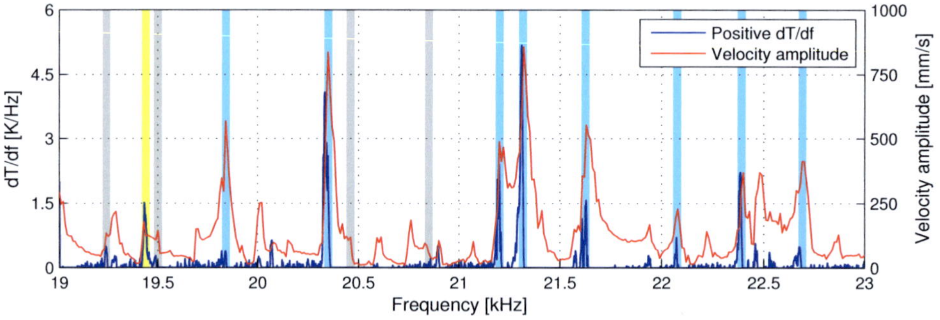

(c) Histories of $\mathrm{d}T/\mathrm{d}f$ and mode III differential velocity amplitude of crack sides 1 and 2

Figure 5.19: Histories of $\mathrm{d}T/\mathrm{d}f$ and mode III velocity amplitudes; specimen 06-12-A; blue: large diff. amplitude and heating, crack sides in-phase; yellow: mode II is determinative of the diff. amplitude and heating; light grey: low diff. amplitude, no heating.

5.4 Three-dimensional scanning laser vibrometry

The final attempt to establish a qualitative relation between spatial crack face displacements and defect heating is based on three-dimensional scanning laser vibrometry. This section describes all steps from the testing procedure and data processing to the composition of 3D mode shapes. Lastly, the theoretical contact model is considered in order to predict the actual defect heating based on the vibrometric measurements.

5.4.1 Testing procedure and data processing

To allow for surface laser scanning, two specimens with a thickness of 6 and 10 mm have been equipped with a uniform grid of retroreflective stickers as shown in Figure 5.1. Along the crack, the film patches of about $6 \times 6\,\mathrm{mm}^2$ were attached as close as possible to both sides of the crack path to effectively measure the differential movements.

3D scanning was achieved by three successive raster scans using the test setup described in Section 5.1.1, where the laser sensor's angle of incidence was changed after each pass. Prior to any automatic movement of the positioning stages the two-dimensional grid point positions are manually approached and stored by means of a proprietary software that controls both linear axes, the ultrasound system as well as the data acquisition. To ensure that the vibrometer sensitivity was adequately chosen, the first automated scan was performed with the laser sensor pointing perpendicular at the plate surface because the pure out-of-plane displacement is expected to have the greatest magnitude. The following scans were done with the laser sensor being inclined and rotated through 30 to 45°, respectively, as shown in Figure 5.1b. In Section 5.1.3 the 3M self-adhesive film was analysed and proven to allow for laser vibrometry using angles of incidence up to 45°.

During each scan the positions of all grid points are approached in the same pre-defined order. Once, a grid point is reached by the positioning stages, the ultrasound system is activated and set to a constant power level. Then, the desired frequencies – primarily resonance frequencies found by preceding sweeps – are tuned in one after another. For each grid point and frequency 540 samples are acquired at a sampling rate of 0.5 MS/s. Again, prior to each data acquisition a programmed delay of 200 ms is incorporated to ensure that the plate reached a steady state vibration. Since the subsequent composition of the 3D grid point displacements requires information about the phase lag of each recorded time history with respect to the ultrasound excitation, a second channel providing a reference signal is synchronously acquired besides of the vibrometer velocity output.

This temporal reference can be distorted to some extent as long as it is clearly periodically and has an invariant phasing with respect to the driving transducer. The toroidal coil installed inside the Branson ultrasound generator offers a periodic voltage signal that sufficiently represents the shape of the harmonic excitation. In case of the TIB amplifier, the high-frequency input signal itself serves as an ideal reference and was looped through to the corresponding analogue input of the data acquisition device.

In total, a complete 3D scan contains a set of $6n$ time histories per frequency with n being the number of grid points. The collected raw data was subjected to a standardised procedure including basic signal conditioning tasks:

- Scaling of the velocity proportional analogue voltage signal according to the used vibrometer measurement range and numerical integration to obtain displacement histories
- High-pass filtering of all acquired signals using a 5th order Butterworth filter with a cut-off frequency of 10 kHz
- Subtraction of a fitted linear detrend function from each displacement history in order to reset the zero position and to compensate for a possible drift
- Subtraction of the mean value from each reference signal and smoothing of the data

Prior to any further processing the data recorded at each excitation frequency is separated and treated individually in the following. Up to this point, the single-point velocity/displacement measurements, together with their corresponding reference records, are not temporally related at all. Their phasing with respect to each other is not known, yet. In order to establish the mutual phase relations, one single reference stream arbitrarily chosen from the complete data set serves as the master reference. Next, the individual phase shift between each reference signal and the master reference signal is determined:

- Computation of the cross-correlation between master reference and each other reference signal based on the original sampling considering a maximum shift of ± 18 time steps covering a full period of $360°$ for all excitation frequencies in the range of 15 to 25 kHz
- Spline interpolation of the sample cross-correlation coefficient function onto a refined grid with a spacing of 1/10 of the original sampling time step
- Determination of the refined phase lag or lead indicated by the position of the maximum correlation coefficient

Each displacement history is then shifted according to the phasing found for its corresponding reference signal. "Moving" a time-discrete signal with sub-sample accuracy requires an interpolation scheme. Here, the method used by Medgenberg [Med08] was adopted. After computation of the new sampling points in time according to the desired time shift, cubic polynomials are fitted to four consecutive data points $\vec{y}^T = (y_1\ y_2\ y_3\ y_4)$ each, where the target point in time t_{int} whose displacement value y_{int} is to be interpolated always lies in between the second and third regarded sample. Stepping forward through the original samples, the considered times t_1, t_2, t_{int}, t_3 and t_4 are mapped to an arbitrarily chosen but equidistant relative time scale (for example using $t_{1,\text{rel}} = -1.5$, $t_{2,\text{rel}} = -0.5 < t_{\text{int,rel}} < t_{3,\text{rel}} = 0.5$ and $t_{4,\text{rel}} = 1.5$). Normalisation of the target time allows using a constant interpolation coefficient matrix $\mathbf{M}_\text{c}$. Each cubic polynomial is determined by its coefficients $\vec{c}^T = (c_1\ c_2\ c_3\ c_4)$ which are retrieved through

$$\vec{c} = \begin{pmatrix} c_1 \\ c_2 \\ c_3 \\ c_4 \end{pmatrix} = \mathbf{M}_\text{c}^{-1} \cdot \vec{y} = \begin{pmatrix} 1 & t_{1,\text{rel}} & t_{1,\text{rel}}^2 & t_{1,\text{rel}}^3 \\ 1 & t_{2,\text{rel}} & t_{2,\text{rel}}^2 & t_{2,\text{rel}}^3 \\ 1 & t_{3,\text{rel}} & t_{3,\text{rel}}^2 & t_{3,\text{rel}}^3 \\ 1 & t_{4,\text{rel}} & t_{4,\text{rel}}^2 & t_{4,\text{rel}}^3 \end{pmatrix} \cdot \begin{pmatrix} y_1 \\ y_2 \\ y_3 \\ y_4 \end{pmatrix} \quad . \tag{5.23}$$

Then, the interpolated displacement values

$$y_{\text{int}} = \sum_{i=1}^{4} c_i \cdot t_{\text{int,rel}}^{i-1} \tag{5.24}$$

are saved as the time-shifted displacement histories. Finally, all single-point displacement records have a common phasing with respect to the ultrasound excitation. Their associated reference signals would now be perfectly in-phase if they had been shifted in the same way. This test was performed several times for verification purposes.

Clearly, the quality of the phase compensation dramatically increased due to the sub-sample interpolation scheme. A certain interval (18 samples) at the beginning and at the end of all records has been discarded for all measurements due to shifting of the signals. Furthermore, no special interpolation rules were used for the first and last instance of t_{int}, when lying between t_1 and t_2 or t_3 and t_4, respectively. Therefore, another two original samples are not considered during the proposed data processing.

5.4.2 Extraction of vibration mode shapes

In order to reconstruct the surface vibration based on the displacements of the measurement grid, the spatio-temporal relations between the three independent measurement passes per point need to be established. After synchronisation of all displacement histories as described in the previous section, a coordinate transformation is required to compute the grid point positions in an orthogonal coordinate system aligned with the plate specimen.

The displacement vector $\vec{d}$ with components u, v, w in a local Cartesian coordinate system can be obtained for each point at any time by the trigonometric relations solely depending on the chosen measurement directions A, B and C. Figure 5.20 illustrates the local system for one grid point viewed from the right side and the top.

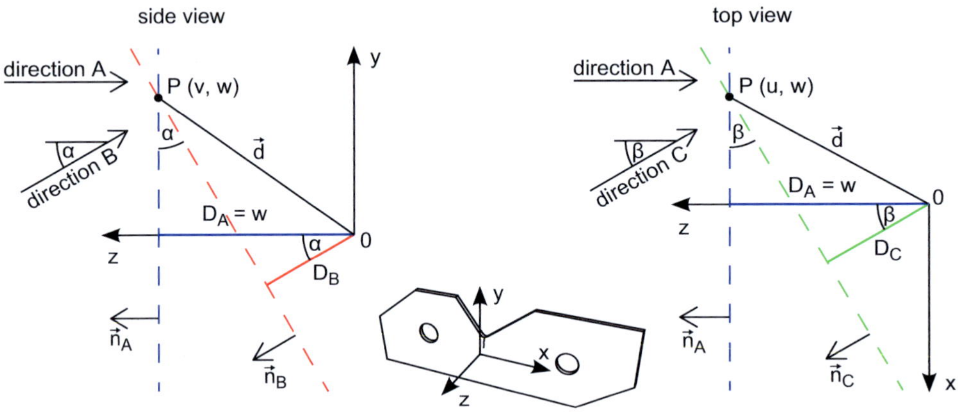

Figure 5.20: Trigonometric relations between laser measurement directions A, B, C and grid point displacements in Cartesian coordinates.

The system's origin corresponds to the zero position of the oscillating point P. Laser measurement direction A is perpendicular to the plate, B and C are rotated by the elevation angle α and the azimuth angle β. Therefore, w is equal to D_A, the displacement measured in direction A. The in-plane components u and v are found using the following relations:

$$u = \frac{-D_A}{\tan\beta} + \frac{D_C}{\cos\beta \tan\beta} \quad , \qquad v = \frac{D_A}{\tan\alpha} - \frac{D_B}{\cos\alpha \tan\alpha} \tag{5.25}$$

Making use of the identity $\sin\varphi = \cos\varphi \cdot \tan\varphi$, the displacement vector of a point P is then assembled as:

$$\vec{d} = \begin{pmatrix} u \\ v \\ w \end{pmatrix} = \begin{pmatrix} \frac{-D_A}{\tan\beta} + \frac{D_C}{\sin\beta} \\ \frac{D_A}{\tan\alpha} - \frac{D_B}{\sin\alpha} \\ D_A \end{pmatrix} \tag{5.26}$$

The current position of P can also be thought of as the intersection of the three planes which are defined by the surface normals $\vec{n}_A$, $\vec{n}_B$ and $\vec{n}_C$ with components according to the measurement directions A, B and C. These are expressed in the Cartesian coordinate system:

$$\vec{n}_A = \begin{pmatrix} 0 \\ 0 \\ 1 \end{pmatrix} \quad , \qquad \vec{n}_B = \begin{pmatrix} 0 \\ -\tan\alpha \\ 1 \end{pmatrix} \quad , \qquad \vec{n}_C = \begin{pmatrix} \tan\beta \\ 0 \\ 1 \end{pmatrix} \tag{5.27}$$

If the planes are given in normal form, the distances to the origin correspond to the displacements D_A, D_B and D_C. The normalised form of the planes based on

$$\vec{d} \cdot \vec{n}_{i,0} - D_i = 0 \quad , \qquad \vec{n}_{i,0} = \frac{\vec{n}_i}{|\vec{n}_i|} \quad , \qquad i = A, B, C \tag{5.28}$$

results in a linear system of equations which is solved for $\vec{d}$ at each point in time:

$$\mathbf{N}_0 \cdot \vec{d} = \begin{pmatrix} \vec{n}_{A,0}^T \\ \vec{n}_{B,0}^T \\ \vec{n}_{C,0}^T \end{pmatrix} \cdot \begin{pmatrix} u \\ v \\ w \end{pmatrix} = \begin{pmatrix} D_A \\ D_B \\ D_C \end{pmatrix} \tag{5.29}$$

$$\vec{d} = \mathbf{N}_0^{-1} \cdot \vec{D} = \begin{pmatrix} n_{A,0,u} & n_{A,0,v} & n_{A,0,w} \\ n_{B,0,u} & n_{B,0,v} & n_{B,0,w} \\ n_{C,0,u} & n_{C,0,v} & n_{C,0,w} \end{pmatrix}^{-1} \cdot \begin{pmatrix} D_A \\ D_B \\ D_C \end{pmatrix} \tag{5.30}$$

Finally, each grid point's 3D dynamic displacement, which has been centred at the zero position of the local system, is added to its static two dimensional coordinates that are already defined in the global coordinate system when programming the scan routine. No further transform is required if both systems, the global and each local one, have the same orientation. After meshing the grid points using MATLAB's built in implementation of the Delaunay triangulation and discarding small or sharp-ended facets, an animated visualisation of the surface vibration is achieved by cycling through all the individual frames at each instance of time.

5.4.3 Prediction of defect heating based on crack face displacements

The vibrothermographic experiments described before aimed at the analysis of the relation between heat dissipation and oscillation amplitude of the crack faces, measured in just one direction. Relative displacements according to mode II and III have been investigated, where the in-plane movement according to crack mode I has not been taken into account, yet. Within this passage it is attempted to find a correlation between the three-dimensional crack face movements and the resulting defect heating. Therefore, the two plate specimens 06-12 and 06-24 were excited by means of the Branson and the TIB ultrasound equipment. At first, several basic sweep runs were performed with both systems in order to identify the ten strongest resonance frequencies within the standard frequency range of 19 to 23 kHz. One-point laser vibrometry in the out-of-plane direction together with a DFT-based data processing was used for this task.

In the next step, 3D scanning laser vibrometry measurements were conducted as described in the previous sections. Only the conspicuous resonance frequencies have been considered when scanning the grid points covering the crack region of each specimen. The experiments on both plates with the two ultrasound systems were repeated several times to prove a sufficient reproducibility of the scanning procedure. Here, the TIB system was operated at full power and the Branson generator was run at relative power levels between 9 and 12 %. Evaluation of the measured surface vibrations allowed for the assignment of corresponding mode shapes that have been excited by the high-power weld generator as well as by means of the TIB linear amplifier. Especially the out-of-plane component is useful to identify common modes since its displacement amplitude is significantly greater than those of the in-plane displacements. Note that the analysed frequencies slightly differ, depending on the ultrasound system. On average, the resonance frequencies found by use of the Branson generator are about 35 to 100 Hz higher than those found with the TIB system. This discrepancy was expected since the analysis of the frequency accuracy comparing actual and nominal frequencies revealed a mean deviation of about 96 Hz in case of the Branson system.

Comparison of the thermal responses during sweep testing confirmed that the low-power excitation achieved by the TIB system does not result in any crack heating within a reasonable time. Instead, Figure 5.21a exemplarily shows the spatial distribution of the absolute temperature of plate 06-24 obtained by means of the Branson generator. The corresponding temperature history ΔT as well as the evolution of the (positive) first derivative $\mathrm{d}T/\mathrm{d}f$ is depicted in Figure 5.21b. The marked frequencies indicate only such resonances whose mode shapes could be identified using both excitation devices. Figure 5.21b clarifies that these are not necessarily the most efficient frequencies with regard to crack heating. However, a certain coincidence of the selected resonance frequencies with those that are heating relevant manifests without any doubt.

The detailed analysis of the reconstructed surface vibrations achieved by means of the Branson generator unfortunately revealed that a proper composition of the in-plane displacements is hardly possible. Contrary to the directly measured out-of-plane component, which proved a sufficient reproducibility, the in-plane measurements are not reliable. Here, many of the scanned grid points do not oscillate harmonically at all and do not seem to move similar to their neighbouring points. The installed toroidal coil that provides the temporal reference signal could be eliminated as a potential error source by using the velocity output of a second vibrometer as an excitation reference, instead.

It was found from further analysis of the raw data records that the Branson system is not able to accurately hold its ultrasound frequency constant over the scan duration. Even small frequency variations lead to short-term but significant mismatching of the phasing. The presented data-processing allows only for a compensation of the average phase lag or lead. Regarding the in-plane components, which are composed from two individual scans considering different angles of incidence, a temporary phase mismatch can lead to extreme errors of the computed grid point displacements. A time-resolved phase recognition using short-time Fourier transform or wavelet analysis is beyond the scope of this work but could be useful in approaching this issue despite of the complicated data processing.

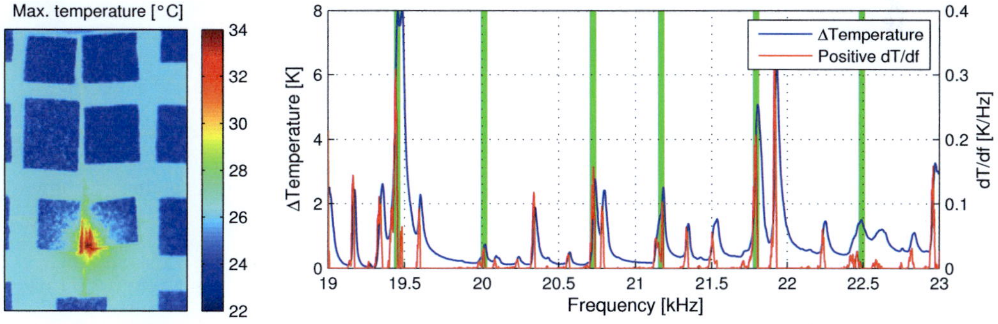

(a) Max. temperature (b) Hot spot temperature increase and positive dT/df history

Figure 5.21: Spatial temperature distribution and time history of the thermal response due to sweep excitation; specimen 06-24-B.

In case of the TIB equipment, no problems have been perceived either during the scan procedure or at the post-processing stage including the 3D displacement composition. As a result, perfectly smooth vibration modes have been reconstructed mainly due to the ideal frequency and amplitude accuracy of the TIB ultrasound system.

Figures 5.22 and 5.23 exemplarily depict two of the six mode shapes which could be identified on specimen 06-24 using both excitation devices. Due to the described problems related with the frequency accuracy of the Branson weld generator, Figures 5.22a and 5.23a only show the maximum out-of-plane deflection achieved. Instead, Figures 5.22b to 5.22d and 5.23b to 5.23d illustrate the maximum displacements according to mode III, I and II, respectively, obtained by using the TIB system. Note the strong coincidence of both measured out-of-plane mode shapes and the different orders of magnitude regarding the maximum deflection. Other mode shapes of specimen 06-24 can be found in Appendix A.8.

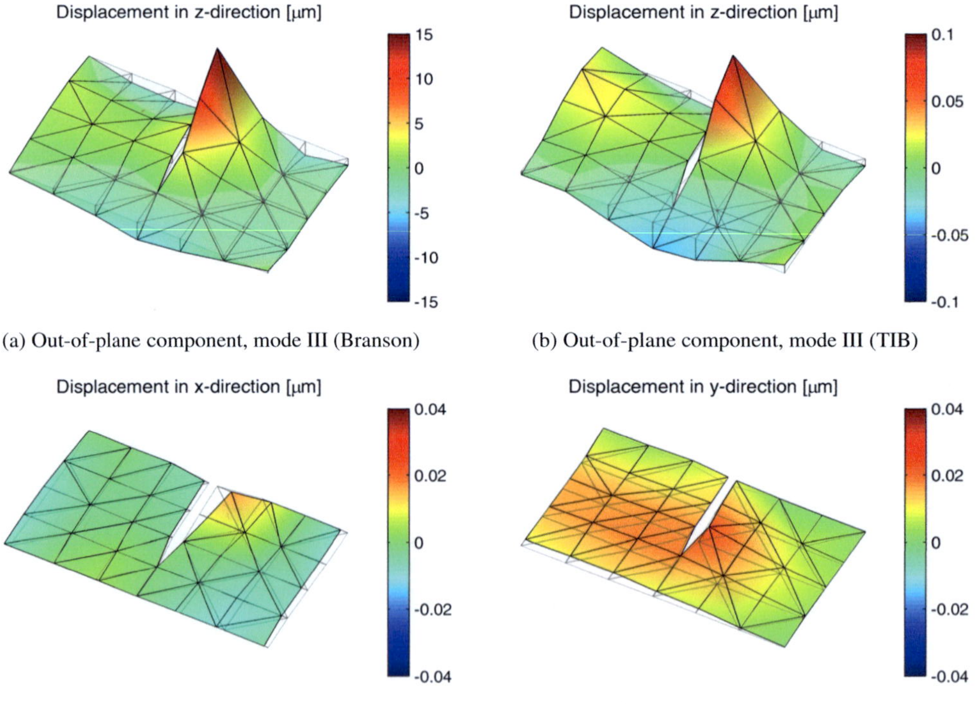

(a) Out-of-plane component, mode III (Branson)

(b) Out-of-plane component, mode III (TIB)

(c) In-plane component, mode I (TIB)

(d) In-plane component, mode II (TIB)

Figure 5.22: Displacement components of the mode shape of specimen 06-24-B at 20.67 kHz.

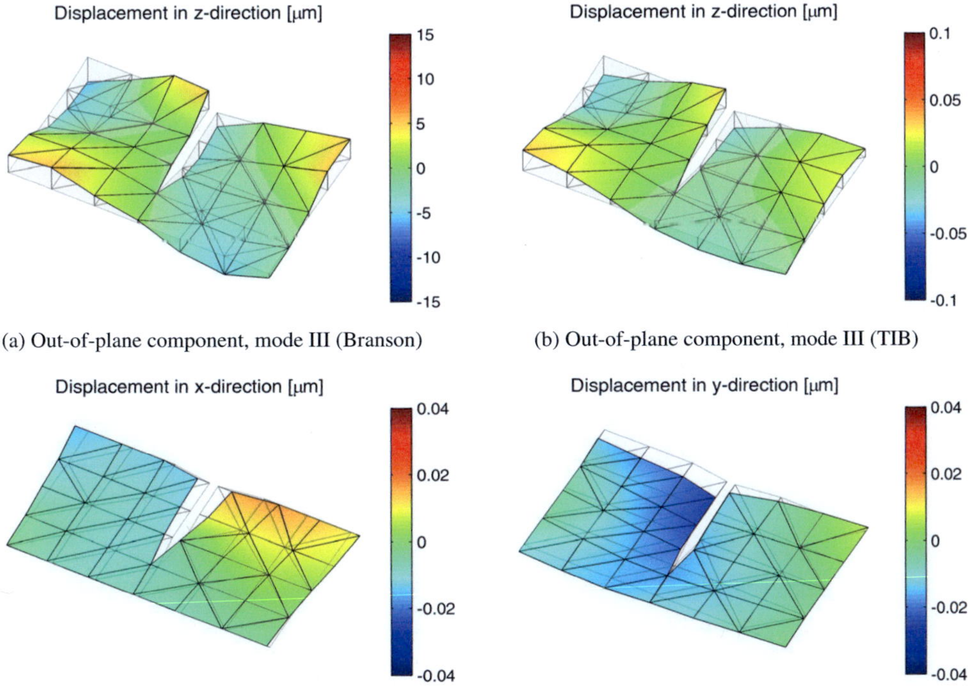

(a) Out-of-plane component, mode III (Branson)

(b) Out-of-plane component, mode III (TIB)

(c) In-plane component, mode I (TIB)

(d) In-plane component, mode II (TIB)

Figure 5.23: Displacement components of the mode shape of specimen 06-24-B at 21.115 kHz.

In the following it is attempted to predict the heat dissipation at the crack faces based on the differential displacements of two grid points lying face-to-face on each side of the crack path. In order to overcome the faulty in-plane measurements when using the Branson generator, it is assumed that all components of each grid displacement history achieved by excitation with the TIB amplifier can be scaled proportionally to the vibration level achieved by means of the Branson system. Thus, scaling is done for each frequency based on the out-of-plane amplitudes, which were proven to be reproducibly measured. This way, the displacements resulting from low-power excitation are multiplied by the average scale factor of the two analysed grid points. Such a linear scaling operation clearly requires linearity of the mechanical system, which cannot be be proven explicitly. However, this kind of simplification can be justified in case of a governing compressive prestress with small variations of the crack face loading. Note that in Section 5.3.1 the correlation of heating (and thus the occurrence of resonance frequencies) and the total harmonic distortion measured according to modes II and III did not show any peaks. Thus, there is no clear indication of non-linear displacements with respect to the harmonic excitation. In fact, this result was found for plate 06-12, a specimen whose thermal hot spot achieved during sweep and mono-frequency excitation is close to the notch bottom. Here, a non-linear contact behaviour is even more likely than in case of specimen 06-24, whose thermal hot spot is located very close to the crack tip, where compressive residual stresses usually act on the crack faces. This corresponds to a positive approach of the mean surfaces z according to Figure 3.10.

In the next step, the differential mode I displacement history (corresponding to the global x direction) is computed. Since the actual residual stress state of the crack region is not known, a variable approach of the crack faces is considered in order to account for potential compressive prestressing. Based on the Hagman model introduced in Chapter 3 an estimate of the static normal pressure acting on a fictitious contact area of $100\,\mathrm{mm}^2$ is made. Therefore, equation 3.47 is applied using the same constants E^*, C and R as assumed in Section 3.2.1. The positive approach is added to the differential grid point movement.

Finally, the tangential movements are evaluated by applying Coulomb's law of dry friction to the differential displacement histories according to mode II and III (corresponding to the global y and z directions). For this purpose the two out-of-plane components are vectorially added to yield the total tangential displacement, and the resulting velocity history is found by numerical differentiation. The absolute value of the shear stress acting on the crack faces is determined by the friction coefficient and the normal pressure modulation, which is taken explicitly into account. Micro-slip is not considered at this point. For each investigated eigenfrequency the theoretical power dissipation is computed by multiplying the absolute values of shear stress and tangential velocity. Figure 5.24a shows the predicted temporal mean of the heating power depending on the normal approach of the crack faces for specimen 06-24. Normalisation of the data with respect to the results achieved at the frequency of 19.45 kHz clarifies that the relative power dissipation converges quickly if an approach greater than 10 µm is considered (Fig. 5.24b). Based on the chosen model parameters this approach corresponds to a macroscopic pressure of about 0.62 MPa, which is very likely to be exceeded by residual stresses acting on the crack faces close to the crack tip.

In order to compare the predicted heat dissipation with the actual heating of the plate specimen 06-24, the thermal quantities ΔT and $\mathrm{d}T/\mathrm{d}f$ achieved by ultrasound sweep thermography are related to their maximum values as well. All relevant results are organised in Table 5.3. Focusing on the relative quantities, a certain coincidence of actual heating intensity and predicted power dissipation is found. In case of the plate specimen 06-24 having its thermal hot spot close to the crack tip, the proposed friction modelling based on linearly scaled displacements leads to a very good prediction of the relative extent of frictional heating. Clearly, the assumptions regarding linearity of the structural response and the governing compressive residual stress need to be further analysed in future works.

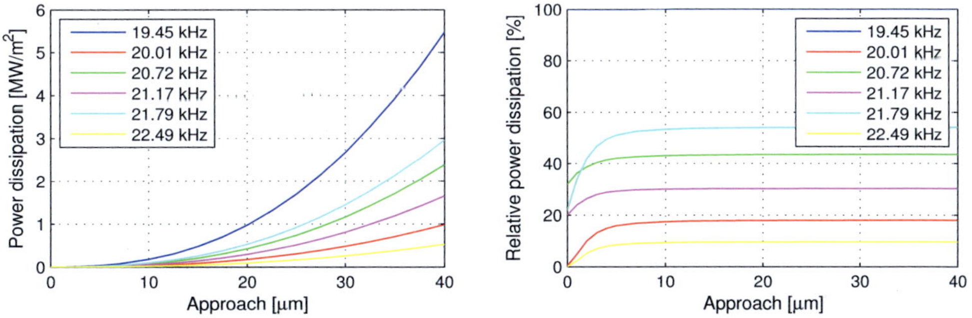

(a) Absolute power dissipation

(b) Relative power dissipation

Figure 5.24: Predicted power dissipation of specimen 06-24-B based on vibration amplitudes considering different crack face approaches and a fictitious contact area.

Since the prediction of the most heating relevant resonances failed for specimen 06-12, the author believes that in this case the contact acoustic non-linearity, in particular the clapping according to mode I, and the absence of a prestress impede the applicability of the described experimental approach including its numerous simplifications. However, the presented investigations reveal that under certain conditions of the crack region a clear correlation of the 3D crack face displacements and the resulting heating can be established when basic concepts of contact mechanics are explicitly taken into account besides of the macroscopic friction laws.

Table 5.3: Achieved defect heating and predicted power dissipation of specimen 06-24.

Frequency [kHz]		mean amp. ratio [-]	ΔT [K]/[%]		dT/df [K/Hz]/[%]		rel. predicted power[1] [%]
Branson	TIB		absolute	relative	absolute	relative	
19.450	19.420	447	7.9	100	0.31	100	100
20.010	19.965	119	0.7	9.2	0.03	10.0	18.0
20.720	20.670	565	2.8	34.9	0.16	50.6	43.5
21.170	21.115	355	2.5	31.5	0.10	32.9	30.3
21.790	21.760	419	5.1	63.9	0.21	66.5	54.1
22.490	22.420	96	1.5	18.8	0.03	8.4	9.6

[1]Values given for an approach of 40 µm

6 Numerical analysis – Simulation of ultrasound excited thermography

This final chapter focuses on the numerical simulation of vibrothermographic testing. A three-step approach based on commercial finite element codes is proposed to simulate the high-frequency mechanical excitation, the frictional heat generation and the transient heat diffusion. Realistic distributions of the crack face loading in terms of contact pressure, friction stress and heat flux are achieved by sophisticated FE modelling techniques, considering the random asperity distribution of the crack faces. Finally, the results of numerous simulation passes representing a complete ultrasound sweep are compared with experimental results.

6.1 Simulation approach

The high-frequency mechanical excitation of a steel plate specimen, its transient oscillation and the local heating of the crack region is simulated based on a three-step approach using ANSYS LS-DYNA (ANSYS Multiphysics 12.0 and LS-DYNA 9.71). The multi-purpose finite element software ANSYS allows for the development of highly customised simulation procedures and provides explicit time integration via the LS-DYNA interface. ANSYS offers multiphysics simulation to solve strongly coupled thermomechanical, electromagnetic or fluid-structure interaction problems. The mutual dependencies between two or more fields are taken into account by sequential coupling where the involved disciplines are computationally treated as individual single-physics, or by a direct coupled-field solution where each finite element has all necessary degrees of freedom. In the latter case the physical interaction of the domains is explicitly implemented.

However, directly coupled FE analyses might become inefficient if the physical phenomena are interacting only in one direction. Then, a clear advantage of decoupling the problems is to be seen in the improved adaptability, especially when it comes to extremely different time scales of the simulation disciplines. Regarding vibrothermographic experiments, the cycle duration t_{cycle} at 20 kHz is 0.05 ms but each frequency step of the commonly used sweep excitation takes about 100 ms, which corresponds to the thermal simulation time. Hence, a reasonable time step for the mechanical simulation is a few thousand times smaller than an appropriate time step used for the thermal computation. In this case, the direct coupling would be extremely time consuming, since the highly dynamic mechanical problem predefines the time step and the thermal problem controls the total simulation time.

Instead, when assuming a steady-state oscillation, the heat generation averaged over a few vibration periods can be used to perform a long-term thermal analysis. At this point, the influence of the temperature field on the displacement field is fully neglected. To realise a straightforward simulation scheme, three distinct finite element models of the plate specimen have been created:

- The global mechanical model of the cracked plate with a rather coarse mesh of the crack region (Figs. 6.2, 6.3a and 6.6)
- The mechanical submodel of the crack region with a fine mesh of the crack faces (Figs. 6.3b, 6.5a) where the node numbers of the outer boundaries correspond to the node numbers of the global model
- The thermal submodel of the crack region having exactly the same mesh and node numbers as the mechanical submodel

At first, the purely mechanical part of the experiment is simulated. The coupled system of ultrasonic transducer and steel plate needs several vibration periods to turn from a transient to a steady-state oscillation. The result output interval t_{output} is chosen to be $t_{cycle}/20$ so the internal computation time step Δt is equal to or smaller than $t_{cycle}/20$. Thus, a complete simulation of a 0.1 s excitation would require at least 40000 time steps to be calculated. Since the computational cost is high using a detailed three-dimensional model of the cracked plate, a different approach is pursued. It is assumed that the plate reaches a steady state after 100 sinusoidal excitation periods. Therefore, the first 100 periods are simulated using the complete mechanical model without saving any results. After continuation of the explicit finite element simulation the nodal displacements of five additional periods are stored.

Next, the mechanical submodel with a finer mesh of the crack region is run for five periods. The nodal displacements obtained from the explicit run are transferred to the outer regions of the submodel and directly used as boundary conditions. This way, the partial model is forced to move like the global model. The reactions of the inner region of the model, which contains the crack faces and the crack tip, can arise freely in the transient analysis. The crack faces are equipped with surface-to-surface contact and target elements, respectively. ANSYS directly outputs the friction stress τ and sliding distance s for each contact element based on the Coulomb friction law. These outputs are systematically stored and used to calculate the resulting surface heat flux q at the contacting crack faces due to dry friction on a per-element basis. Corresponding to equations (3.36) and (3.41) the surface element heat load q is given for every time step i as:

$$q_i = \zeta \cdot \frac{|\tau_i|}{2} \cdot \frac{|s_{i+1} - s_i|}{\Delta t} \tag{6.1}$$

Here, ζ determines the ratio of dissipated work converted to heat and work spent for wearing off the mating surfaces. Within a first approach it is supposed that all mechanical work done at the interface is dissipated to heat ($\zeta = 1$). The division by two takes into account that the generated heat spreads out in two opposing crack faces.

Since the submodel is subjected to inertia forces as well, only the last three of five simulated periods are monitored. With the assumption of a steady-state oscillation the frictional heat load history q_i is averaged over time.

Because most of the performed experiments indicated that plasticity at the crack tip does not play a major role concerning heating of the crack, the material is estimated linear elastic in the simulations. Thus, clapping and rubbing of the crack faces represent the only source of non-linear response which is taken into account at this time.

In the last step the calculated heat loads are transferred to the thermal model, which has the same node and element numbering as the structural model. Again, a transient calculation is performed, where the output interval and the desired duration of heat generation can be chosen independently of the structural analysis. The thermal simulation finally results in the temporally and spatially resolved heat diffusion. After evaluation of the temperature field at the plate surface, a direct comparison between experimental and numerical results can be drawn.

Table 6.1 gives a summary of the three-step simulation approach. The output quantities of one step represent the input quantities of the subsequent step.

Table 6.1: Overview of the proposed three-step simulation approach.

Step	1	2	3
Discipline	Mechanical (explicit dynamics)	Mechanical (implicit dynamics)	Thermal
Model	Global model	Partial model	Partial model
Loading/ input	100 + 5 displacement-controlled sine periods at the horn position	Nodal displacements for five periods calculated in step one	Averaged heat flux at crack faces calculated in step two
Time step	$\Delta t \leq t_{\text{cycle}}/20$	$\Delta t \leq t_{\text{cycle}}/20$	$\Delta t = 0.05\,\text{s}$
Sim. time	$t_{\text{final}} = 105\,t_{\text{cycle}}$	$t_{\text{final}} = 5\,t_{\text{cycle}}$	$t_{\text{final}} = 1\,\text{s}$
Results/ output	Nodal displacements of the crack region for the last five periods	Element results at crack faces: contact pressure, friction stress, velocity, sliding, heat flux	Transient temperature distribution over the crack region

The conceptual decoupling of the structural and the thermal simulation was verified by means of a simple model of a steel block sinusoidally sliding on another one while being subjected to a constant normal pressure of 4.6 MPa. To avoid unwanted oscillations in normal direction, the sliding amplitude of 20 µm is forced by displacement control and the contact pressure is adjusted by a given approach of the two blocks. Static and dynamic friction coefficients are chosen to be 0.1 and the thermal material properties are taken from Peil and Wichers [PW04, PW05].

Figure 6.1 shows the (predefined) sliding distance and velocity, the friction stress, the heat flux and the resulting temperature increase at the centre of the interface. A comparison is drawn between the results obtained from a thermo-structural model using ANSYS coupled-field elements, from a set of decoupled models and from the decoupled models using averaged heat loads. Note that the velocity is simply derived from the sliding distances at two consecutive points in time. Furthermore, no temperature dependent material properties were considered within the coupled simulation. Hence, "manual" computation of q_i according to equation (6.1) exactly matches the actual heat flux through the contact elements internally computed by ANSYS using coupled-field elements.

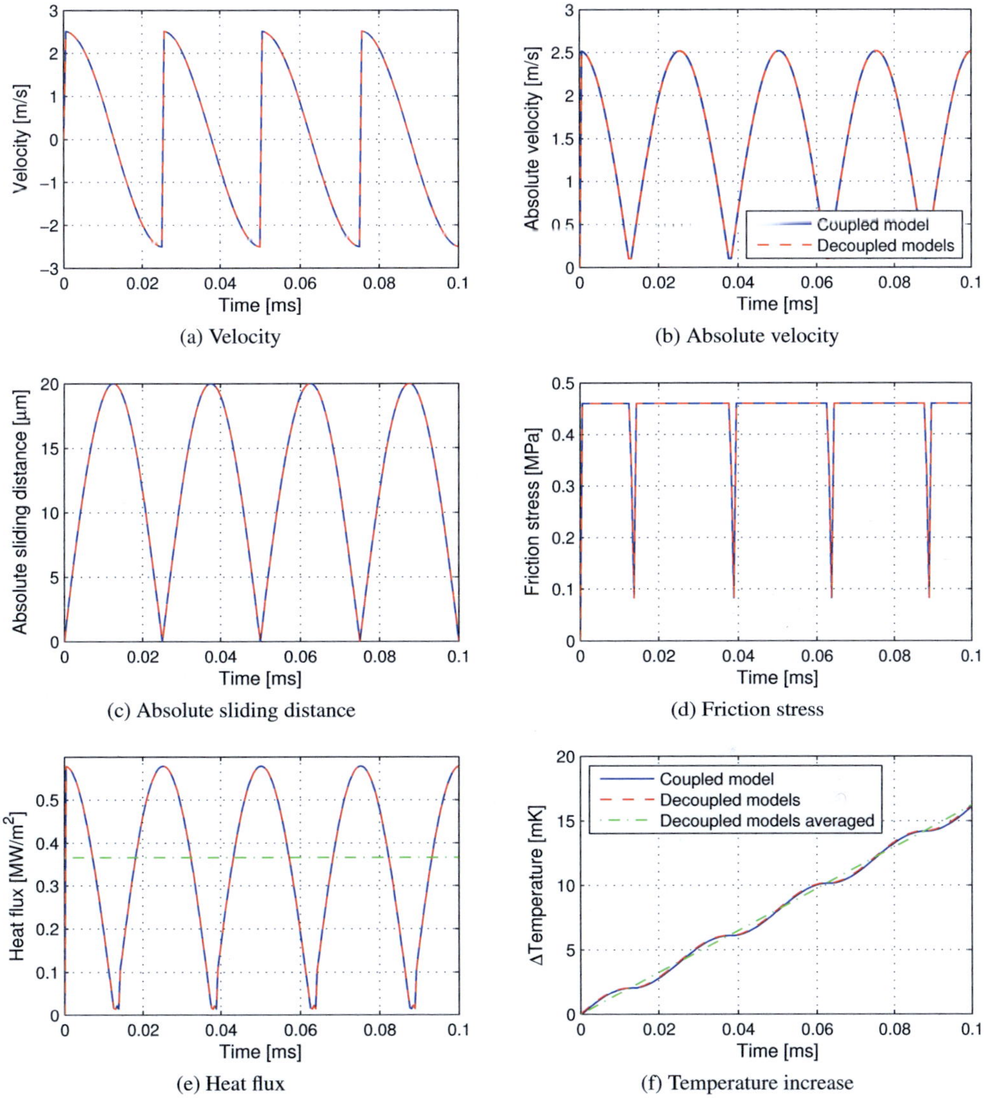

(a) Velocity

(b) Absolute velocity

(c) Absolute sliding distance

(d) Friction stress

(e) Heat flux

(f) Temperature increase

Figure 6.1: Verification of the decoupling approach for the computation of dry frictional heating using a test model of two blocks being pressed against each other and forced to slide sinusoidally; comparison of the velocity (a, b), the sliding distance (c), the friction stress (d), the heat flux (e) and the temperature increase (f) using ANSYS coupled-field elements and consecutive single-physics models.

6.2 Finite element modelling of plate specimens

Coupled ultrasonic transducer and steel plate are forming a rather complex oscillating system. The converter with its piezo stack inside is not modelled within the scope of this work. From the author's point of view it is very expensive to realistically simulate the mass distribution and the complex vibrational characteristics of an ultrasonic transducer. Moreover, the non-linear hammering of the converter tip against the structure, which is included in the works of Han et al. (see for example [HIN$^+$05]), is irrelevant here. Thus, the section focuses on the detailed modelling of the cracked plate specimen and the boundary conditions.

6.2.1 Global and partial finite element models

As a first approach, only the cracked steel plate is three-dimensionally modelled using ANSYS LS-DYNA explicit dynamic elements SOLID164 (8-node brick solid). Clearly, neglecting the transducer mass will result in a different frequency response compared to the coupled system. On the other hand, it cannot be expected anyway that the FE model exactly matches eigenfrequencies of the real plate at ultrasonic frequencies without iteratively adjusting material properties or geometry.

Within an explicit mechanical simulation the element having the smallest characteristic length is critical for determination of the internal time step size. Such being the case and due to the fact that the complete structural model should only be able to reproduce the global dynamics of the plate, a moderate element size is chosen as depicted in Figure 6.2. In thickness direction the plate is built up by four brick element layers. Radial meshing starting from the holes allows to completely avoid any degenerated brick elements. The relatively fine resolved edges of the holes are advantageous for future simulation of the prestressed specimen. It has been found that the element size is suitable for simulating vibrational modes in the 20 kHz regime. Here, the crack region is not refined and the through-thickness crack is realised as a wedge-shaped cut with a length of 24 mm. When using the LS-DYNA solver, the definition of sliding interfaces does not require specialised contact and target elements as it is known from an implicit FE analysis.

The structural and thermal submodels are based on the global model. Fundamental differences can only be found in the discretisation of the crack region. In order to properly resolve the crack face interaction, this region is strongly refined. Considering a 24 mm deep crack, the crack faces are represented by more than 16×12 segments instead of 5×4 facets in the global model. Since plasticity is not taken into account at this time, the crack front zone is modelled by dragged singular elements which are arranged fan-like around the crack front. This type of discretisation of the crack front allows for calculation of the stress intensity factors (SIF) along the thickness direction (see for example [IW03] or [Kun08] for details on numerical analysis of crack tip/front loading and computational fracture mechanics). The evaluation of SIF under static or dynamic loading is not undertaken in this work because none of the vibrothermographic experiments resulted in a significant propagation of the initial fatigue crack. However, this interesting aspect should be taken into account within future works since other research groups observed crack growth due to ultrasound excitation.

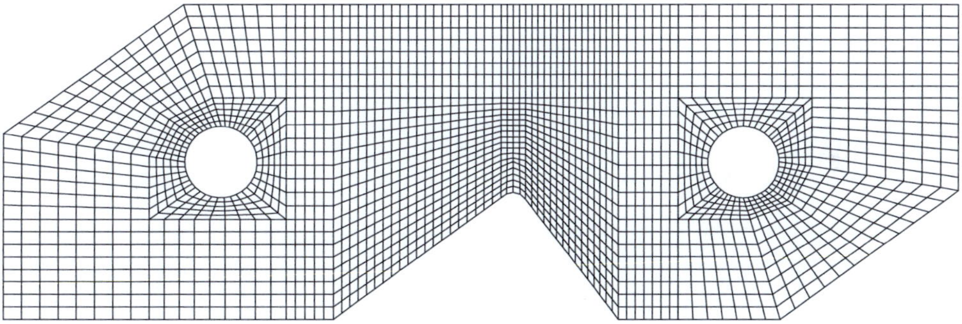

Figure 6.2: Global finite element model of a 10 mm thick steel plate specimen.

In order to transfer nodal displacements from the global model to the partial structural model, a certain region or at least a path of both models ideally matches each other's node locations and numbers. Otherwise, an additional interpolation step becomes necessary to map the global displacements onto the nodes of the submodel. Figure 6.3 shows the central region of the global model and the partial model identifying their common elements and their different inner crack regions (highlighted bright).

Both structural models are highly parametrised allowing to adjust the crack orientation, its position offset from the notch bottom and of course the crack opening at the beginning and at 90 % of the crack depth. The crack path is generally assumed to be straight although experiments revealed that they are often kinking to some extent. For building the mechanical implicit model, ANSYS elements SOLID185 (8-node brick solid), CONTA173/174 (4/8-node surface-to-surface contact) and TARGE170 (target segment) are used to model the outer region and the crack faces. The quarter-point crack tip elements are based on quadratic elements SOLID186 (20-node brick solid), and the transition zone is filled with elements of type SOLID92 (10-node tetrahedral solid). Mid-points of higher-order elements are automatically omitted when connected to linear shaped elements.

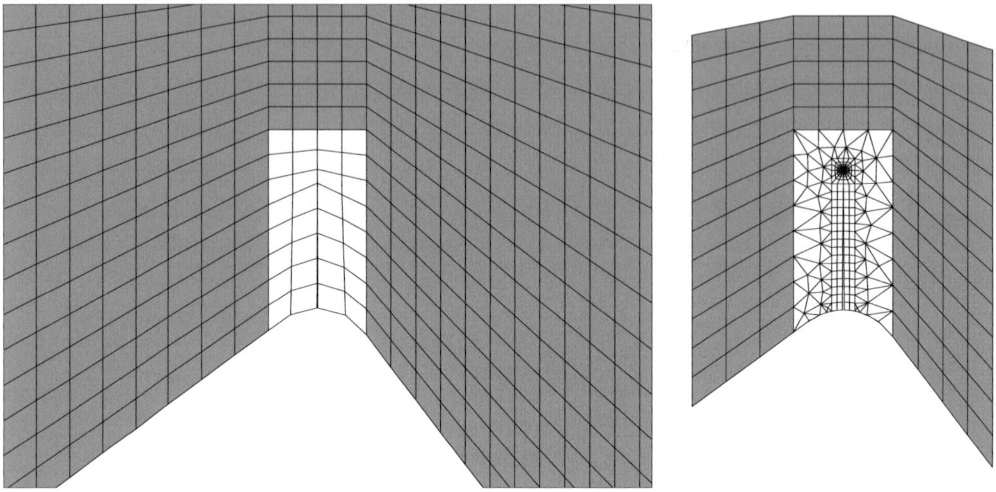

(a) Notch region of the global model (b) Submodel

Figure 6.3: Comparison of the global and partial FE model illustrating the common mesh (dark) and the different discretisations of the crack region (bright).

The thermal model fully corresponds to the mechanical submodel. Only the element types have been switched from structural to thermal. ANSYS elements SOLID70 (8-node brick solid), SOLID90 (20-node brick or 10-node tetrahedral solid), CONTA173/174 (4/8-node surface-to-surface contact) and TARGE170 (target segment) are chosen. The heat fluxes which are calculated for each contact element are automatically applied to both of the crack faces so that symmetrical heat diffusion is achieved. To make sure that the boundaries of the submodel do not artificially affect the heat conduction in case of a long-term simulation (e.g. 1 s), the partial model is chosen to be much bigger than necessary for transferring the nodal displacements.

Table 6.2: Mechanical and thermal properties of steel S355J2+N used for finite element analysis of a steel plate specimen.

E [GPa]	ν [−]	ρ [kg/m^3]	C_p [J/(kg K)]	k [W/(m K)]
210	0.277[1]	7837[1]	461[1]	40.4[1]

[1]Values of ν, ρ, C_p and k according to Peil and Wichers [PW04, PW05]

Throughout the numerical investigations, mechanical and thermal material properties are chosen according to the publications of Peil and Wichers [PW04, PW05] except from the Young's modulus that was set to the commonly used value of 210 GPa. Table 6.2 summarises the relevant material parameters for steel S355J2+N at room temperature.

6.2.2 Modelling of crack face asperities

In order to achieve a more realistic crack face modelling, an asperity distribution of the mating faces is implemented by randomly moving the nodes of both faces in space orthogonally to their planes. This is not a standard routine and care has to be taken since existing elements are deliberately distorted, which might result in a poor element quality. However, node locations can be altered to some fraction of the element edge length without encountering any problems.

Thompson proposed four different methods of generating rough surfaces in ANSYS and analysed the contact behaviour when pressed against a perfectly flat and rigid surface [Tho06]. Figures 6.4a and 6.4b illustrate two of the methods which have been considered for implementation. The top surfaces of both blocks are generated by the same pattern of asperities. The block on the left was built from the "bottom up" with free tetrahedral meshing afterwards, and the one on the right is constructed "top down" using a simple block volume, mapped brick meshing and subsequent node modification.

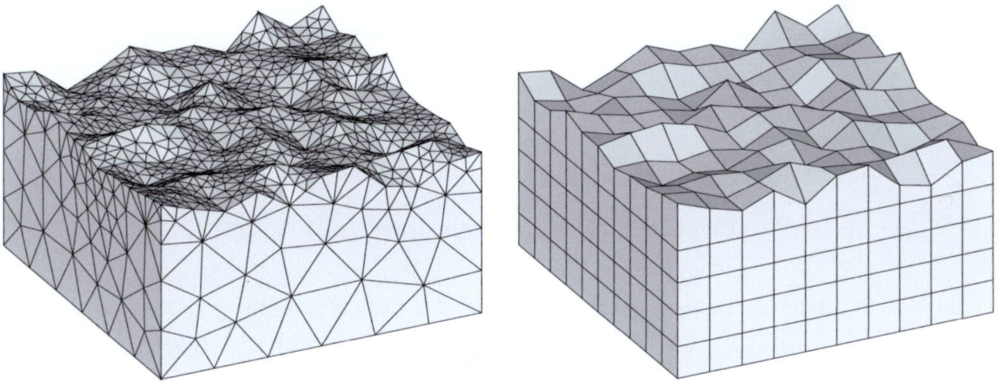

(a) Rough surface using "bottom up" modelling

(b) Rough surface using "top down" modelling

Figure 6.4: Two different methods of generating rough surfaces using the finite element software ANSYS as proposed by Thompson [Tho06].

The second version was used for the plate models because the structural error in case of contact is almost as low as with the other version and because "bottom up" modelling is rather expensive when all other volumetric components are generated by geometric primitives using the "top down" modelling technique. Moreover, the number of contact elements is much smaller, which reduces the computation time. Figure 6.5a illustrates the detailed crack region of the submodels. Note that the maximum asperity height as well as the crack opening are exaggerated for better visibility. Figure 6.5b shows the corresponding random distribution of asperities generated by ANSYS using a Gaussian distribution with a standard deviation of 0.33. Approaching the crack front, the asperities are omitted in order to comply with the straight front and to prevent distortion of the singular crack tip elements. The asperity map can be freely scaled in magnitude to simulate various surface roughness levels.

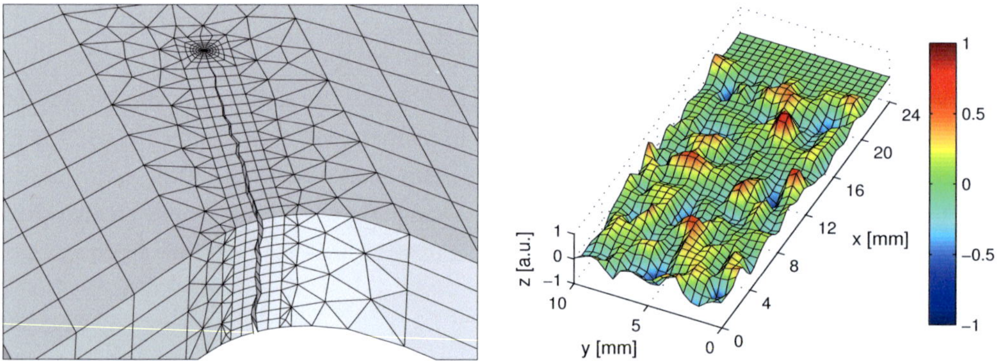

(a) Partial model with rough crack faces (b) Unscaled random distribution of asperities

Figure 6.5: Finite element realisation of rough crack faces using "top down" modelling and random node repositioning.

6.2.3 Mechanical excitation and support conditions

Within a first approach the mechanical loading of the steel plate is realised by harmonically moving the surface nodes inside a circular area with a diameter of 40 mm (corresponding to the transducer tip) in the out-of-plane direction. The excitation is simulated at coupling position 1, located in a distance of about 75 mm from the lower left corner of the plate geometry according to Figure 4.1. A displacement amplitude of 15 µm is chosen.

During the experiments described in Chapter 4 the plate was vertically supported at two and clamped at four positions, respectively. Very soft foam was placed beneath each of the aluminium clamps. The effective Young's modulus E of the compressed foam was estimated to be 5 MPa, which is an average value for polystyrene. In the explicit FE model spring-damper elements (COMBI165) are orthogonally attached to the plate at the support positions and constrained at their free node. The element length l corresponds to the thickness of the foam block and the spring stiffness k_s is evaluated based on its section area A of $15 \times 15\,\mathrm{mm}^2$ and the modulus of elasticity.

$$k_s = \frac{EA}{l} = 112.5\,\mathrm{N/mm} \tag{6.2}$$

Structural damping of the plate and the supports is not considered, yet. In case of matching a resonance frequency, the plate deflection is limited due to the displacement controlled excitation. Moreover, gravitational forces are not included in the simulation because the effects on the stress and displacement fields are considered to be negligible. Figure 6.6 clarifies the position of the spring supports and the harmonic loading of the plate model.

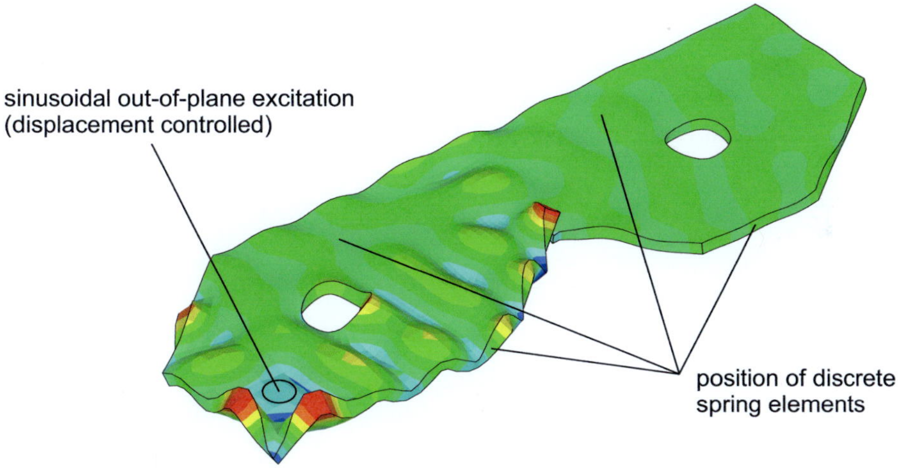

Figure 6.6: Snapshot of the transient wave propagation in a steel plate specimen after several excitation periods at 21.08 kHz; the deflection is scaled up by a factor of 1000 and the out-of-plane displacement component is mapped to colour for better visibility.

The mechanical submodel of the crack region does not require any additional supports since all nodes of three outer edges are used to transfer the motion of the global model to it and therefore represent the displacement boundary conditions.

6.2.4 Thermal boundary conditions

The plate specimen is permanently exchanging heat with its surrounding. Thus, in a steady state without excitation the plate's temperature matches the ambient temperature. Mechanical energy induced by the ultrasound system heats up the crack region. When a temperature gradient develops, heat conduction immediately sets in and the transient heat diffusion starts. Simultaneously, the warmed up plate boundaries are exposed to the encompassing air. Thus, energy losses due to free convection and radiation occur. Recalling equations (3.128) and (3.129) gives the total heat flux at the boundary surfaces that attenuates the crack heating to some unknown extent.

$$q_{\mathrm{tot}} = q_{\mathrm{c}} + q_{\mathrm{r}} = \alpha_{\mathrm{c}} \left(T_{\mathrm{obj}} - T_{\mathrm{fluid}} \right) + \epsilon_{\mathrm{obj}} \, \sigma \left(T_{\mathrm{obj}}^4 - T_{\mathrm{env}}^4 \right) \tag{6.3}$$

With the assumption that the surrounding fluid has approximately the same temperature as the laboratory environment ($T_{\mathrm{fluid}} \approx T_{\mathrm{env}}$) the radiative and convective part of the heat transfer can be summarised. Then, equation (6.3) can be formally rearranged so that it corresponds to Newton's law of cooling.

$$q_{\mathrm{tot}} = \alpha_{\mathrm{tot}} \left(T_{\mathrm{obj}} - T_{\mathrm{env}} \right) = \left(\alpha_{\mathrm{c}} + \alpha_{\mathrm{r}} \right) \left(T_{\mathrm{obj}} - T_{\mathrm{env}} \right) \tag{6.4}$$

Here, the introduced radiative heat transfer coefficient α_r is defined through (see [BS06]):

$$\alpha_{\mathrm{r}} = \epsilon_{\mathrm{obj}} \, \sigma \frac{T_{\mathrm{obj}}^4 - T_{\mathrm{env}}^4}{T_{\mathrm{obj}} - T_{\mathrm{env}}} = \epsilon_{\mathrm{obj}} \, \sigma \left(T_{\mathrm{obj}}^2 + T_{\mathrm{env}}^2 \right) \left(T_{\mathrm{obj}} + T_{\mathrm{env}} \right) \tag{6.5}$$

This expression is advantageous since it enables a quick estimation of the radiative heat loss and allows for direct comparison with convective heat transfer. Clearly, α_{r} and α_{c} are depending on the actual temperature difference and flow velocity fields, respectively. The presented relations are valid in the case of a steady state so that simplified estimations of the maximum heat loss during dynamic processes are possible.

In order to quantitatively assess the influence of radiation and convection on the crack heating preliminary simulations are conducted. Typical scenarios are considered with a constant room temperature of 20 °C and a local object temperature of 40 to 120 °C. Determination of the mean film coefficient α_{c} of a vertical plate, corresponding to the experimental work, requires an estimate for the mean Nusselt number Nu. A prediction which is widely accepted was reported by Churchill and Chu [BS06, HMS04, VDI06]:

$$Nu = \left(0.825 + \frac{0.387\,Ra^{1/6}}{\left[1 + \left(\frac{0.492}{Pr}\right)^{9/16}\right]^{8/27}} \right)^2 \tag{6.6}$$

This equation describes a correlation of measurements on laminar and turbulent flows. Lienhard [LILV03] gives an alternative formula based on the Squire-Eckert result for the local heat transfer from a vertical wall during laminar natural convection:

$$Nu = 0.678\,Ra^{1/4} \left(\frac{Pr}{0.952 + Pr} \right)^{1/4} \tag{6.7}$$

A third equation valid for laminar flow is cited by Hering et al. [HMS04]:

$$Nu = 0.53\,(Gr \cdot Pr)^{1/4} \tag{6.8}$$

Calculation of the Nusselt number requires various thermophysical properties of the surrounding fluid. Here, the Prandtl and Rayleigh numbers Pr and Ra are to be evaluated at the mean temperature $\overline{T} = \left(T_{\text{obj}} + T_{\text{env}}\right)/2$, whereas the volume expansion coefficient β should be evaluated at T_{env}. Table A.4 in Appendix A.9 summarises values of the thermal conductivity k, heat capacity C_p, density ρ and the dynamic viscosity η of air at different temperatures. The thermal diffusivity α, the kinematic viscosity υ and the dimensionless numbers Pr, Gr and Ra are then calculated according to Section 3.3.3.

In the following, a 10 mm thick plate specimen with a 24 mm deep crack is considered and assumed to be held vertically. A uniform heat load is applied on both crack faces for 1 s. Its actual value was iteratively determined so that maximum temperature increases of 20, 40, 60, 80 and 100 K are achieved at the plate surface at half the crack depth. The characteristic length l is chosen to be 50 mm, which is a trade-off between the maximum path length of heat generation and the actual lateral dimension of the plate at the notch (131 mm). Table A.5 in Appendix A.9 presents the estimated Nusselt numbers based on equations (6.6), (6.7) and (6.8) for the different object temperatures. The resulting heat transfer coefficients α_{c} and α_{r} are derived from equations (3.130) and (6.5), respectively. A comparison of the different sets for Nu and α_{c} proves a reasonable consistency. It can be concluded that the three approaches reveal very similar results, so the average values of α_{c} are used in the numerical simulation.

Table 6.3 summarises the heat transfer coefficients depending on the temperature difference $\Delta T = T_{\mathrm{obj}} - T_{\mathrm{env}}$. Since the corresponding values for α_{c} and α_{r} are close together, it is expected that convective and radiative heat transfer will have similar effects on the temperature development.

Table 6.3: Radiative and convective heat transfer coefficients.

ΔT	[K]	20	40	60	80	100
α_{c} (avg.)	[W/(m^2 K)]	6.09	7.23	7.99	8.57	9.05
α_{r}		6.33	6.99	7.72	8.51	9.36

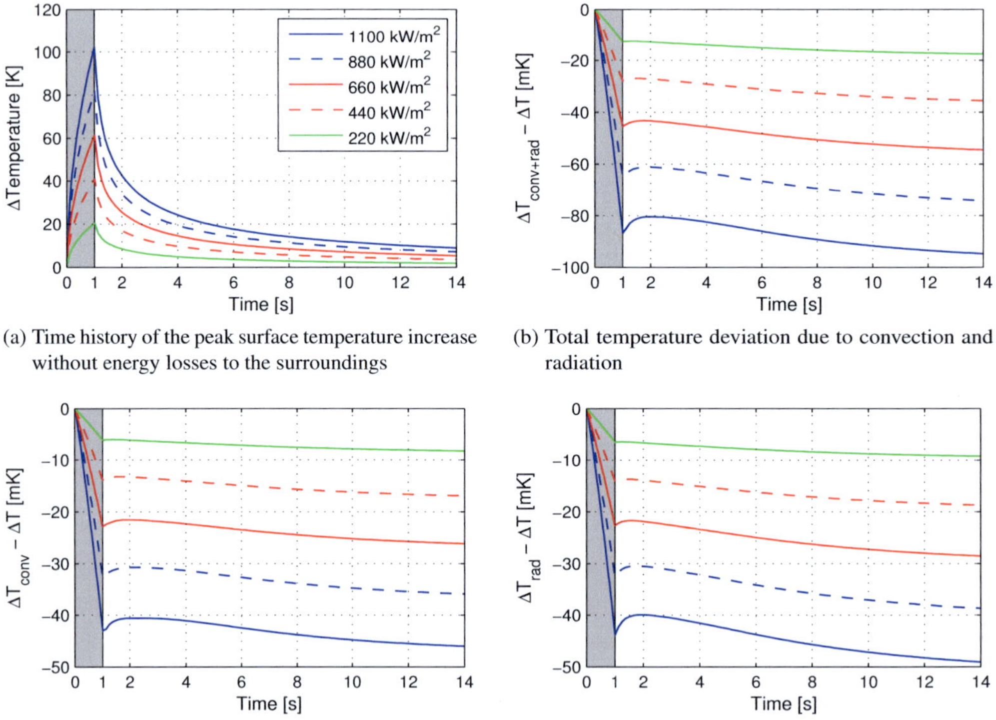

(a) Time history of the peak surface temperature increase without energy losses to the surroundings

(b) Total temperature deviation due to convection and radiation

(c) Temperature deviation due to convection

(d) Temperature deviation due to radiation

Figure 6.7: Hot spot temperature history of the plate surface due to different simulated heat loads uniformly applied for 1 s on both crack faces (a); calculated temperature deviations due to energy losses to the surroundings by convection and radiation (b), convection only (c) and radiation only (d).

Starting from an initial state at 20 °C, the transient temperature field due to the applied heat flux is simulated for the case without energy losses to the surrounding, first. Then, convective and radiative heat transfer have been included separately and in combination by means of surface effect elements SURF152, which are attached to the solid elements. Within the simulation approach using ANSYS the radiative heat flux is updated in each time step. It solely depends on the actual temperature difference ΔT, the Stefan-Boltzmann constant and the emissivity that was set to its maximum value of 1.0. In contrast, the convective part depends on ΔT but also on the film coefficient, which is temperature or rather situation dependent as shown before. α_c was therefore defined as a temperature dependent parameter. In addition to the estimated values given in Table 6.3, a value of 5.0 W/(m^2 K) was extrapolated for $\Delta T = 0$ K. ANSYS internally interpolates between the existent input data.

Figure 6.7a depicts the hot spot temperature history and Figures 6.7b to 6.7d illustrate the reduction in temperature due to heat losses at the plate surface. It is found that both convective and radiative heat transfer have a minor impact on the temperature evolution. Even if the crack region is heated up to 120 °C, the reduction of the peak temperature at $t = 1$ s is less than 0.1 K. As it was expected, the same quantitative effect can be attributed to both heat loss mechanisms since convection (Fig. 6.7c) and radiation (Fig. 6.7d) lead to very similar temperature decays. Due to the fact that the surface area which is affected by the crack heating is relatively small and because of the rather low heat transfer coefficients, neglecting all convective and radiative energy losses in the simulation seems to be fully justified.

6.3 Batch processing of simulation passes

A complete simulation run consisting of the global dynamic simulation, the refined structural analysis and the thermal simulation is only valid for one set of input parameters such as excitation amplitude, ultrasound frequency, crack opening and asperity distribution, etc. Within a first study the frequency is varied from 19 to 23 kHz corresponding to the performed ultrasound sweep experiments described in Section 4.4.6. In order to save computation time, the frequency step size is chosen to be 20 Hz instead of 10 Hz. The thermal model is run up to 1 s simulation time with averaged heat fluxes applied to the crack faces. Static and dynamic friction coefficients are set to 0.3. The maximum asperity height is chosen to be 50 µm and the maximum crack opening is 25 µm as measured on specimen 10-24-II (compare Table 4.6). Running the 200 sets of simulations in a batch mode takes about 24 hours on a standard PC (dual core CPU, 4 GB RAM). Most of the computation time (about 65 %) is required for solving the implicit structural problem.

6.4 Numerical results in comparison with experiments

For each time step of the implicit structural and thermal simulation passes various results are saved for further analysis. At first, the temperature increase is compared to the conducted experiment on specimen 10-24-II. The ultrasonic sweep was done by incrementally raising the frequency in steps of 10 Hz. Theoretically, each frequency step was hold for 166 ms (66.7 s/401), whereas a short time is required for the ultrasound system to tune in the frequency set point and for the vibrating system to reach a steady state. Thus, the simulated temperature distribution at 150 ms is evaluated. It is found that the proposed modelling of the crack faces leads to spatially restricted hot spots, which heat up strongest at 21.08 kHz. In Figure 6.8 the distribution of the maximum temperature increase and the maximum value $\mathrm{d}T/\mathrm{d}f$ found during sweep thermography of plate 10-24-II is compared with the simulated surface temperature distribution after 150 ms excitation. The positions of the thermal hot spot with respect to the complete crack path are in a very good agreement and the maximum temperatures are of the same order of magnitude. Clearly, the experimental sweep is done continuously. Thus, the total energy dissipated during the complete sweep spreads and results in the diffuse thermal signature shown in Figure 6.8a. As described earlier, imaging of the positive values of the first derivative with respect to frequency enhances the defect contrast.

To compare the frequency dependence of the thermal response, the infrared data along the oblique crack path A-B indicated in Figure 6.8a and the FEA results at the corresponding nodes of the

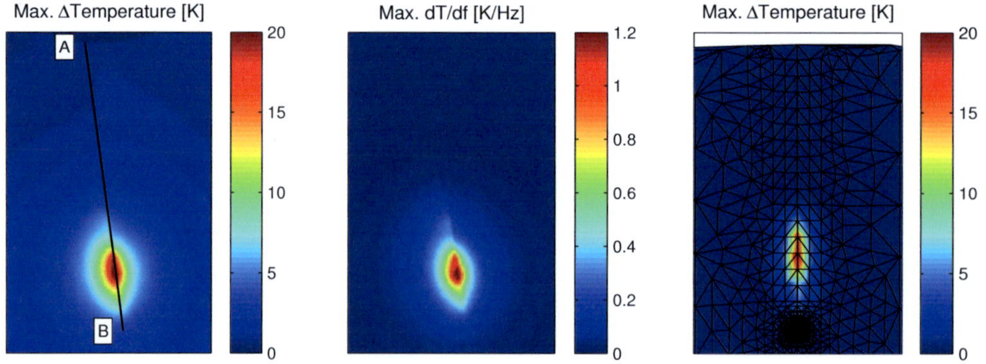

(a) Maximum ΔT and path A-B; specimen 10-24-II

(b) Maximum positive $\mathrm{d}T/\mathrm{d}f$; specimen 10-24-II

(c) Maximum ΔT; simulation

Figure 6.8: Comparison of the thermal hot spot location and intensity found during sweep excitation of specimen 10-24-II (a, b) and after 150 ms simulated excitation at 21.08 kHz (c).

modelled crack face ($x = 0 \ldots 24$ mm, $y = 10$ mm) are further analysed. Figures 6.9a and 6.9b depict the experimental temperature difference over position and frequency and the maximum temperature increase achieved along the whole crack path. Analogously, Figure 6.10 illustrates the 200 data sets of the thermal model after 150 ms simulated mono-frequency excitation. A direct comparison with Figure 6.9 reveals that the strong frequency dependence of the crack heating is reasonably reproduced. Experimental and numerical histories of max. ΔT even show a rough correlation regarding the heating relevant frequencies. The ΔT distribution along the crack path is not just scaled to different maximum values but modified in shape with varying frequency in both cases.

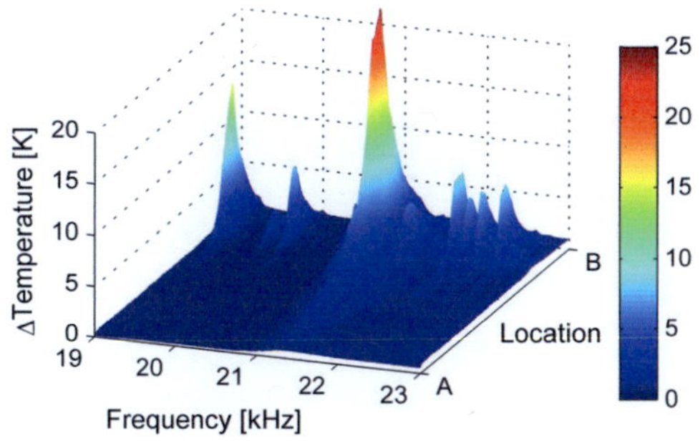

(a) Frequency dependent temperature increase along the crack path A-B

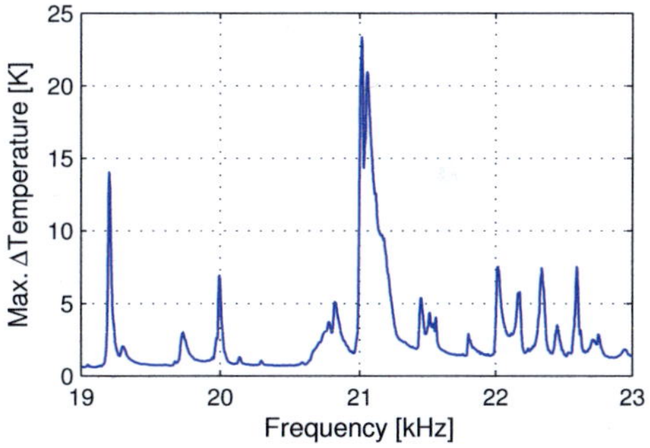

(b) Maximum values of ΔT along the crack path A-B over frequency

Figure 6.9: Thermal results of the ultrasound sweep on plate 10-24-II-A (compare Section 4.4.6).

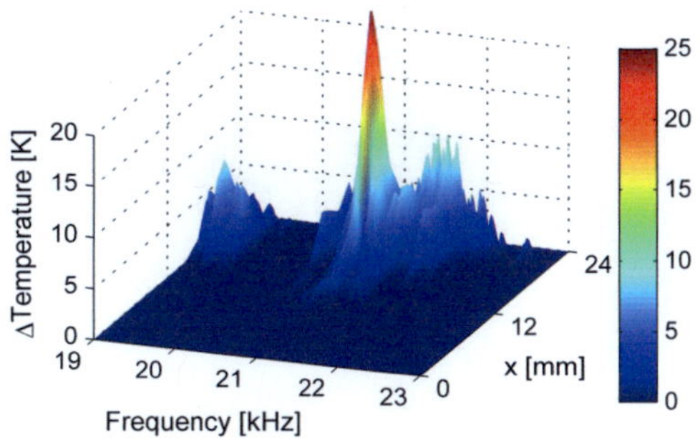

(a) Frequency dependent temperature increase along the crack path $x = 0 \ldots 24$ mm, $y = 10$ mm

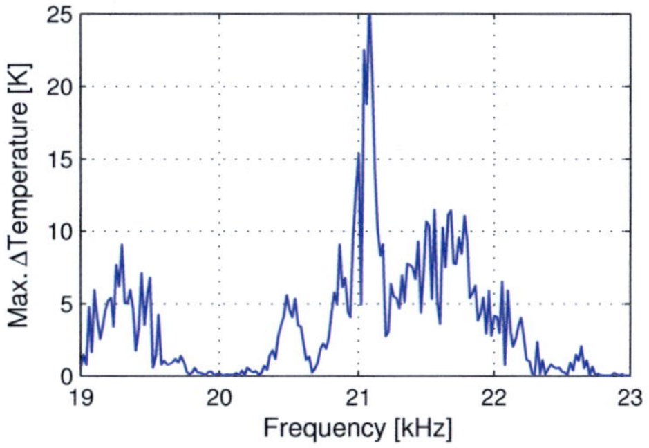

(b) Maximum values of ΔT along the crack path $x = 0 \ldots 24$ mm, $y = 10$ mm over frequency

Figure 6.10: Thermal results of 200 simulation passes of mono-frequency excitation after 150 ms.

Figure 6.11 exemplarily presents the numerical crack face loading for the ultrasound frequency of 21.08 kHz. All parameters are based on the contact elements used for the implicit structural simulation run, regarding only the last periods that also allow averaging of the heat flux values. Since all quantities are determined on a per-element basis and saved together with each element centroid, the values are interpolated onto a uniform grid for demonstration purposes. Figures 6.11a to 6.11d map the maximum values of the velocity, the contact pressure, the friction stress and the heat flux. Because the models do not include plasticity, yet, the stresses might exceed the yield limit of 360 MPa. This is the case for the contact pressure very close to the crack front (lower right corner). Velocity values are only defined and not equal to zero in case of a sliding contact.

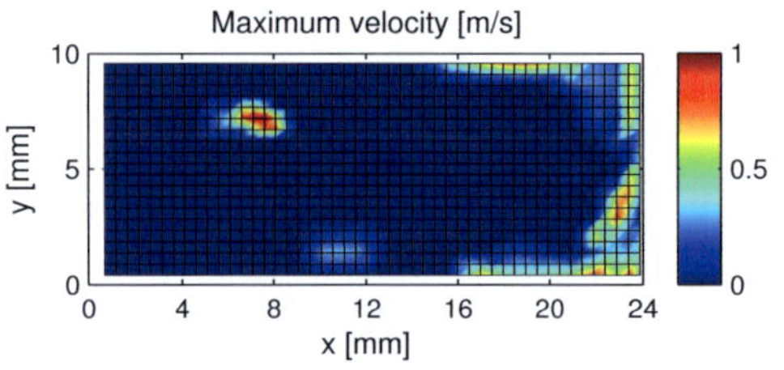

(a) Maximum relative velocity

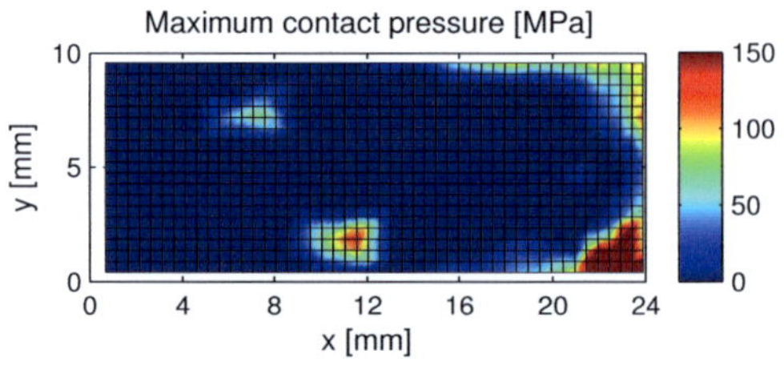

(b) Maximum contact pressure

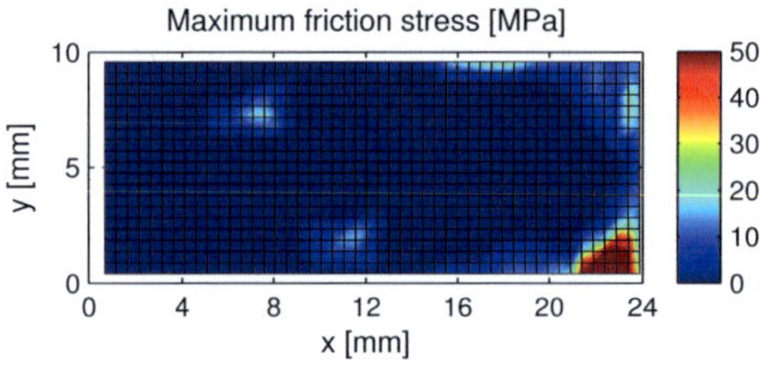

(c) Maximum friction stress

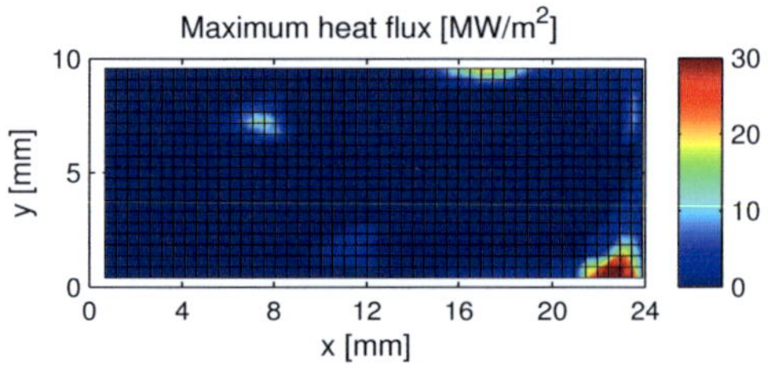

(d) Maximum heat flux

Figure 6.11: Simulated spatial distributions of heating relevant quantities over the crack face; maximum values of the relative velocity (a), contact pressure (b), friction stress (c) and heat flux (d) at 21.08 kHz.

The mappings in Figure 6.11 reveal that only a few contact locations exist and that most of the crack face area is rather inactive referring to frictional heat generation. This effect clearly results from the modelled crack face asperity and proves that realistic crack face loading is possible by means of the simulation approach. Such non-uniform distributions of heat generation (and wear) have been found for all investigated specimens in Section 4.5.

It should be noted that the proposed simulation approach can be further improved since several aspects have not been considered, yet. The following issues should be addressed in the future to provide even more realistic modelling of ultrasound excited thermography testing:

- Switching to force controlled excitation, considering non-constant loading amplitudes
- Adjustment of Rayleigh damping coefficients in the explicit FE model
- Adjustment of the dynamic and thermal response of the model by means of synchronous (scanning 3D) laser vibrometry and infrared thermography
- Frequency analysis of the experimental and numerical crack face displacement components and adjustment of the numerical contact stiffness
- Implementation of a more sophisticated friction law if identified experimentally
- Consideration of the residual stress state
- Implementation of cyclic plasticity

Except for the implementation of cyclic plasticity all of the above points require highly accurate oscillation measurements of the specimen. If the crack face morphology could be precisely measured without destroying the specimen, advanced numerical simulations might be useful to identify the underlying friction law. This could be achieved when the actual displacement field is transferred to the finite element model and a user defined friction law can be adjusted by means of an optimisation procedure.

7 Conclusion

7.1 Summary and results

The purpose of this work is to assess the use of ultrasound excited thermography as a non-destructive test method for detecting surface breaking fatigue cracks in structural steel members. Motivated by a lack of fast and imaging crack detection methods on the one hand and the increasing acceptance and technological progress of active thermography techniques on the other hand, this work presents the current status of vibrothermography applied to thick-walled steel members.

Chapter 2 gives a brief introduction to the use of infrared technology in the field of non-destructive testing and presents the general concept and interdisciplinarity of ultrasound excited thermography. From a comprehensive literature survey it is found that vibrothermography was never applied to steel structures in the sense of an in-situ crack detection. Instead, most practical development efforts focus on rather specialised applications in the field of mechanical engineering, e.g. maintenance of gas turbine blades or inspection of aerospace composite structures. Here, considerable progress has been achieved, such as the implementation in series production and process control, usage as standard test method, as alternative to or substitution for dye penetrant testing, the prototype development of hand-held excitation devices, etc. Current basic research includes the analysis of vibration data, numerical analysis and tribological evaluation of crack faces in order to clarify the heating mechanisms. Finding reasonable probability of detection (POD) curves to rate the method in a common way turns out extremely difficult due to the numerous involved dependencies of defect heating and the non-standardised excitation and data processing.

Chapter 3 outlines the involved technical subdisciplines following the logical order of the three scopes, namely energy input, energy conversion and energy detection. Regarding the generation of structure borne sound in plate-like components of about 10 mm thickness, it is found that the classical plate theory is sufficient for an estimation of out-of-plane vibration wavelengths and the corresponding eigenfrequencies. The potential heating mechanisms dry friction, material damping and crack tip plasticity are discussed in detail. Focusing on frictional heating, the theoretical power dissipation of a steel-steel tribological pairing is analysed regarding Coulomb friction, variation of the normal contact and micro-slip behaviour according to the model of Hagman. A parametric study revealed that various relations between heating power and tangential displacement, such as linear functions or power laws with different exponents, can be expected, depending on the extent and type of normal contact modulation and the transition from micro-slip to gross-slip.

Next, the heat transfer mechanisms conduction, radiation and convection are reviewed. With regard to the common excitation methods using burst and modulated ultrasound insonification, essential dependencies of the surface temperature response on the area, depth and power of a buried heat source are exemplarily presented by means of a simplified thermal finite element model. The influence of the base material on the peak temperature and the attenuation due to a protective coating are estimated. As an outcome, most plastic coatings will not prevent vibrothermographic crack detection if low modulation frequencies or longer burst durations are chosen. Instead, the impact of the modulation frequency on the absolute temperature amplitude itself and the defect distance to the surface become decisive for a successful crack detection. Finally, the basic procedure for remote temperature measurements by means of a focal plane array infrared camera is outlined and the most accepted approaches for vibrothermographic crack detection are reviewed.

Chapter 4 presents experimental investigations towards the practical use of ultrasound excited thermography. Starting with the manufacture and preparation of raw plate specimens, the chapter introduces the local strain approach, which is successfully used to estimate the appropriate loading in order to generate fatigue cracks at a specified number of cycles, mostly in the HCF regime. A limitation of the considered defect types, namely corner flaws and through-thickness cracks, allows for a comprehensive analysis of technically relevant effects that occur during practical testing of plate-like components. Firstly, it is demonstrated by means of a modified 2.2 kW ultrasonic welding generator and a highly sensitive infrared camera that standing wave patterns show up even in a 10 mm thick steel plate, which might lead to misinterpretations regarding the presence of defects. While proving that the used test stand allows for reproducible thermal results even with imprecisely placed specimens, the dramatic influence of the actual excitation frequency and the coupling position on the crack heating is clarified. For all subsequent tests using a mono-frequency burst or frequency sweep excitation in the range of 19 to 23 kHz, a reasonable burst duration and a compromise between frequency step width and total sweep duration was found, respectively. Already at this point, the variation of the excitation power indicates divergent functional relations of the relative power and the crack heating when testing different specimens. The impact of the initial plate temperature on the heating relevant frequencies manifests itself clearly. Routine testing of all available plate specimens with thicknesses of 6, 8, 10, 12 and 14 mm and crack depths of 1, 3, 6, 12, 24 and 40 mm revealed that medium-sized flaws (12 to 24 mm) are better detectable than small and large ones. With increasing plate thickness the absolute heating decreases and the crack depth leading to the strongest thermal response shifts towards greater values. Regarding the unloaded specimens, crack detection limits below 1 to 6 mm are found, depending on the plate thickness. An extrapolation of the achieved generator power with screw-coupled transducer indicates that even plates thicker than 14 mm can be excited by the used equipment. Ultrasound excitation of statically prestressed plates puts the previously found detection limits into perspective since the actual preload, or more specifically the local gap between crack faces along the crack path, substantially influences the potential maximum crack heating. With increasing tensile preload the thermal hot spot moves towards the crack tip. In some cases, the crack tip remained detectable even at highest loading, which is advantageous as the full crack depth would be directly unveiled.

In order to investigate the effects of repeated ultrasound insonification, up to 1700 sweep excitations were performed on a single specimen. The maximum heating generally decreases exponentially with the number of tests and in some cases the hot spot locations steadily moved towards the notch bottom, away from the crack tip. After cutting the specimens into pieces, visual inspection and confocal laser microscopy revealed significant signs of wear on the crack faces, mostly restricted to a small near-surface region that exactly corresponds to the thermally detected hot spot location. These worn areas show oxidation particles but also metallic bright spots, which have a reduced roughness compared to the regular fracture surface. Instead of a propagating initial fatigue crack a new surface crack was generated in one of the six examined 10 mm thick plates. Therefore, the method is not truly non-destructive if excessively applied.

A simulated in-situ application of vibrothermography on a 3.2 m long hot-rolled girder featuring multiple cracks brought promising results. Using a screw-coupled transducer, six of eight cracks were detected independent of an artificially increased structural damping, which was realised by attached sandbags. The crack showing the strongest thermal response could be securely detected by means of a portable excitation device with a magnetic coupling. However, the defect heating showed a strong scatter with a ratio of 11:1 between maximum and minimum heating level. The girder's web area that was used for placing the exciter is to be understood as an influence field.

In order to relate the mechanical displacement of the structure to the defect heating, Chapter 5 introduces a fibre-optic laser vibrometer, which allows for non-contacting single-point and two-point differential velocity measurements. At first, a detailed characterisation of the freely vibrating transducer was conducted. Severe harmonic distortions and a small offset error regarding the linearity of the amplitude response could be attributed to the ultrasonic welding generator. Differential vibrometry of crack faces confirmed that significant crack heating only occurs if eigenfrequencies are excited. Here, out-of-plane (crack opening mode III) velocity amplitude peaks showed a stronger correlation with the temperature increase compared to in-plane shear (mode II) measurements. The integral distortion measure THD did not reveal any correlation with the heating although a stronger occurrence of the contact acoustic non-linearity could be expected especially in case of resonant excitation. Apparently, the displacements did not reach the level at which they get seriously affected by the non-linearity. The temperature dependency of the specimen's eigenfrequencies was confirmed. Furthermore, it was clarified that an excitation of about $\pm 25\,\text{Hz}$ around the exact natural frequencies in the $20\,\text{kHz}$ regime leads to a resonant vibration.

The question concerning the relation between vibration intensity and defect heating was pursued but the dependency could not be limited to a single characteristic. Instead, a variety of reasonable functional relations are found from experiments on different plates. One of the data sets can be interpreted as a power law with exponent 2.65, which corresponds to a pure micro-slip behaviour, and another one appears as a straight line that deviates at some point from the micro-slip curve, indicating the transition to Coulomb friction. A third function clearly manifests a large gap between the opposing crack faces. The defect did not heat up until a threshold vibration level was exceeded. Here, a power law with exponent 3.5 is found, which corresponds to the Coulomb friction with a normal contact modulation. In another case, the relation between velocity amplitude and crack heating appeared almost perfectly linear, indicating pure Coulomb friction without any normal contact modulation or with contact modulation superimposed by a strong static compression.

Ultrasound sweep thermography accompanied by single-point and two-point differential velocity measurements of the crack faces showed that most out-of-plane differential amplitude peaks match a heating relevant frequency. However, considering only mode II and III displacements is not sufficient for predicting the extent of defect heating. Because some frequencies showed large differential movement but no heating at all, three-dimensional scanning vibrometry was employed to incorporate the effects of in-plane (mode I) clapping displacements and the relative phase shift of all displacement components. Mounted on a two-axis positioning system, the laser sensor was moved parallel to the plate specimen allowing to target a complete grid pattern defined around the crack region. The grid displacement components in the plate coordinate system are computed based on three independent scans with varying laser direction. Acquisition of a temporal reference signal provided by the generator allows to determine the phasing of each raw velocity signal with respect to the excitation and thus reconstruction of 3D vibration mode shapes. Due to the harmonic distortion and the poor frequency stability of the high-power generator, both in-plane components could not be measured with adequate accuracy. Therefore, a low-power linear amplifier, which was found to provide an ideal response behaviour, was utilised. The perfectly smooth mode shapes achieved with the alternative system have been scaled up in magnitude so that the out-of-plane displacement amplitudes match the level that was obtained using the high-power unit. Evaluation of the crack face displacements in the time domain considering Coulomb friction and a normal contact modulation yields a reasonable prediction of the relative power dissipation at different resonant frequencies. The prognosis of the defect heating failed on another specimen with the thermal hot spot located much closer to the notch bottom. It is presumed that the linear scaling of crack face displacements is invalid as they are affected by the contact non-linearity in case of high-power ultrasound excitation.

Finally, Chapter 6 presents a finite element modelling scheme using ANSYS LS-DYNA to simulate the vibrothermographic testing of a 10 mm thick plate specimen. In a three-step analysis the highly dynamic excitation and vibration of the plate, the contact problem at the crack faces, the localised heat generation as well as the transient heat diffusion are approached. 200 individual simulation passes with increasing excitation frequency are performed in a batch mode in order to approximately represent a complete ultrasound sweep thermography test. Assuming that an oscillatory steady state is reached after a certain number of excitation periods for each frequency, the frictional heat generation is computed on a refined mechanical sub-model of the crack region. As a main improvement, the modelling of rough crack faces is introduced at this point. A subsequent thermal simulation with the temporally averaged heat flux applied to the crack face elements reproduces the transient heat diffusion, which is evaluated after a certain simulation time. Concerning the thermal response of the crack region, the numerical results are in a good agreement with experimental data. Randomly distributed asperities lead to spatially restricted sliding contacts and a realistic non-uniform frictional heating. The pronounced frequency dependence of the crack detectability clearly manifests in the simulations.

7.2 Outlook and future work

The present work contributes to the implementation of ultrasound excited thermography in the field of non-destructive testing of large-scale steel structures. Experimental investigations proved the technique's potential to be further developed towards a fail-safe, imaging NDT method with a significant range of operation and intuitive result interpretation. Further efforts are required in order to overcome the extremely restricted conditions which are to be met by the crack and component geometry and/or the excitation in order to achieve a significant thermal response. Three main issues emerged that need to be addressed in the future to finally judge on the practical use and reliability of the method and to specify a reasonable probability of detection:

- Since the coupling position has a major impact on the achievable defect heating, a compensation by advanced excitation devices is required. The proposed coupling of the exciter by means of an electromagnet was proven viable in case of ferritic steel components. A more efficient excitation unit should allow for a quick relocation of the transducer tip in a certain radius without fully detaching and re-coupling the whole device.
- Strongly gaping or compressed cracks can not be detected with the employed equipment. This will not change unless greater vibration amplitudes are realised. A superimposed low-frequency excitation that facilitates large amplitude/long wavelength vibrations of the structure might improve the thermographic crack detection if the gap between crack faces can be closed and opened periodically. This would also lead to a certain modulation, which can be used to enhance the signal-to-noise ratio similar as with the lock-in technique.
- Based on laboratory experiments, it seems to be hardly possible to simulate the impact of connected structural members on the vibration attenuation and the defect heating. Thus, in-situ testing of large-scale structures is required to identify the practical range of operation.

Regarding the prevailing heating mechanisms in vibrothermography applied to metals, the present work clearly motivates further investigations. To gain deeper insight into the involved mechanics, three-dimensional laser vibrometry and a high-power transducer with an accurate amplitude linearity, low distortion and good frequency stability is required. The following issues should be addressed in the future for a better understanding of ultrasound excited thermography:

- In-depth analysis of the residual stress state/crack closure stresses
- Explicit analysis of the contact acoustic non-linearity
- Consideration of tribologic models for the generation of frictional heat or flash temperatures

The conducted numerical simulation of vibrothermographic testing indicates that detailed finite element modelling might be further enhanced to predict the thermal response of arbitrarily shaped crack-like flaws in the future. After a more sophisticated adjustment of the model parameters based on experimental results, the simulation can contribute to the assessment of detection limits, restrictions and vibrational conditions which have to be met in order to successfully detect cracks in complex steel structures using ultrasound excited thermography.

References

[AAK91] M. F. Ashby, J. Abulawi, and H. S. Kong. Temperature maps for frictional heating in dry sliding. *Tribology Transactions*, 34(4):577–587, 1991.

[Abr98] O. V. Abramov. *High-intensity ultrasonics*. Gordon & Breach Science Publishers, Amsterdam, 1998.

[ADC94] B. Armstrong-Hélouvry, P. Dupont, and C. Canudas de Wit. A survey of models, analysis tools and compensation methods for the control of machines with friction. *Automatica*, 30(7):1083–1138, 1994.

[Ana93] L. Anand. A constitutive model for interface friction. *Computational Mechanics*, 12(4):197–213, 1993.

[BACM06] T. J. Barden, D. P. Almond, P. Cawley, and M. Morbidini. Advances in thermosonics for detecting impact damage in CFRP composites. *AIP Conference Proceedings (Quantitative Nondestructive Evaluation)*, 820:550–557, 2006.

[BAM$^+$04] T. J. Barden, D. P. Almond, M. Morbidini, P. Duffour, and P. Cawley. A quantitative investigation of thermosonics. In *Proceedings of the 7th Conference on Quantitative Infrared Thermography, Brussels, Belgium*, pages H.12.1–6, 2004.

[BDZ01] G. Busse, A. Dillenz, and T. Zweschper. Defect-selective imaging of aerospace structures with elastic-wave-activated thermography. *Proceedings of SPIE (Thermosense XXIII)*, 4360:580–586, 2001.

[Bel68] M. H. Belgen. Infrared radiometric stress instrumentation application range study. NASA contractor report CR-1067, National Aeronautics and Space Administration, Washington, D.C., 1968.

[Ber06] D. Bergmannshoff. *Das Instabilitätsverhalten zug-/scherbeanspruchter Risse bei Variation des Belastungspfades*. Dissertation, Ruhr-Universität Bochum, 2006.

[BGHR13] L. Bienkowski, M. Goldammer, C. Homma, and M. Rothenfusser. Real-View-Induktionsthermografie. In *DGZfP-Berichtsband 143-CD: Thermografie-Kolloquium, Leinfelden-Echterdingen, Germany*, 2013.

[BGPH10] G. Bolu, A. Gachagan, G. Pierce, and G. Harvey. Reliable thermosonic inspection of aero engine turbine blades. *Insight - Non-Destructive Testing & Condition Monitoring*, 52(9):488–493, 2010.

[BGPT10] G. Bolu, A. Gachagan, G. Pierce, and Barden T. A comparison of methods used to predict the vibrational energy required for a reliable thermosonic inspection. In *Proceedings of the 49th Annual Conference of The British Institute of Non-Destructive Testing, Cardiff, UK*, 2010.

[BHT73] M. B. Bever, D. L. Holt, and A. L. Titchener. *The Stored Energy of Cold Work*. Pergamon Press, Oxford, 1973.

[Bil93] M. Bily. *Cyclic deformation and fatigue of metals*. Elsevier, Amsterdam, 1993.

[BMR78] O. Buck, W. L. Morris, and J. M. Richardson. Acoustic harmonic generation at unbonded interfaces and fatigue cracks. *Applied Physics Letters*, 33(5):371–373, 1978.

[Bod11] M. D. Bode. Sonic IR for engine disk crack detection & evaluation efforts using end-of-life turbine disks. Presentation held at the Air Transport Association Non-Destructive Testing Forum, Charlotte, NC, USA, 2011.

[BP00] J. S. Bendat and A. G. Piersol. *Random Data: Analysis and Measurement Procedures*. John Wiley & Sons, New York, 3rd edition, 2000.

[BP07] W. Böge and W. Plaßmann, editors. *Vieweg Handbuch Elektrotechnik*. Vieweg + Teubner, Wiesbaden, 4th edition, 2007.

[Bra04] Branson. *Ultraschallgenerator 2000b/bdc Betriebsanleitung*. Branson Ultraschall, Dietzenbach, 2nd edition, 2004.

[Bra05] Branson. *Ultraschall für das Kunststoff-Fügen: Grundlagen und praktische Anwendung*. Branson Ultraschall, Dietzenbach, 2005.

[Bri97] E. O. Brigham. *FFT-Anwendungen*. Oldenbourg, München, 1997.

[BS87] C. Boller and T. Seeger. *Materials data for cyclic loading: Part A: Unalloyed steels*. Elsevier, Amsterdam, 1987.

[BS06] H. D. Baehr and K. Stephan. *Wärme- und Stoffübertragung*. Springer, Berlin, 5th edition, 2006.

[BSMM97] I. N. Bronstein, K. A. Semendjajew, G. Musiol, and H. Mühlig. *Taschenbuch der Mathematik*. Harri Deutsch, Thun, 3rd edition, 1997.

[BT09] P. N. Bogdanovich and D. V. Tkachuk. Thermal and thermomechanical phenomena in sliding contact. *Journal of Friction and Wear*, 30(3):153–163, 2009.

[BWK92] G. Busse, D. Wu, and W. Karpen. Thermal wave imaging with phase sensitive modulated thermography. *Journal of Applied Physics*, 71(8):3962–3965, 1992.

[BZH03] K. S. Bhalla, A. T. Zehnder, and X. Han. Thermomechanics of slow stable crack growth: closing the loop between experiments and computational modeling. *Engineering Fracture Mechanics*, 70(17):2439–2458, 2003.

[CB76] G. M. Carlomagno and P. G. Berardi. Unsteady thermotopography in non-destructive testing. In *Proceedings of the 3rd Biannual Exchange, St. Louis, MO, USA*, pages 33–39, 1976.

[CH10] H. Czichos and K.-H. Habig. *Tribologie-Handbuch*. Vieweg + Teubner, Wiesbaden, 3rd edition, 2010.

[CJ59] H. S. Carslaw and J. C. Jaeger. *Conduction of heat in solids*. Oxford University Press, Oxford, 2nd edition, 1959.

[CKLR07] J. C. Chen, J. Kephart, K. Lick, and W. T. Riddell. Crack growth induced by sonic IR inspection. *Nondestructive Testing and Evaluation*, 22(2-3):83–92, 2007.

[CKR06] J. C. Chen, J. Kephart, and W. T. Riddell. A parametric study of crack propagation during sonic IR inspection. *AIP Conference Proceedings (Quantitative Nondestructive Evaluation)*, 820:1577–1584, 2006.

[CPE57] J. S. Courtney-Pratt and E. Eisner. The effect of a tangential force on the contact of metallic bodies. *Proceedings of the Royal Society of London, Series A: Mathematical and Physical Sciences*, 238(1215):529–550, 1957.

[CRC04] CRC Industries Europe BVBA. Kontakt Chemie Graphit 33 technical data sheet. URL: http://www.crceurope.com/wwwcrc/tds/TKC3%20Graphit33.pdf (last accessed: June 14, 2014), 2004.

[CT65] J. W. Cooley and J. W. Tukey. An algorithm for the machine calculation of complex Fourier series. *Mathematics of Computation*, 19(90):297–301, 1965.

[DA02] T. Dirgantara and M. H. Aliabadi. Stress intensity factors for cracks in thin plates. *Engineering Fracture Mechanics*, 69(13):1465–1486, 2002.

[DBW99] A. Dillenz, G. Busse, and D. Wu. Ultrasound lockin thermography: feasibilities and limitations. *Proceedings of SPIE (Diagnostic Imaging Technologies and Industrial Applications)*, 3827:10–15, 1999.

[dCMM98] J. R. M. d'Almeida, N. Cella, S. N. Monteiro, and L. C. M. Miranda. Thermal diffusivity of an epoxy system as a function of the hardener content. *Journal of Applied Polymer Science*, 69(7):1335–1341, 1998.

[DH05] H. Dresig and F. Holzweißig. *Maschinendynamik*. Springer, Berlin, 6th edition, 2005.

[Dil01] A. Dillenz. Ultraschall-Burst-Phasen-Thermografie. *MP Materials Testing*, 43:30–34, 2001.

[DO89] W. DuMouchel and F. O'Brien. Integrating a robust option into a multiple regression computing environment. In *Proceedings of the 21st Symposium on the Interface, Orlando, FL, USA*, pages 297–302, 1989.

[Dom98] H. Domininghaus. *Die Kunststoffe und ihre Eigenschaften*. Springer, Berlin, 5th edition, 1998.

[DR03] W. R. Davis and M. B. Rankin. Laser vibrometry and strain gage measurements of thermosonic activation. *AIP Conference Proceedings (Quantitative Nondestructive Evaluation)*, 657:492–499, 2003.

[Dru54] D. C. Drucker. Coulomb friction, plasticity, and limit loads. *ASME Journal of Applied Mechanics*, 21:71–74, 1954.

[DSZ05] A. Dillenz, C. Spiessberger, and T. Zweschper. Vollautomatisierte Ultraschallthermografie in der industriellen Serienfertigung. In *DGZfP-Berichtsband 94-CD: DGZfP-Jahrestagung, Rostock, Germany*, 2005.

[DZB00] A. Dillenz, T. Zweschper, and G. Busse. Phase angle thermography with ultrasound burst excitation. In *Proceedings of the 5th Conference on Quantitative Infrared Thermography, Reims, France*, pages 247–252, 2000.

[DZRB03] A. Dillenz, T. Zweschper, G. Riegert, and G. Busse. Progress in phase angle thermography. *Review of Scientific Instruments*, 74(1):417–419, 2003.

[Ede09] Edevis GmbH. Active thermography methods overview. URL: http://www.edevis.de/thermography/thermography_overview_de.php (last accessed: August 1, 2009), 2009.

[Elb71] W. Elber. The significance of fatigue crack closure. In M. S. Rosenfeld, editor, *Damage Tolerance in Aircraft Structures*, pages 230–442. American Society for Testing and Materials, Philadelphia, 1971.

[FHO+00] L. D. Favro, X. Han, Z. Ouyang, G. Sun, H. Sui, and R. L. Thomas. Infrared imaging of defects heated by a sonic pulse. *Review of Scientific Instruments*, 71(6):2418–2421, 2000.

[FHO+01] L. D. Favro, X. Han, Z. Ouyang, G. Sun, and R. L. Thomas. Sonic IR imaging of cracks and delaminations. *Analytical Sciences*, 17:451–453, 2001.

[FLI05a] FLIR. *ThermaCAM Phoenix DTS Camera System user manual.* FLIR Systems Inc., 2005.

[FLI05b] FLIR. *ThermaCAM RCal IR calibration tool user manual.* FLIR Systems Inc., 2005.

[FR06] N. A. Fouad and T. Richter. *Leitfaden Thermografie im Bauwesen: Theorie, Anwendungsgebiete, praktische Umsetzung*. Fraunhofer IRB-Verlag, Stuttgart, 2006.

[Fre10] M. Frenz. *Lebensdauerermittlung stählerner Tragwerke unter Berücksichtigung von Systemumlagerungen*. Dissertation, Technische Universität Braunschweig, 2010.

[FTH+01] L. D. Favro, R. L. Thomas, X. Han, Z. Ouyang, G. Newaz, and D. Gentile. Sonic infrared imaging of fatigue cracks. *International Journal of Fatigue*, 23(Suppl. 1):471–476, 2001.

[Gei10] J. Geisler. *Numerische und experimentelle Untersuchungen zum dynamischen Verhalten von Strukturen mit Fügestellen*. Dissertation, Universität Erlangen-Nürnberg, 2010.

[GM01] C. Groth and G. Müller. *FEM für Praktiker - Band 3: Temperaturfelder*. Expert, Renningen, 4th edition, 2001.

[GMFM09] M. Genest, F. Mabrouki, A. Fahr, and N. Mrad. Investigation of alternative excitation sources for vibrothermography. *Proceedings of SPIE (Nondestructive Characterization for Composite Materials, Aerospace Engineering, Civil Infrastructure, and Homeland Security)*, 7294:72940Y, 2009.

[GMVC99] F. Galmiche, X. Maldague, S. Valler, and J.-P. Couturier. Pulsed phased thermography with the wavelet transform. *AIP Conference Proceedings (Quantitative Nondestructive Evaluation)*, 509:609–616, 1999.

[GS07] D. Gross and T. Seelig. *Bruchmechanik.* Springer, Berlin, 4th edition, 2007.

[GSB07] A. Gleiter, C. Spiessberger, and G. Busse. Improved ultrasound activated thermography using frequency analysis. *Quantitative Infrared Thermography Journal*, 4(2):155–164, 2007.

[Hag93] L. Hagman. *Micro-slip and surface deformation.* Licentiate thesis, Royal Institute of Technology Stockholm, 1993.

[Hai02] Erwin Haibach. *Betriebsfestigkeit.* Springer, Berlin, 2nd edition, 2002.

[Har78] F. J. Harris. On the use of windows for harmonic analysis with the discrete Fourier transform. *Proceedings of the IEEE*, 66(1):51–83, 1978.

[Has06] D. Hasenberg. *Zerstörungsfreie Prüfung geklebter Verbindungen mittels Ultraschall angeregter Thermografie.* Dissertation, Technische Universität Braunschweig, 2006.

[HBB08] B. Hosten, C. Bacon, and C. Biateau. Finite element modeling of the temperature rise due to the propagation of ultrasonic waves in viscoelastic materials and experimental validation. *The Journal of the Acoustical Society of America*, 124(6):3491–3496, 2008.

[HC91] N. Harwood and W. M. Cummings, editors. *Thermoelastic Stress Analysis.* Adam Hilger, Bristol, 1991.

[HDZ03] D. Hasenberg, K. Dilger, and T. Zweschper. Charakterisierung von Klebverbindungen mittels ultraschallangeregter Thermografie. In *DGZfP-Berichtsband 86-CD: Thermografie-Kolloquium, Stuttgart, Germany*, 2003.

[He10] Q. He. *Develop sonic infrared imaging NDE for local damage assessment in civil structures.* PhD thesis, Wayne State University, Detroit, 2010.

[HFNT06] X. Han, L. D. Favro, G. Newaz, and R. L. Thomas. Recent developments in IR imaging with chaotic sound excitation for NDI of aircraft structures. In *Proceedings of the 9th joint FAA/DoD/NASA Conference on Aging Aircraft, Atlanta, GA, USA*, 2006.

[HFOT01] X. Han, L. Favro, Z. Ouyang, and R. L. Thomas. Thermosonics: detecting cracks and adhesion defects using ultrasonic excitation and infrared imaging. *The Journal of Adhesion*, 76(2):151–162, 2001.

[HFOT02] X. Han, L. D. Favro, Z. Ouyang, and R. L. Thomas. Recent developments in thermosonic crack detection. *AIP Conference Proceedings (Quantitative Nondestructive Evaluation)*, 615:552–557, 2002.

[HFT01] X. Han, L. D. Favro, and R. L. Thomas. Thermosonic imaging of cracks: applications to teeth. *Proceedings of SPIE (Hybrid and Novel Imaging and New Optical Instrumentation for Biomedical Applications)*, 4434:70–73, 2001.

[HHW+07] W. Hassan, C. Homma, Z. Wen, F. Vensel, and B. Hogan. Detection of tight fatigue cracks at the root of dampers in fan blades using sonic IR inspection: a feasibility demonstration. *AIP Conference Proceedings (Quantitative Nondestructive Evaluation)*, 894:455–462, 2007.

[HIN+05] X. Han, S. Islam, G. Newaz, L. D. Favro, and R. L. Thomas. Finite-element modeling of acoustic chaos to sonic infrared imaging. *Journal of Applied Physics*, 98(1):014907, 2005.

[HIN+06] X. Han, S. Islam, G. Newaz, L. D. Favro, and R. L. Thomas. Finite element modeling of the heating of cracks during sonic infrared imaging. *Journal of Applied Physics*, 99(7):074905, 2006.

[HLZT04] X. Han, V. Loggins, Z. Zeng, and R. L. Thomas. Mechanical model for the generation of acoustic chaos in sonic infrared imaging. *Applied Physics Letters*, 85(8):1332–1334, 2004.

[HMS04] E. Hering, R. Martin, and M. Stohrer. *Physik für Ingenieure*. Springer, Berlin, 9th edition, 2004.

[Hol07] S. D. Holland. First measurements from a new broadband vibrothermography measurement system. *AIP Conference Proceedings (Quantitative Nondestructive Evaluation)*, 894:478–483, 2007.

[Hom07] C. Homma. *Untersuchungen zu Mechanismus und technischer Umsetzung der akustischen Thermographie*. Dissertation, Universität des Saarlandes, Saarbrücken, 2007.

[HR10] S. D. Holland and J. Renshaw. Physics-based image enhancement for infrared thermography. *NDT & E International*, 43(5):440–445, 2010.

[HRBS06] C. Homma, M. Rothenfusser, J. Baumann, and R. Shannon. Study of the heat generation mechanism in acoustic thermography. *AIP Conference Proceedings (Quantitative Nondestructive Evaluation)*, 820:566–573, 2006.

[HRRR00] J. Hodowany, G. Ravichandran, A. J. Rosakis, and P. Rosakis. Partition of plastic work into heat and stored energy in metals. *Experimental Mechanics*, 40(2):113–123, 2000.

[HRS79] E. G. Henneke, K. L. Reifsnider, and W. W. Stinchcomb. Thermography - an NDI method for damage detection. *Journal of Metals*, 31(9):11–15, 1979.

[HUO+11] S. D. Holland, C. Uhl, Z. Ouyang, T. Bantel, M. Li, W. Q. Meeker, J. Lively, L. Brasche, and D. Eisenmann. Quantifying the vibrothermographic effect. *NDT & E International*, 44(8):775–782, 2011.

[HUR08] S. D. Holland, C. Uhl, and J. Renshaw. Toward a viable strategy for estimating vibrothermographic probability of detection. *AIP Conference Proceedings (Quantitative Nondestructive Evaluation)*, 975:491–497, 2008.

[HW77] P. W. Holland and R. E. Welsch. Robust regression using iteratively reweighted least-squares. *Communications in Statistics - Theory and Methods*, 6(9):813 – 827, 1977.

[HW90] K. Herrmann and L. Walther. *Wissensspeicher Infrarottechnik*. Fachbuchverlag, Leipzig, 1990.

[HZL+04a] X. Han, Z. Zeng, W. Li, S. Islam, J. Lu, V. Loggins, L. D. Favro, G. M. Newaz, and R. L. Thomas. Importance of acoustic chaos in sonic IR imaging NDE. *AIP Conference Proceedings (Quantitative Nondestructive Evaluation)*, 700:496–500, 2004.

[HZL+04b] X. Han, Z. Zeng, W. Li, S. Islam, J. Lu, V. Loggins, E. Yitamben, L. D. Favro, G. Newaz, and R. L. Thomas. Acoustic chaos for enhanced detectability of cracks by sonic infrared imaging. *Journal of Applied Physics*, 95(7):3792–3797, 2004.

[IBS09] IBS Magnet. Electromagnets EM8 and EM15 technical data sheet. URL: http://www.ibs-magnet.de/products/e_haftmagnete/hamaelt.php (last accessed: June 9, 2014), 2009.

[IC05] C. Ibarra-Castanedo. *Quantitative subsurface defect evaluation by pulsed phase thermography: depth retrieval with the phase*. PhD thesis, Université Laval, Québec City, 2005.

[ICGS+07] C. Ibarra-Castanedo, M. Genest, G. Stéphane, J.-M. Piau, X. P. V. Maldague, and A. Bendada. Inspection of aerospace materials by pulsed thermography, lock-in thermography, and vibrothermography: a comparative study. *Proceedings of SPIE (Thermosense XXIX)*, 6541:654116, 2007.

[ICSK+08] C. Ibarra-Castanedo, M. Susa, M. Klein, M. Grenier, J.-M. Piau, W. Ben Larby, A. Bendada, and X. P. V. Maldague. Infrared thermography: principle and applications to aircraft materials. In *Proceedings of the 1st International Symposium on NDT in Aerospace, Fürth, Germany*, 2008.

[Iow13] Iowa State University. Thermography research group. URL: https://thermal.cnde.iastate.edu/research.xml (last accessed: June 9, 2014), 2013.

[IRH97] L. Issler, H. Ruoß, and P. Häfele. *Festigkeitslehre - Grundlagen*. Springer, Berlin, 2nd edition, 1997.

[IW03] A. R. Ingraffea and P. A. Wawrzynek. Finite element methods for linear elastic fracture mechanics. In I. Milne, R. O. Ritchie, and B. L. Karihaloo, editors, *Comprehensive structural integrity*, pages 1–88. Elsevier Pergamon, Amsterdam, 2003.

[Jan07] Janos Technology, Inc. MWIR Asio 50 mm lens technical data sheet. URL: http://www.janostech.com/products-services/thermal-imaging-lenses/asio-lenses.html (last accessed: June 9, 2014), 2007.

[Kat00] K. Kato. Wear in relation to friction - a review. *Wear*, 241(2):141–157, 2000.

[KCZ05] J. Kephart, J. Chen, and H. Zhang. Characterization of crack propagation during sonic IR inspection. *Proceedings of SPIE (Thermosense XXVII)*, 5782:234–244, 2005.

[KDDW07] A. M. Korsunskya, D. Dinia, F. P. E. Dunnea, and M. J. Walshb. Comparative assessment of dissipated energy and other fatigue criteria. *International Journal of Fatigue*, 29(9-11):1990–1995, 2007.

[KFA+88] P. K. Kuo, Z. J. Feng, T. Ahmed, L. D. Favro, Thomas R. L., and J. Hartikainen. Parallel thermal wave imaging using a vector lock-in video technique. In P. Hess and J. Pelzl, editors, *Photoacoustic and Photothermal Phenomena*, pages 415–418. Springer, 1988.

[KK86] J. Krautkrämer and H. Krautkrämer. *Werkstoffprüfung mit Ultraschall*. Springer, Berlin, 5th edition, 1986.

[Kra98] J.-C. Krapez. Compared performances of four algorithms used for modulation thermography. In *Proceedings of the 4th Conference on Quantitative Infrared Thermography, Lódź, Poland*, pages 148–153, 1998.

[KSA06] F. G. Kollmann, T. F. Schösser, and R. Angert. *Praktische Maschinenakustik*. Springer, Berlin, 2006.

[Kun08] M. Kuna. *Numerische Beanspruchungsanalyse von Rissen*. Vieweg + Teubner, Wiesbaden, 2008.

[Lei69] A. W. Leissa. *Vibration of plates*. Special Publication SP-160, Office of Technology Utilization, National Aeronautics and Space Administration, Washington D.C., 1969.

[Lei73] A. W. Leissa. *Vibration of shells*. Special Publication SP-288, Office of Technology Utilization, National Aeronautics and Space Administration, Washington D.C., 1973.

[LHM10a] M. Li, S. D. Holland, and W. Q. Meeker. Automatic crack detection algorithm for vibrothermography sequence-of-images data. *AIP Conference Proceedings (Quantitative Nondestructive Evaluation)*, 1211:1919–1926, 2010.

[LHM10b] M. Li, S. D. Holland, and W. Q. Meeker. Statistical methods for automatic crack detection based on vibrothermography sequence-of-images data. *Applied Stochastic Models in Business and Industry*, 26(5):481–495, 2010.

[LHN+07] J. Lu, X. Han, G. Newaz, L. D. Favro, and R. L. Thomas. Study of the effect of crack closure in sonic infrared imaging. *Nondestructive Testing and Evaluation*, 22(2):127–135, 2007.

[LILV03] J. H. Lienhard IV and J. H. Lienhard V. *A heat transfer textbook*. Phlogiston Press, Cambridge, 3rd edition, 2003.

[LM87] R. W. Lang and J. A. Manson. Crack tip heating in short-fibre composites under fatigue loading conditions. *Journal of Materials Science*, 22(10):3576–3580, 1987.

[MÖ07] M. Möser. *Technische Akustik*. Springer, Berlin, 7th edition, 2007.

[Mal01] X. P. V. Maldague. *Theory and practice of infrared technology for nondestructive testing*. John Wiley & Sons, New York, 2001.

[Mal02] X. P. V. Maldague. Introduction to NDT by active infrared thermography. *Materials Evaluation*, 60(9):1060–1073, 2002.

[MAS$^+$09] A. Mendioroz, E. Apinaniz, A. Salazar, P. Venegas, and I. Saez-Ocariz. Quantitative study of buried heat sources by lock-in vibrothermography: an approach to crack characterization. *Journal of Physics, D: Applied Physics*, 42(5):055502, 2009.

[MC04a] M. A. Meggiolaro and J. T. P. Castro. Statistical evaluation of strain-life fatigue crack initiation predictions. *International Journal of Fatigue*, 26(5):463–476, 2004.

[MC04b] C. Meola and G. M. Carlomagno. Recent advances in the use of infrared thermography. *Measurement Science and Technology*, 15:R27–R58, 2004.

[MC08] M. Morbidini and P. Cawley. The reliable implementation of thermosonics NDT. In *Proceedings of the 17th World Conference on Nondestructive Testing, Shanghai, China*, 2008.

[MC09a] M. Morbidini and P. Cawley. A calibration procedure for sonic infrared nondestructive evaluation. *Journal of Applied Physics*, 106(2):106, 023504, 2009.

[MC09b] M. Morbidini and P. Cawley. The detectability of cracks using sonic IR. *Journal of Applied Physics*, 105:093530, 2009.

[MCB$^+$06] M. Morbidini, P. Cawley, T. Barden, D. Almond, and P. Duffour. The relationship between vibration level and minimum detectable defect size in Sonic-IR inspection. In *Proceedings of the 9th European Conference on Non-Destructive Testing, Berlin, Germany*, page We.3.8.4, 2006.

[MCB$^+$07] M. Morbidini, P. Cawley, T. Barden, D. Almond, and P. Duffour. Prediction of the thermosonic signal from fatigue cracks in metals using vibration damping measurements. *Insight - Non-Destructive Testing & Condition Monitoring*, 49(5):254–263, 2007.

[McK87] J. McKelvie. Consideration of the surface temperature response to cyclic thermoelastic heat generation. *Proceedings of SPIE (Stress Analysis by Thermoelastic Techniques)*, 731:44–53, 1987.

[MDBR03] W. O. Miller, I. M. Darnell, M. W. Burke, and C. L. Robbins. Defining the envelope for sonic IR: detection limits and damage limits. *Proceedings of SPIE (Thermosense XXV)*, 5073:406–416, 2003.

[Med08] J. Medgenberg. *Investigation of localized fatigue properties in unalloyed steels by infrared thermography.* Dissertation, Technische Universität Braunschweig, 2008.

[Meh03] M. Mehdianpour. *Lebensdauervorhersage von ermüdungsbeanspruchten Stahltragwerken mit Hilfe von Monitoring und begleitenden Versuchen.* Dissertation, Technische Universität Braunschweig, 2003.

[MGB11] P. Menner, H. Gerhard, and G. Busse. Lockin-interferometry: principle and applications in NDE. *Journal of Mechanical Engineering*, 57(3):183–191, 2011.

[MGD+81] R. B. Mignogna, R. E. Green, J. C. Duke, E. G. Henneke, and K. L. Reifsnider. Thermographic investigation of high-power ultrasonic heating in materials. *Ultrasonics*, 19(4):159–163, 1981.

[MHIN04] A. Mian, X. Han, S. Islam, and G. Newaz. Fatigue damage detection in graphite/epoxy composites using sonic infrared imaging technique. *Composites Science and Technology*, 64(5):657–666, 2004.

[Mik07] Mikron. *Mikron M345X6 Blackbody Source user manual.* Mikron Infrared, Inc., Oakland, NJ, USA, 2007.

[MK10] M. Möser and W. Kropp. *Körperschall.* Springer, Berlin, 3rd edition, 2010.

[MM96] X. P. V. Maldague and S. Marinetti. Pulse phase infrared thermography. *Journal of Applied Physics*, 79(5):2694–2698, 1996.

[MNH+04] A. Mian, G. Newaz, X. Han, T. Mahmood, and C. Saha. Response of sub-surface fatigue damage under sonic load - a computational study. *Composites Science and Technology*, 64(9):1115–1122, 2004.

[Mor07] M. Morbidini. *A comparison of the vibro-modulation and thermosonic NDT techniques.* PhD thesis, University of London (Imperial College), 2007.

[MR09] B. Muralikrishnan and J. Raja. *Computational surface and roundness metrology*. Springer, London, 2009.

[MRF10] R. Montanini, G. L. Rossi, and F. Freni. Ultrasound lockin thermography as a quantitative technique for quality control assessment of cast iron turbine components. In *Proceedings of the 10th Conference on Quantitative Infrared Thermography, Québec City, Canada*, pages 753–759, 2010.

[MRKS02] N. G. H. Meyendorf, H. Rösner, V. Kramb, and S. Sathish. Thermo-acoustic fatigue characterization. *Ultrasonics*, 40(1):427–434, 2002.

[MTGF09] F. Mabrouki, M. Thomas, M. Genest, and A. Fahr. Frictional heating model for efficient use of vibrothermography. *NDT & E International*, 42(5):345–352, 2009.

[MTGF10] F. Mabrouki, M. Thomas, M. Genest, and A. Fahr. Numerical modeling of vibrothermography based on plastic deformation. *NDT & E International*, 43(6):476–483, 2010.

[NW99] H. Naubereit and J. Weihert. *Einführung in die Ermüdungsfestigkeit*. Carl Hanser, München, 1999.

[OFTH02] Z. Ouyang, L. D. Favro, R. L. Thomas, and X. Han. Theoretical modeling of thermosonic imaging of cracks. *AIP Conference Proceedings (Quantitative Nondestructive Evaluation)*, 615:577–581, 2002.

[OH97] U. Olofsson and L. Hagman. A model for micro-slip between flat surfaces based on deformation of ellipsoidal elastic bodies. *Tribology International*, 30(8):599–603, 1997.

[Olo95] U. Olofsson. Cyclic micro-slip under unlubricated conditions. *Tribology International*, 28(4):207–217, 1995.

[OSB99] A. V. Oppenheim, R. W. Schafer, and J. R. Buck. *Discrete-time signal processing*. Prentice Hall, Upper Saddle River, New Jersey, 2nd edition, 1999.

[OWG$^+$07] R. Olbrycht, B. Wiecek, G. Gralewicz, T. Swiatczak, and G. Owczarek. Comparison of Fourier and wavelet analyses for defect detection in lock-in and pulse phase thermography. *Quantitative Infrared Thermography Journal*, 4(2):219–232, 2007.

[PA09] U. Polimeno and D. P. Almond. A compact thermosonic inspection system for the inspection of composites. In *Proceedings of the 17th International Conference on Composite Materials, Edinburgh, UK*, 2009.

[Pap01] L. Papula. *Mathematik für Ingenieure und Naturwissenschaftler - Band 3*. Vieweg, Braunschweig, 4th edition, 2001.

[PBML08] J.-M. Piau, A. Bendada, X. P. V. Maldague, and J.-G. Legoux. Nondestructive testing of open microscopic cracks in plasma-sprayed-coatings using ultrasound excited vibrothermography. *Nondestructive Testing and Evaluation*, 23(2):109–120, 2008.

[Pet96] C. Petersen. *Dynamik der Baukonstruktionen*. Vieweg, Braunschweig, 1996.

[Pfl06] K. Pfleiderer. *Frequenzkonversion aufgrund nichtlinearer akustischer Phänomene: Grundlagen und Anwendung zur defektselektiven zerstörungsfreien Prüfung*. Dissertation, Universität Stuttgart, 2006.

[PI 08] PI miCos GmbH. Linear stage LS-110 technical data sheet. URL: http://www.pimicos.com/web2/en/1,4,110,ls110.html (last accessed: January 17, 2012), 2008.

[Pol09] Polytec. *Vibrometer Controller OVF-5000 Handbuch*. Polytec GmbH, Waldbronn, 2009.

[Pop09] V. Popov. *Kontaktmechanik und Reibung*. Springer, Berlin, 2009.

[PSA$^+$10] L. J. Pieczonka, W. J. Staszewski, F. Aymerich, T. Uhl, and M. Szwedo. Numerical simulations for impact damage detection in composites using vibrothermography. *IOP Conference Series: Materials Science and Engineering*, 10:012062, 2010.

[PU08] R. Plum and T. Ummenhofer. Ultrasound excited thermography of load bearing members used in constructional steelwork. In *Proceedings of the 9th Conference on Quantitative Infrared Thermography, Kraków, Poland*, pages 461–467, 2008.

[PU09] R. Plum and T. Ummenhofer. Ultrasound excited thermography of thick-walled steel load bearing members. *Quantitative Infrared Thermography Journal*, 6(1):79–100, 2009.

[PW04] U. Peil and M. Wichers. Schweißen unter Betriebsbeanspruchung, Werkstoffkennwerte für einen S 355 J2G3 unter Temperaturen bis 1200 °C. *Stahlbau*, 73(6):400–416, 2004.

[PW05] U. Peil and M. Wichers. Schweißen unter Betriebsbeanspruchung - Werkstoffkennwerte zur Temperaturfeldberechnung für einen S 355 J2G3. *Stahlbau*, 74(4):249–257, 2005.

[Rad03] Dieter Radaj. *Ermüdungsfestigkeit.* Springer, Berlin, 2nd edition, 2003.

[RCHT11] J. Renshaw, J. Chen, S. D. Holland, and R. B. Thompson. The sources of heat generation in vibrothermography. *NDT & E International*, 44(8):736–739, 2011.

[RDQB10] A. F. Robinson, J. M. Dulieu-Barton, S. Quinn, and R. L. Burguete. Paint coating characterization for thermoelastic stress analysis of metallic materials. *Measurement Science and Technology*, 21(8):085502, 2010.

[RH05] M. Rothenfusser and C. Homma. Acoustic thermography: vibrational modes of cracks and the mechanism of heat generation. *AIP Conference Proceedings (Quantitative Nondestructive Evaluation)*, 760:624–631, 2005.

[RHB09] J. Renshaw, S. D. Holland, and D. J. Barnard. Viscous material-filled synthetic defects for vibrothermography. *NDT & E International*, 42(8):753–756, 2009.

[RHT08] J. Renshaw, S. D. Holland, and R. B. Thompson. Measurement of crack opening stresses and crack closure stress profiles from heat generation in vibrating cracks. *Applied Physics Letters*, 93(8):081914, 2008.

[RHTA11] J. Renshaw, S. D. Holland, R. B. Thompson, and J. Anderegg. Vibration-induced tribological damage to fracture surfaces via vibrothermography. *International Journal of Fatigue*, 33(7):849–857, 2011.

[Rie06] G. Riegert. Lockin and burst-phase induction thermography for NDE. *Quantitative Infrared Thermography Journal*, 3(2):141–154, 2006.

[Row00] M. Rowe. Generate a swept sine test signal. URL: http://www.edn.com/electronics-news/4384064/Generate-a-Swept-Sine-Test-Signal (last accessed: June 9, 2014), 2000.

[RPnt] A. Ruina and R. Pratap. *Introduction to Statics and Dynamics.* Oxford University Press, Oxford, 2009 (Preprint).

[RRHR01] G. Ravichandran, A. J. Rosakis, J. Hodowany, and P. Rosakis. On the conversion of plastic work into heat during high-strain-rate deformation. *AIP Conference Proceedings (Shock Compression of Condensed Matter)*, 620:557–562, 2001.

[RS12] H. A. Richard and M. Sander. *Ermüdungsrisse - Erkennen, sicher beurteilen, vermeiden.* Vieweg + Teubner, Wiesbaden, 2nd edition, 2012.

[RWB96] J. Rantala, D. Wu, and G. Busse. Amplitude-modulated lock-in vibrothermography for NDE of polymers and composites. *Research in Nondestructive Evaluation*, 7(4):215–228, 1996.

[RWSB96] J. Rantala, D. Wu, A. Salerno, and G. Busse. Lock-in thermography with mechanical loss angle heating at ultrasonic frequenices. In *Proceedings of the 3rd Conference on Quantitative Infrared Thermography, Stuttgart, Germany*, pages 389–393, 1996.

[SAL04] S. M. Shepard, T. Ahmed, and J. R. Lhota. Experimental considerations in vibrothermography. *Proceedings of SPIE (Thermosense XXVI)*, 5405:332–335, 2004.

[SCR88] J. O. Street, R. J. Carroll, and D. Ruppert. A note on computing robust regression estimates via iteratively reweighted least squares. *The American Statistician*, 42(2):152–154, 1988.

[SDW$^+$98] A. Salerno, A. Dillenz, D. Wu, J. Rantala, and G. Busse. Progress in ultrasound lockin thermography. In *Proceedings of the 4th Conference on Quantitative Infrared Thermography, Lódź, Poland*, pages 154–160, 1998.

[SKB02] I. Y. Solodov, N. Krohn, and G. Busse. CAN: an example of nonclassical acoustic nonlinearity in solids. *Ultrasonics*, 40(1-8):621–625, 2002.

[SKB10] I. Solodov, N. Krohn, and G. Busse. Nonlinear ultrasonic NDT for early defect recognition and imaging. In *Proceedings of the 10th European Conference on Non-Destructive Testing, Moscow, Russia*, 2010.

[Sne72] I. N. Sneddon. *The linear theory of thermoelasticity*. Springer, Wien, 1972.

[SO96] U. Sellgren and U. Olofsson. A user programmable micro-slip feature. In *Proceedings of the 7th International ANSYS Conference and Exhibition, Pittsburgh, PA, USA*, 1996.

[Sol98] I. Y. Solodov. Ultrasonics of non-linear contacts: propagation, reflection and NDE-applications. *Ultrasonics*, 36(1-5):383–390, 1998.

[SR10] G. Sutter and N. Ranc. Flash temperature measurement during dry friction process at high sliding speed. *Wear*, 268(11-12):1237–1242, 2010.

[Sur98] S. Suresh. *Fatigue of materials*. Cambridge University Press, Cambridge, 2nd edition, 1998.

[SWT70] K. N. Smith, P. Watson, and T. H. Topper. A stress-strain function for the fatigue of metals. *Journal of Materials*, 5(4):767–778, 1970.

[The08] The Mathworks, Inc. MATLAB 7.6 (R2008a) documentation, 2008.

[THFN10] R. L. Thomas, X. Han, L. D. Favro, and G. Newaz. Infrared imaging of defects in materials with chaotic sonic excitation. In *Proceedings of the 15th IEEE International Ultrasonics Symposium, San Diego, CA, USA*, pages 591–594, 2010.

[Tho06] M. K. Thompson. Methods for generating rough surfaces in ANSYS. In *Proceedings of the 13th International ANSYS Conference and Exhibition, Pittsburgh, PA, USA*, 2006.

[TIB10] TIB. *Leistungsverstärker S250K Bedienungsanleitung*. TIB GmbH, Stipshausen, 8th edition, 2010.

[TR04] K. A. Tsoi and N. Rajic. Effect of sonic thermographic inspection on fatigue crack growth in an Al alloy. Technical Report DSTO-TN-0584, Australian Government Department of Defence, 2004.

[TR07] K. A. Tsoi and N. Rajic. Interference fit fastener inspection using sonic thermography. In *Proceedings of the 5th Australasian Congress on Applied Mechanics, Brisbane, Australia*, 2007.

[Val10] Vallen Systeme GmbH. Vallen Dispersion R2010 documentation. URL: http://www.vallen.de (last accessed: April 7, 2013), 2010.

[vB06] P. von Böckh. *Wärmeübertragung*. Springer, Berlin, 2nd edition, 2006.

[VDI06] VDI Verein Deutscher Ingenieure, VDI-Gesellschaft Verfahrenstechnik und Chemieingenieurwesen (GVC), editor. *VDI-Wärmeatlas*. Springer, Berlin, 10th edition, 2006.

[Wan86] R. J. H. Wanhill. Short cracks in aerospace structures. In K. J. Miller and E. R. de los Rios, editors, *The behaviour of short fatigue cracks*, volume 3, pages 27–36. Mechanical Engineering Publications, London, 1986.

[Wau08] J. Wauer. *Kontinuumsschwingungen*. Vieweg + Teubner, Wiesbaden, 2008.

[WB98] D. Wu and G. Busse. Lock-in thermography for nondestructive evaluation of materials. *Revue Générale de Thermique*, 37(8):693–703, 1998.

[WJS87] A. K. Wong, R. Jones, and J. G. Sparrow. Thermoelastic constant or thermoelastic parameter? *Journal of Physics and Chemistry of Solids*, 48(8):749–753, 1987.

[WKN+99] G. Walle, W. Karpen, U. Netzelmann, H. Rösner, and N. G. H. Meyendorf. Zerstörungsfreie Prüfung mittels thermographischer Verfahren. *tm - Technisches Messen*, 66(9):312–321, 1999.

[Wri06] P. Wriggers. *Computational contact mechanics*. Springer, Berlin, 2nd edition, 2006.

[WSD88] A. K. Wong, J. G. Sparrow, and S. A. Dunn. On the revised theory of the thermoelastic effect. *Journal of Physics and Chemistry of Solids*, 49(4):395–400, 1988.

[WVN+09] G. Walle, B. Valeske, U. Netzelmann, S. Bessert, H. Strauß, and F. Wolf. Eine thermische Prüftechnik zur Oberflächenrissprüfung leitfähiger Materialien. *MP Materials Testing*, 51(9):593–602, 2009.

[ZDB00] T. Zweschper, A. Dillenz, and G. Busse. Ultrasound lockin thermography - an NDT method for the inspection of aerospace structures. In *Proceedings of the 5th Conference on Quantitative Infrared Thermography, Reims, France*, pages 212–217, 2000.

[ZDB01] T. Zweschper, A. Dillenz, and G. Busse. Ultrasound lockin thermography - a defect-selective NDT method for the inspection of aerospace structures. *Insight - Non-Destructive Testing & Condition Monitoring*, 43(3):173–179, 2001.

[ZDR+02] T. Zweschper, A. Dillenz, G. Riegert, D. Scherling, and G. Busse. Lockin thermography methods for the NDT of CFRP aircraft components. In *Proceedings of the 8th European Conference on Nondestructive Testing, Barcelona, Spain*, 2002.

[ZDR+03] T. Zweschper, A. Dillenz, G. Riegert, D. Scherling, and G. Busse. Ultrasound excited thermography using frequency modulated elastic waves. *Insight - Non-Destructive Testing & Condition Monitoring*, 45(3):178–182, 2003.

[ZRDB03] T. Zweschper, G. Riegert, A. Dillenz, and G. Busse. Ultrasound burst phase thermography (UBP) for applications in the automotive industry. *AIP Conference Proceedings (Quantitative Nondestructive Evaluation)*, 657:531–536, 2003.

[ZRDB04] T. Zweschper, G. Riegert, A. Dillenz, and G. Busse. Ultrasound excited thermography - advances due to frequency modulated elastic waves. In *Proceedings of the* 7th *Conference on Quantitative Infrared Thermography, Brussels, Belgium*, pages H.8.1–6, 2004.

[ZS06] P. J. Zombo and R. E. Shannon. Advanced NDE systems for flexible operation and maintenance of gas turbine components. In *Proceedings of the 2006 POWER-GEN International Conference, Orlando, FL, USA*, 2006.

[ZZZF06] K. Zheng, H. Zhang, S. Zhang, and L. Fan. A dynamical model of subharmonic generation in ultrasonic infrared thermography. *Ultrasonics*, 44(Suppl. 1):e1343–e1347, 2006.

Appendix

A.1 Competitive active thermography methods

Table A.1: Competitive active thermography methods and their fields of application [Ede09].

Structure	Defect/detail/property	Excitation method		
		Optical[1]	Ultrasonic	Inductive
Joints	Adhesive joints	o	+	+
	Screw joint	+	o	o
	Rivet joint	o	+	o
	Spot/laser welding	o	+	o
	Clinch joint	o	+	o
Layers	Layer thickness	+	–	–
	Adhesive strength	+	–	–
Fibre composites	Separating foils	+	–	o
	Impact damages	+	+	o
	Delaminations	o	+	o
	Porosity	o	–	o
Fibre ceramics	Delaminations	+	o	o
	Cracks	–	+	o
	Inhomogeneities	+	–	–
	Porosity	+	–	–
Metals	Cracks	–	+	+
	Cavities	–	o	–
	Press fits	–	+	–
Fibre metal laminates	Impact damages	o	+	o
	Delaminations	o	+	o

[1]Lock-in and pulse approaches not distinguished + well suited o conditionally suited

A.2 Chemical compositions and mechanical properties of S355J2+N base material

Table A.2: Chemical compositions of S355J2+N according to manufacturer ladle analyses.

Thickness	[mm]		6	8	10	12	14
Carbon (C)	[%]	(≤ 0.2[1])	0.17	0.17	0.14	0.18	0.15
Silicium (Si)	[%]	(≤ 0.55[1])	0.02	0.02	0.02	0.39	0.31
Manganese (Mn)	[%]	(≤ 1.6[1])	1.59	1.55	1.54	1.44	1.41
Phosphor (P)	[%]	(≤ 0.025[1])	0.011	0.018	0.014	0.019	0.014
Sulphur (S)	[%]	(≤ 0.025[1])	0.011	0.018	0.014	0.019	0.014
Nitrogen (N)	[%]		0.004	0.003	0.004	0.006	0.005
Aluminium (Al)	[%]		0.028	0.035	0.049	0.035	0.037
Copper (Cu)	[%]	(≤ 0.55[1])	0.03	0.05	0.01	0.02	0.04
Chromium (Cr)	[%]		0.02	0.05	0.03	0.07	0.06
Nickel (Ni)	[%]		0.03	0.05	0.03	0.02	0.02
Molybdenum (Mo)	[%]		0.001	0.010	0.002	0.005	0.005
Niobium (Nb)	[%]		0.001	0.001	0.020	0.003	0.003
Titanium (Ti)	[%]		0.001	0.001	0.001	0.004	0.003
Vanadium (V)	[%]		0.003	0.004	0.004	0.004	0.004
Carbon equivalent (CEV)	[%]	(≤ 0.45[1])	0.44	0.45	0.41	0.44	0.40

[1]Maximum percentage according to EN 10025-2:2004

Table A.3: Mechanical properties of S355J2+N base material.

	Tensile test			Inspection certificate			
Thickness [mm]	R_{eH} [MPa] (≥ 355[1])	R_m [MPa] ($470-630$[1])	E [GPa]	R_{eH} [MPa] (≥ 355[1])	$R_{p0.2}$ [MPa] (≥ 355[1])	R_m [MPa] ($470-630$[1])	A_5 [%] (≥ 22[1])
6	347	504	216	–	406	570	26
8	375	542	210	371	–	533	29
10	432	551	209	443	–	565	29
12	392	546	213	365	–	538	29
14	373	529	207	384	–	532	29

[1]Minimum or range of nominal values according to EN 10025-2:2004

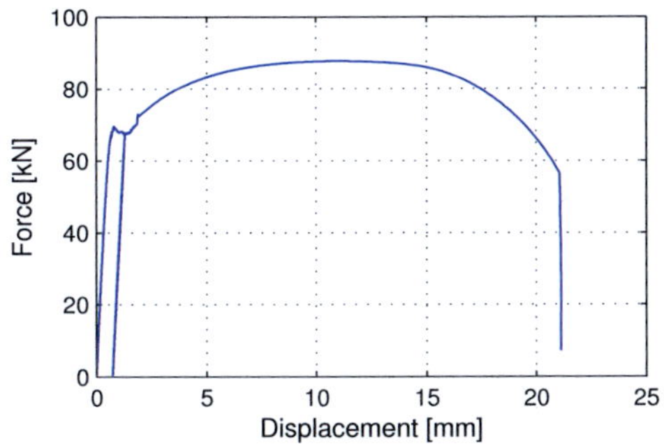

(a) Load-displacement curve (10 mm thick slab)

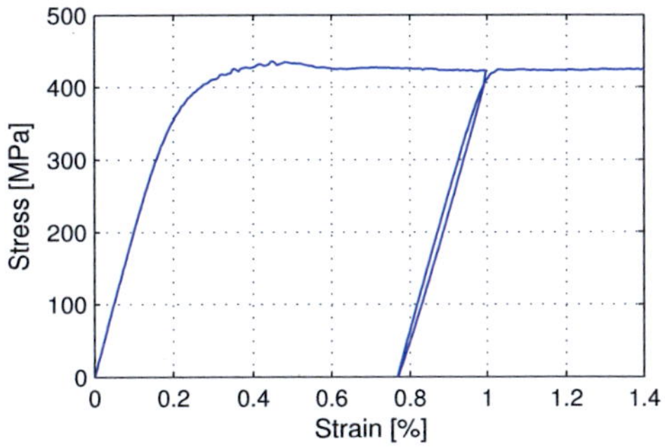

(b) Stress-strain curve (10 mm thick slab)

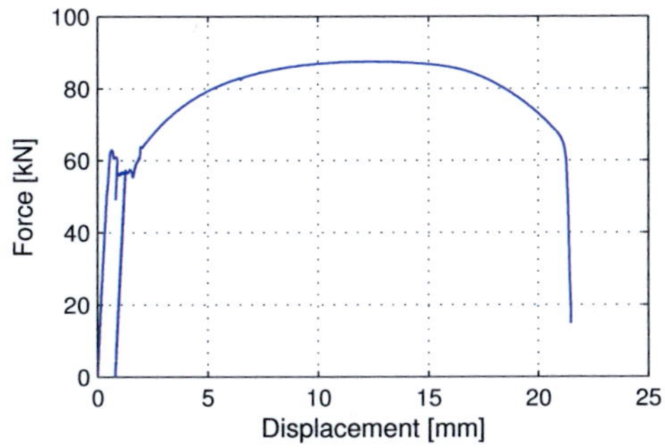

(c) Load-displacement curve (12 mm thick slab)

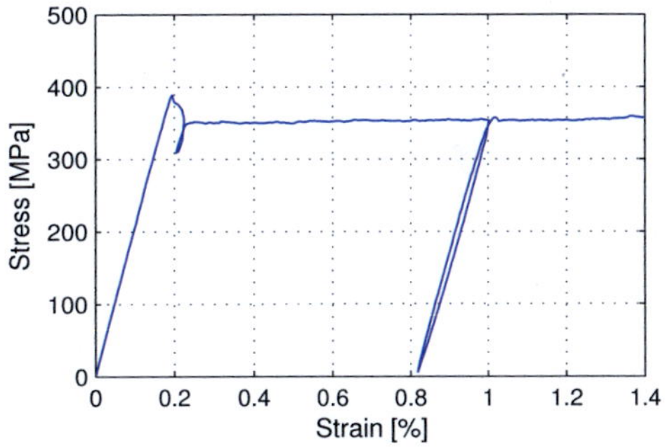

(d) Stress-strain curve (12 mm thick slab)

Figure A.1: Representative load-displacement and stress-strain curves for the delivered 6, 8 and 10 mm (a, b) and for the 12 and 14 mm (c, d) thick S355J2+N base material slabs.

A.3 Thermal response of plates to sweep and burst excitation

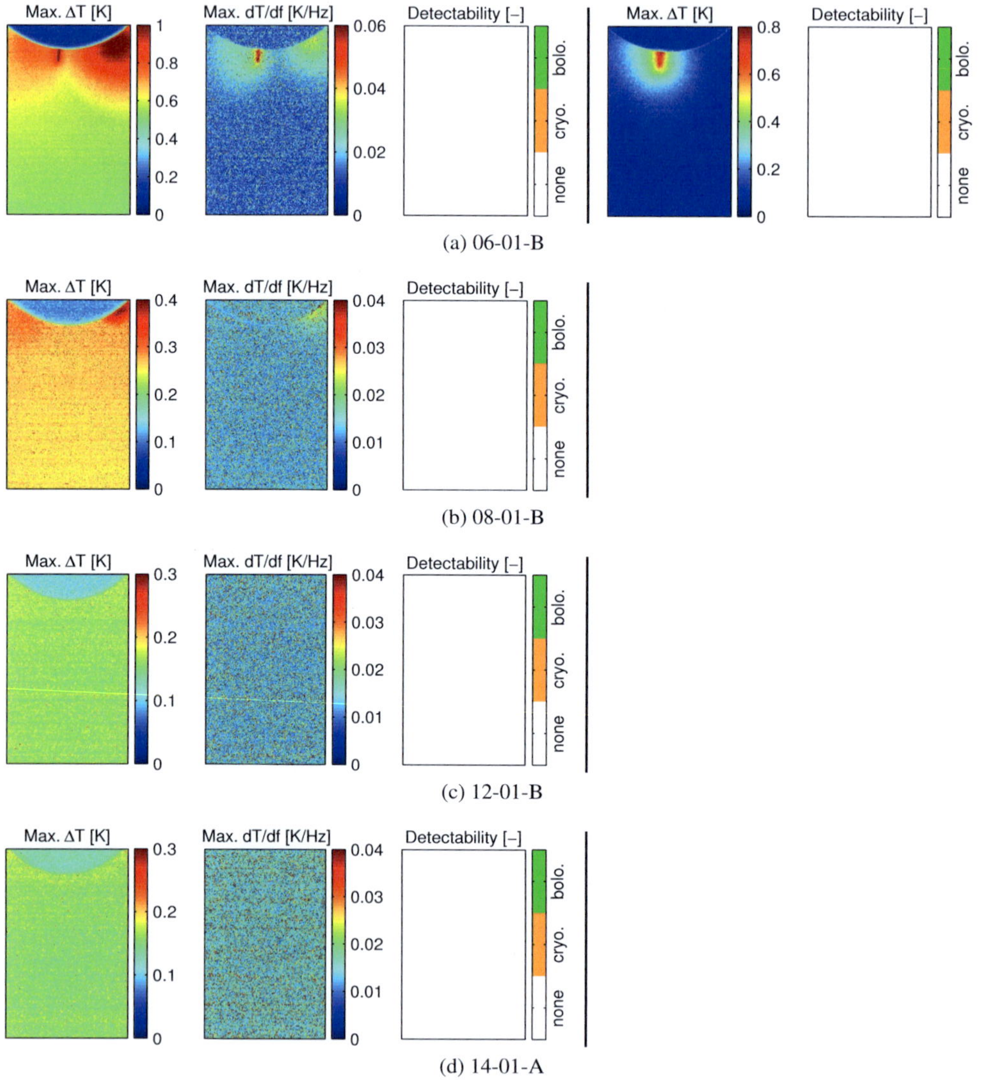

Figure A.2: Detection of 1 mm deep cracks; plate thicknesses 6, 8, 12 and 14 mm; left: strongest sweep excitation; right: strongest mono-frequency excitation.

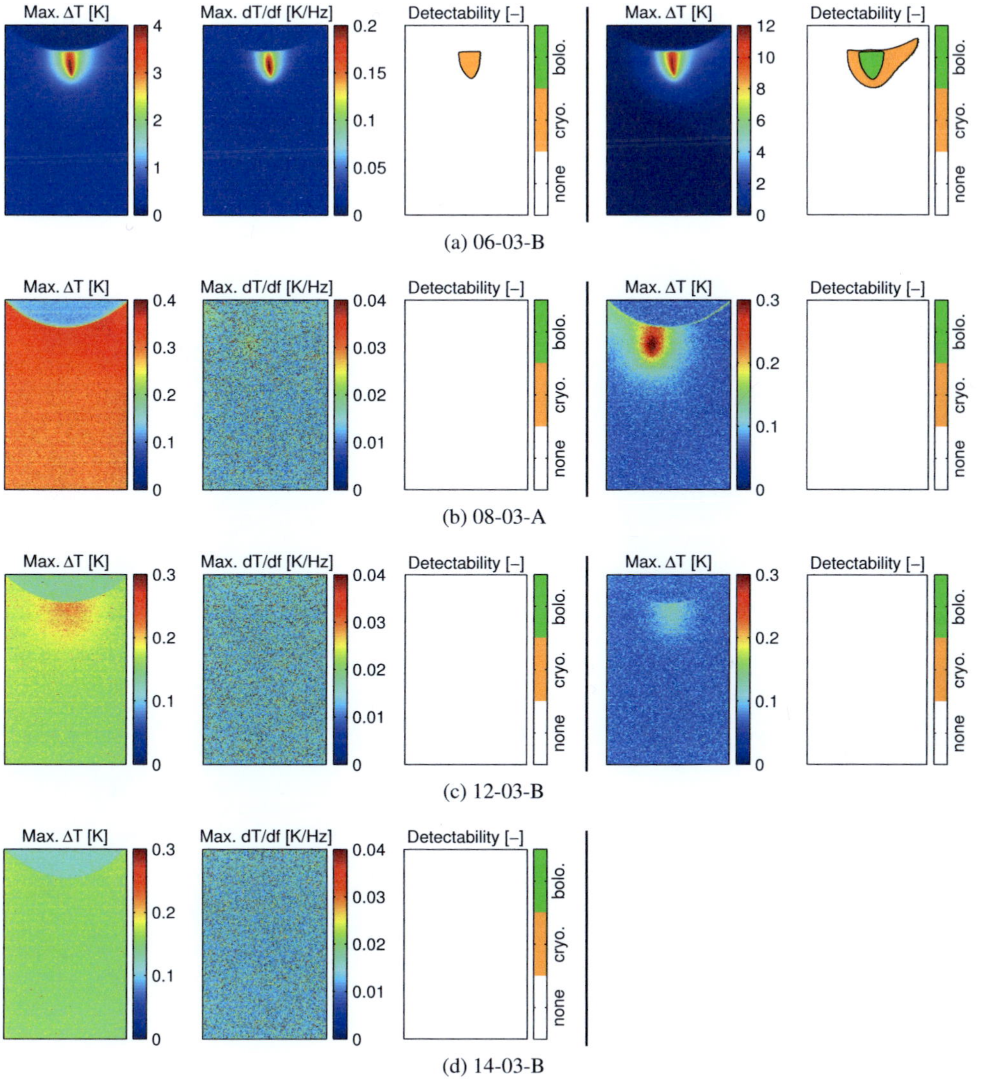

Figure A.3: Detection of 3 mm deep cracks; plate thicknesses 6, 8, 12 and 14 mm; left: strongest sweep excitation; right: strongest mono-frequency excitation.

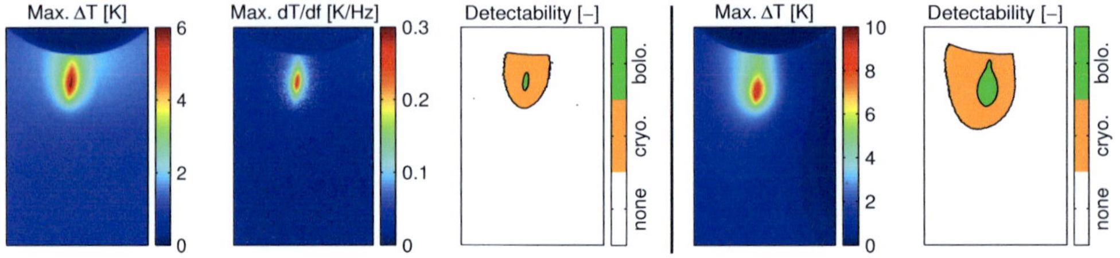

(a) 06-06-B

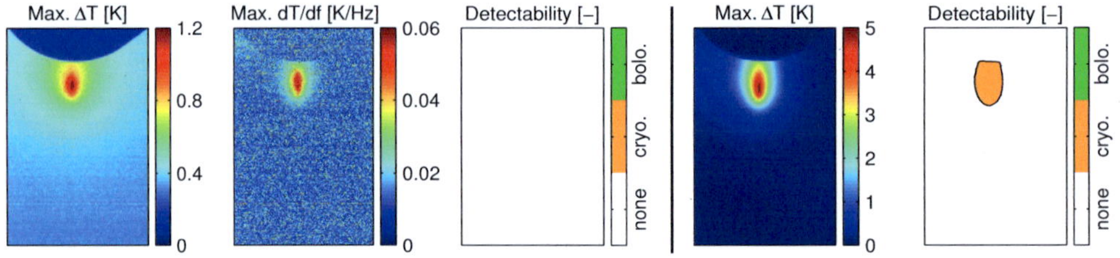

(b) 08-06-A

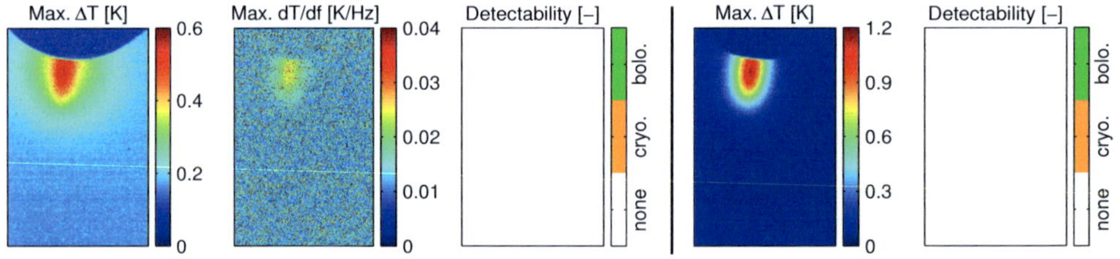

(c) 12-06-A

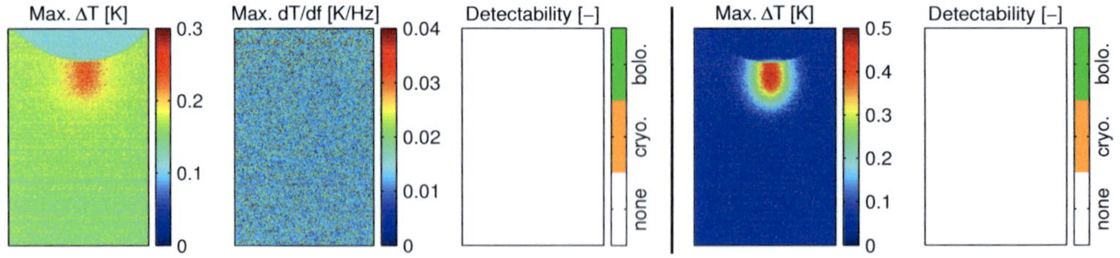

(d) 14-06-B

Figure A.4: Detection of 6 mm deep cracks; plate thicknesses 6, 8, 12 and 14 mm; left: strongest sweep excitation; right: strongest mono-frequency excitation.

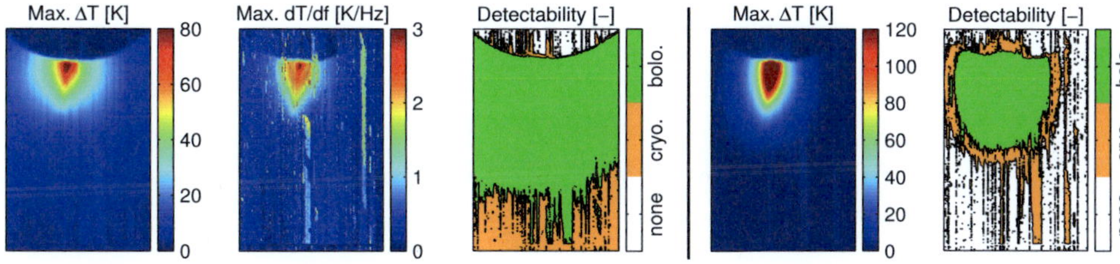

(a) 06-12-A

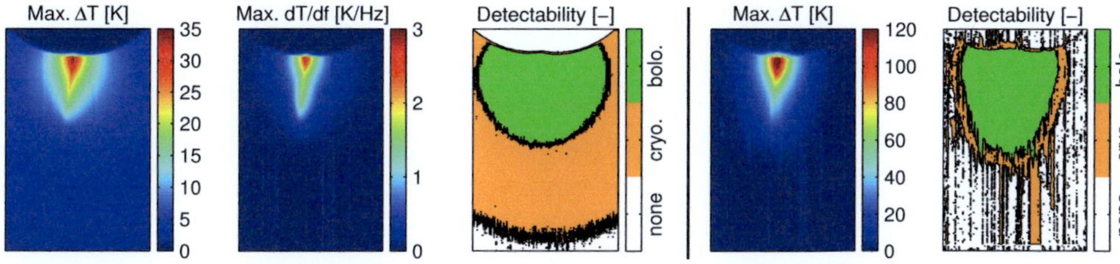

(b) 08-12-B

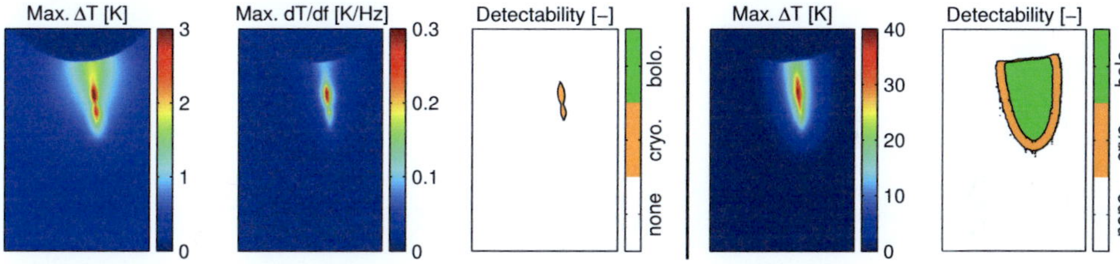

(c) 12-12-B

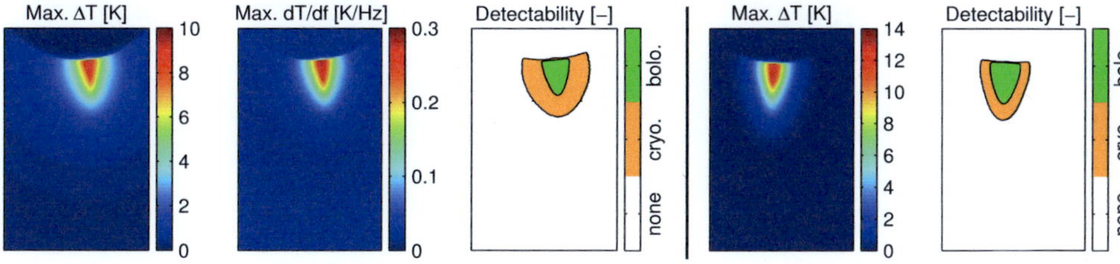

(d) 14-12-B, -A

Figure A.5: Detection of 12 mm deep cracks; plate thicknesses 6, 8, 12 and 14 mm; left: strongest sweep excitation; right: strongest mono-frequency excitation.

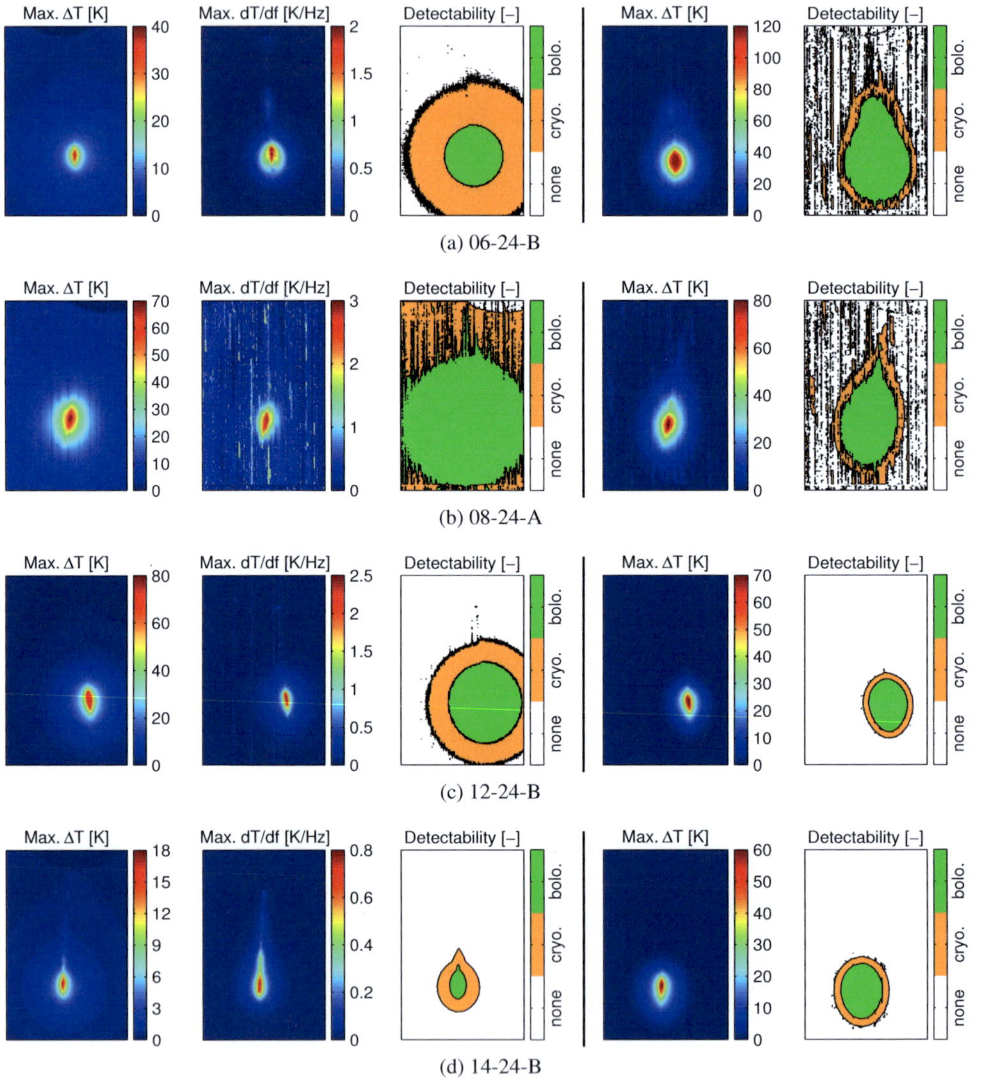

Figure A.6: Detection of 24 mm deep cracks; plate thicknesses 6, 8, 12 and 14 mm; left: strongest sweep excitation; right: strongest mono-frequency excitation.

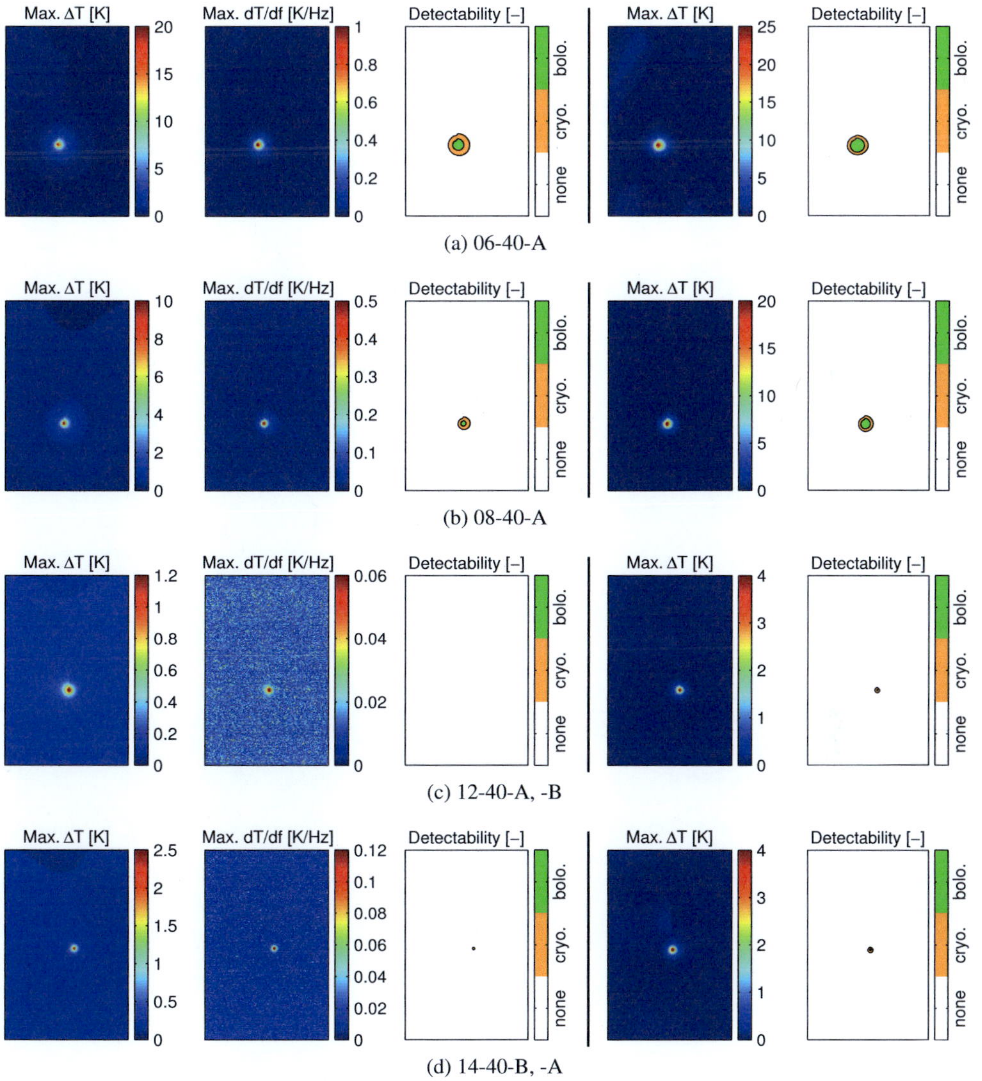

Figure A.7: Detection of 40 mm deep cracks; plate thicknesses 6, 8, 12 and 14 mm; left: strongest sweep excitation; right: strongest mono-frequency excitation.

A.4 Crack mouth opening displacement and defect heating under static prestress

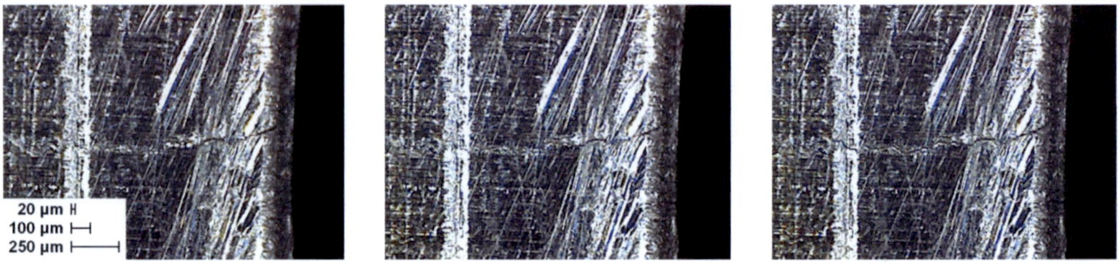

(a) Force: 0 kN (b) Force: 40 kN (c) Force: 80 kN

Figure A.8: Crack mouth opening displacement at varying machine forces; specimen 10-01-II-A.

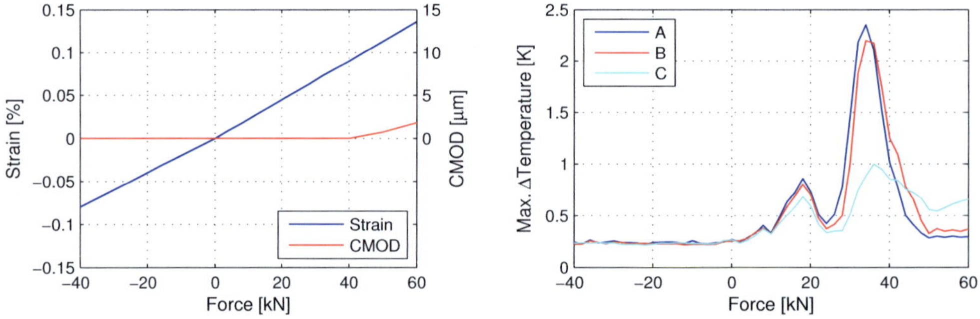

(a) Strain and CMOD (b) Max. ΔT over force

Figure A.9: Axial strain near the fillet edge, crack mouth opening displacement and max. ΔT at points A, B and C achieved at varying machine forces; specimen 10-01-II-A.

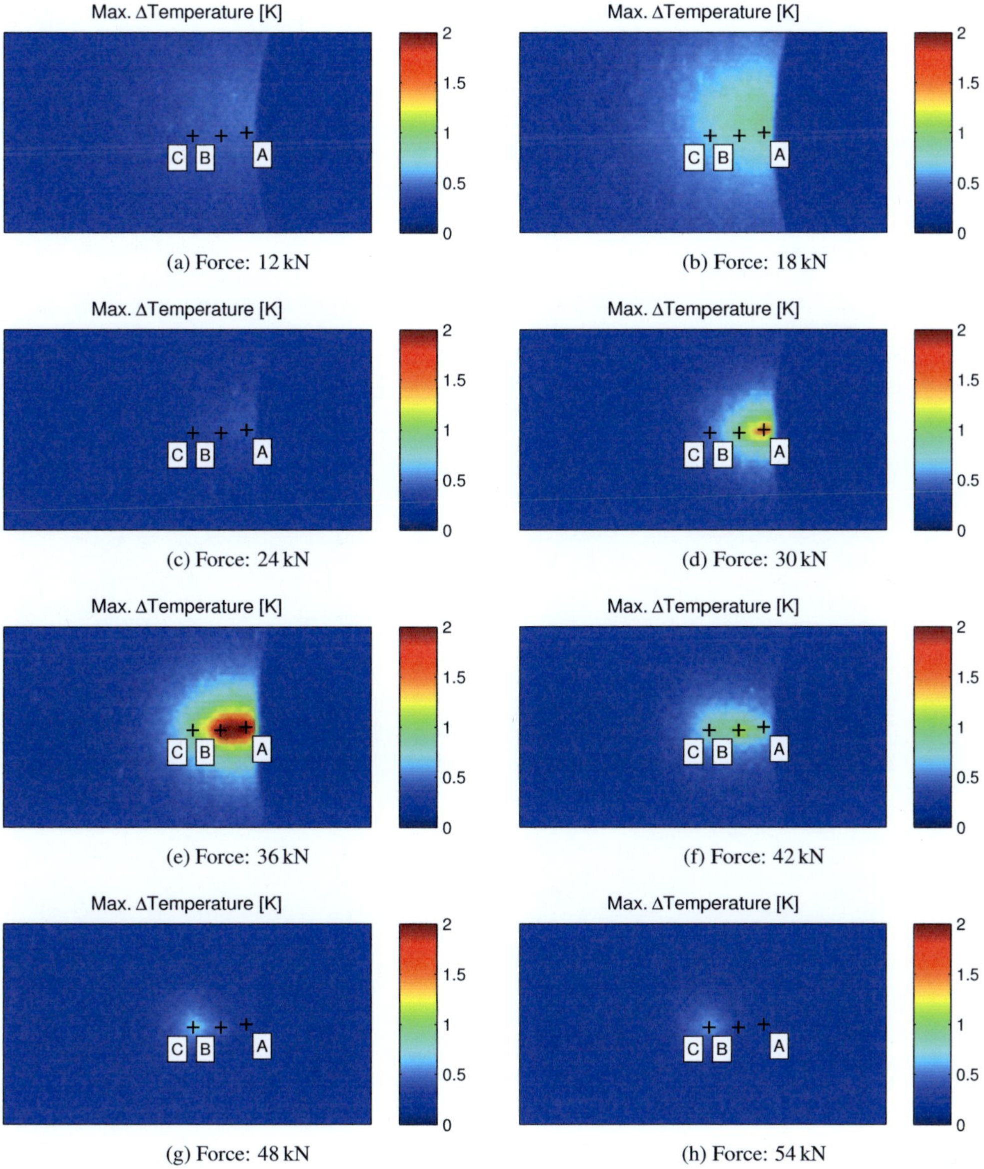

(a) Force: 12 kN
(b) Force: 18 kN
(c) Force: 24 kN
(d) Force: 30 kN
(e) Force: 36 kN
(f) Force: 42 kN
(g) Force: 48 kN
(h) Force: 54 kN

Figure A.10: Distribution of max. ΔT achieved at different preloading states; specimen 10-01-II-A.

(a) Force: 0 kN (b) Force: 40 kN (c) Force: 80 kN

Figure A.11: Crack mouth opening displacement at varying machine forces; specimen 10-03-B.

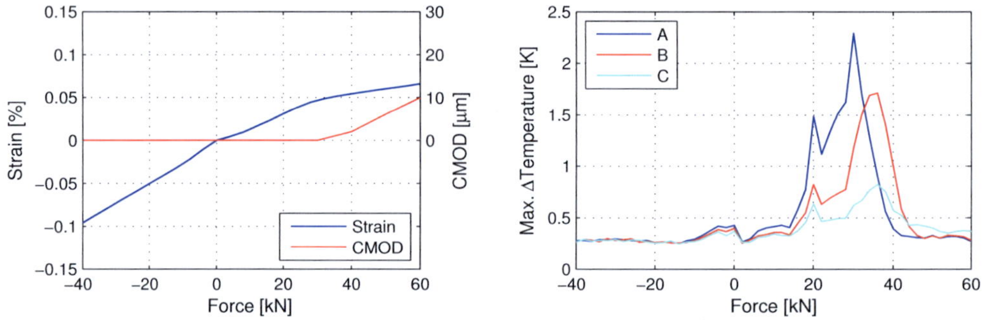

(a) Strain and CMOD (b) Max. ΔT over force

Figure A.12: Axial strain near the fillet edge, crack mouth opening displacement and max. ΔT at points A, B, and C achieved at varying machine forces; specimen 10-03-B.

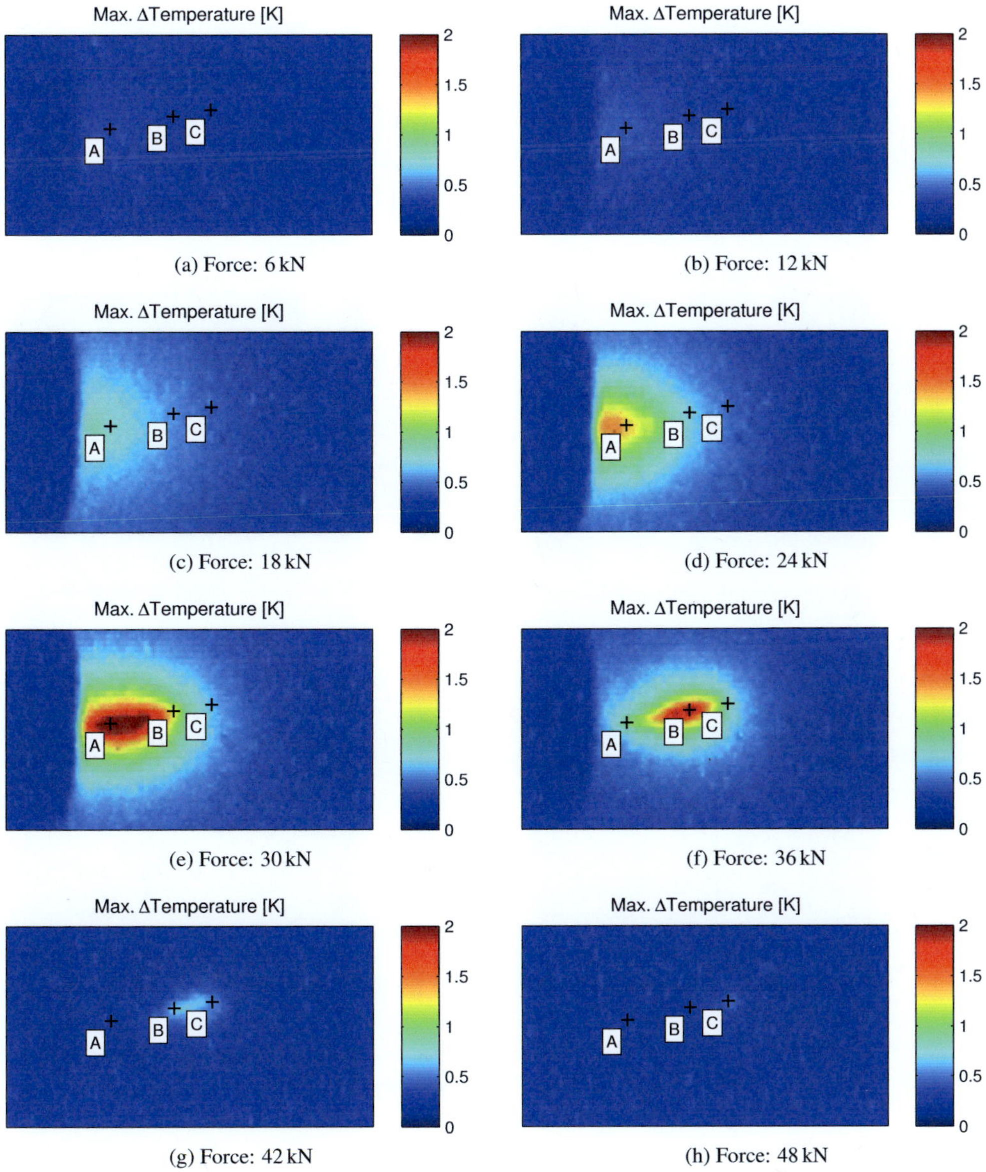

(a) Force: 6 kN
(b) Force: 12 kN
(c) Force: 18 kN
(d) Force: 24 kN
(e) Force: 30 kN
(f) Force: 36 kN
(g) Force: 42 kN
(h) Force: 48 kN

Figure A.13: Distribution of max. ΔT achieved at different preloading states; specimen 10-03-B.

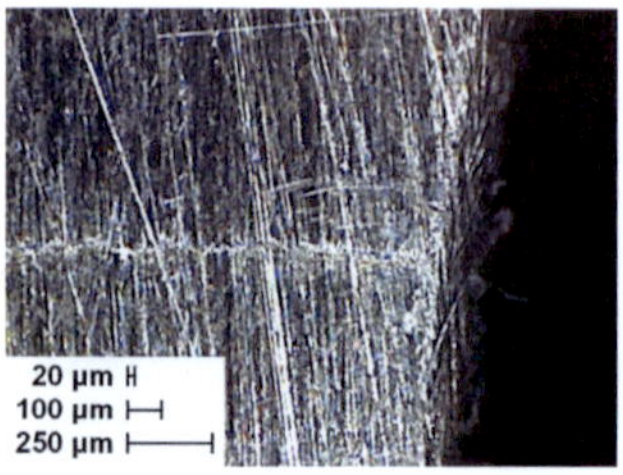

(a) Force: −40 kN

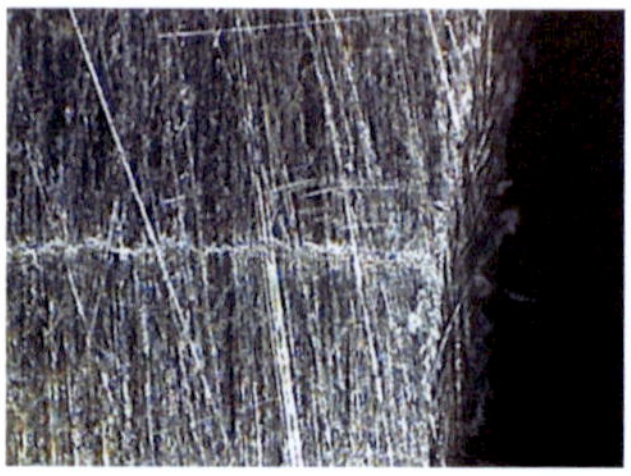

(b) Force: −20 kN

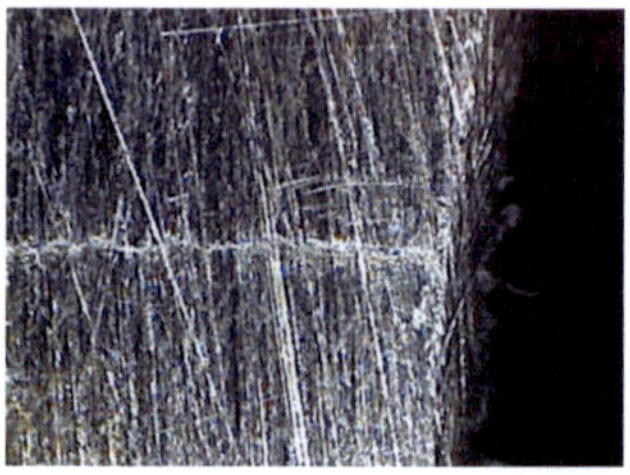

(c) Force: 0 kN

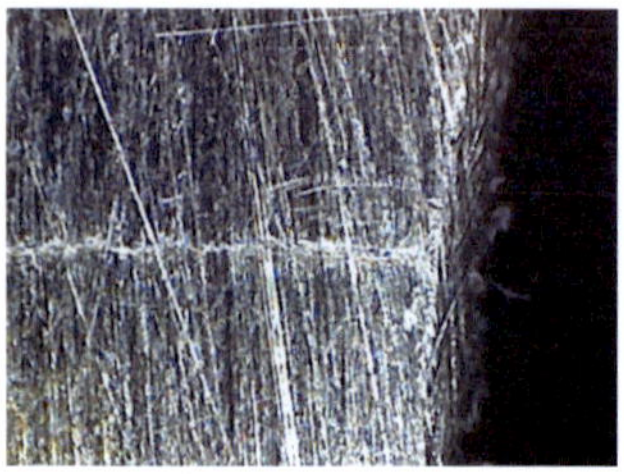

(d) Force: 20 kN

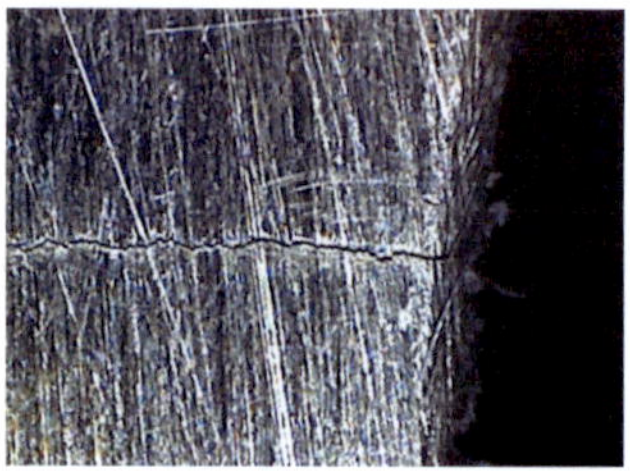

(e) Force: 40 kN

(f) Force: 60 kN

Figure A.14: Crack mouth opening displacement at varying machine forces; specimen 10-06-II-A.

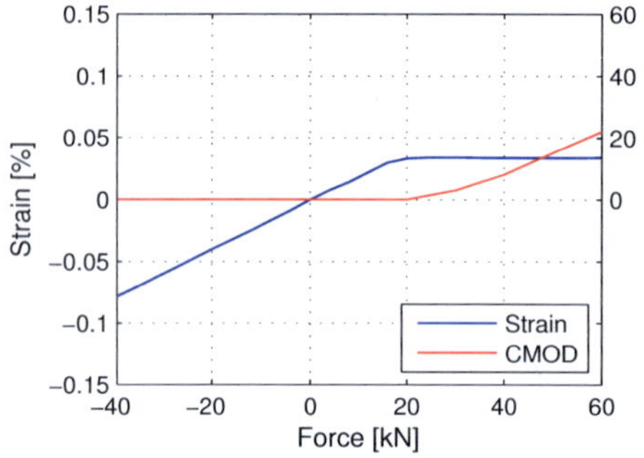

(a) Strain and CMOD

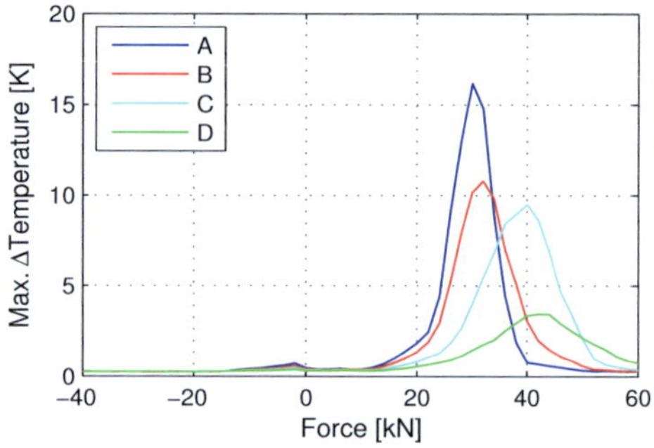

(b) Max. ΔT over force

Figure A.15: Axial strain near the fillet edge, crack mouth opening displacement and max. ΔT at points A, B, C and D achieved at varying machine forces; specimen 10-06-II-A.

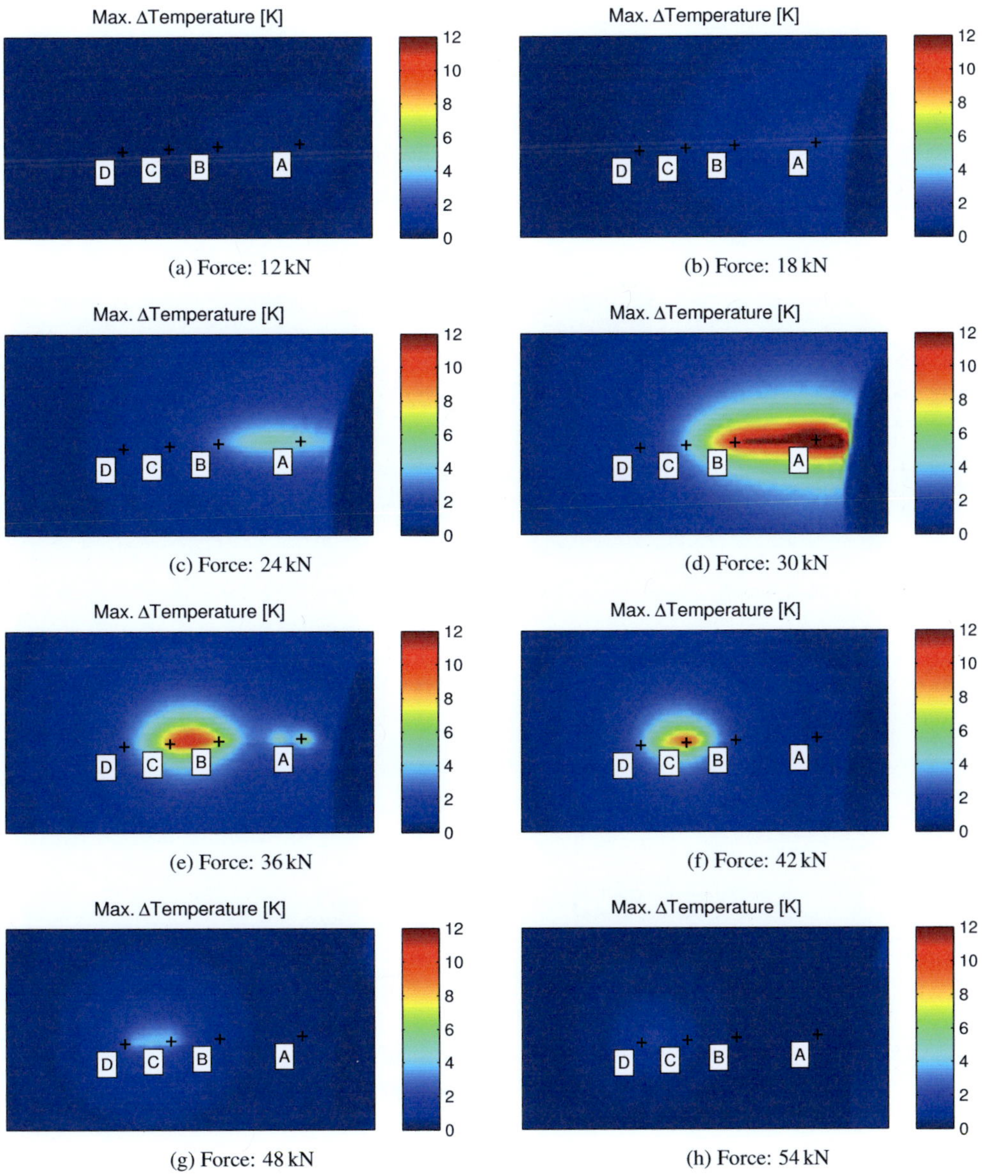

(a) Force: 12 kN
(b) Force: 18 kN
(c) Force: 24 kN
(d) Force: 30 kN
(e) Force: 36 kN
(f) Force: 42 kN
(g) Force: 48 kN
(h) Force: 54 kN

Figure A.16: Distribution of max. ΔT achieved at different preloading states; specimen 10-06-II-A.

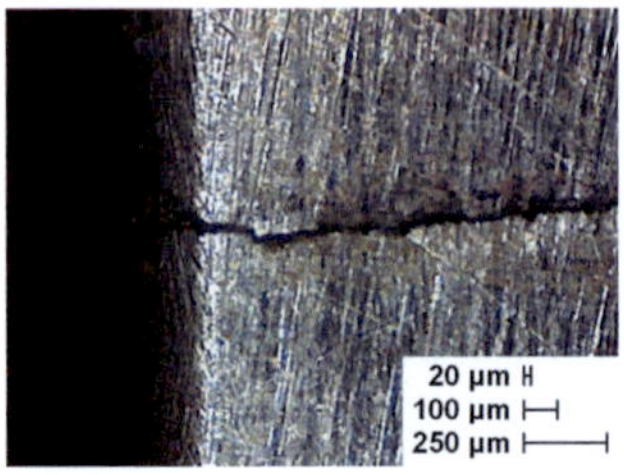

(a) Force: -40 kN

(b) Force: -20 kN

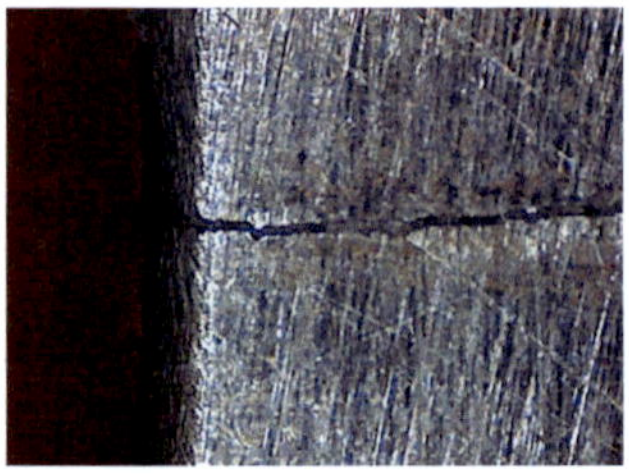

(c) Force: 0 kN

(d) Force: 20 kN

(e) Force: 40 kN

(f) Force: 60 kN

Figure A.17: Crack mouth opening displacement at varying machine forces; specimen 10-12-III-B.

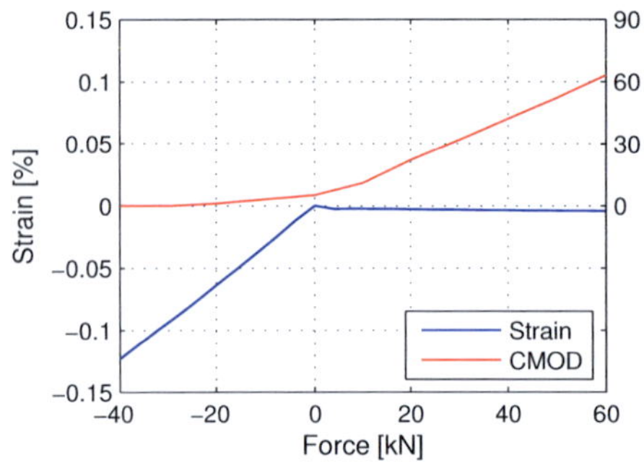

(a) Strain and CMOD

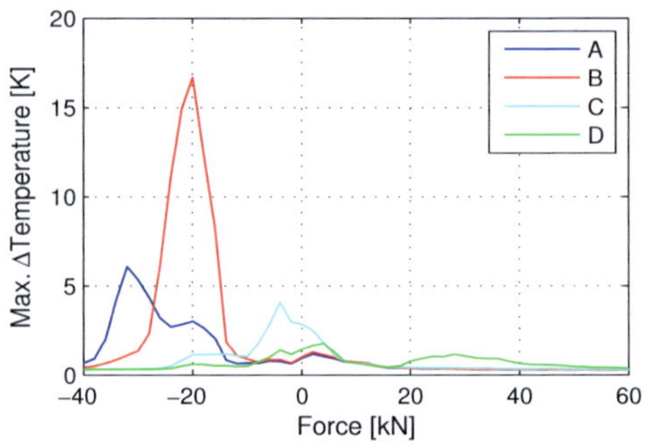

(b) Max. ΔT over force

Figure A.18: Axial strain near the fillet edge, crack mouth opening displacement and max. ΔT at points A, B, C and D achieved at varying machine forces; specimen 10-12-III-B.

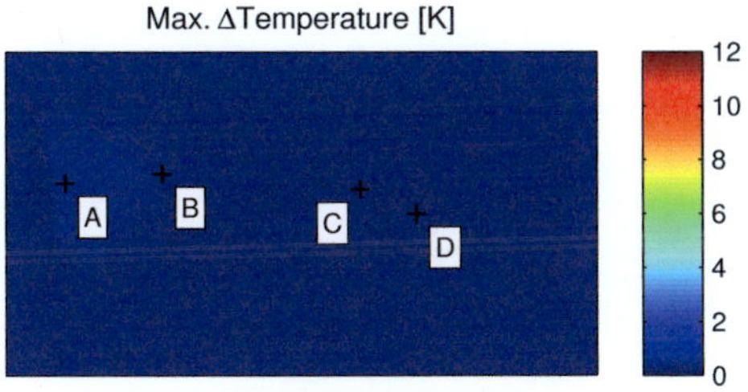

(a) Force: −42 kN

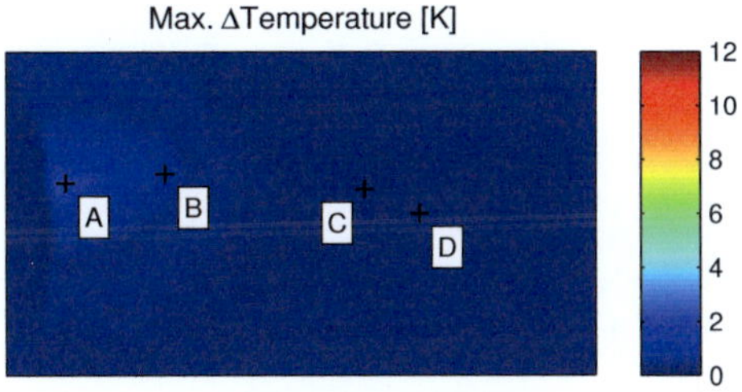

(b) Force: −36 kN

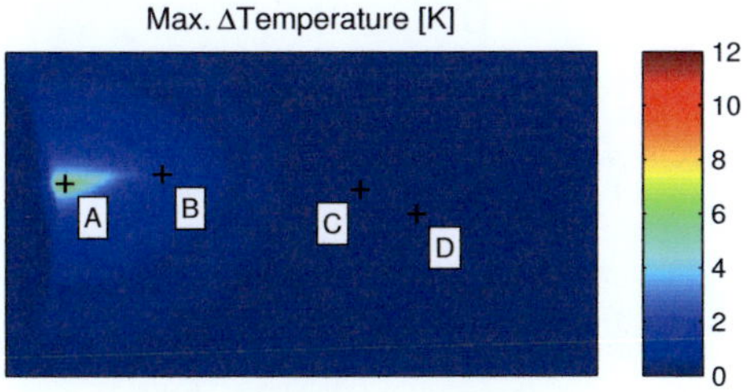

(c) Force: −30 kN

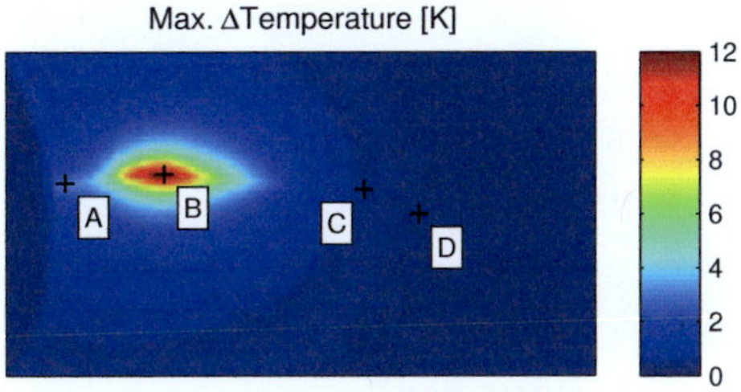

(d) Force: −24 kN

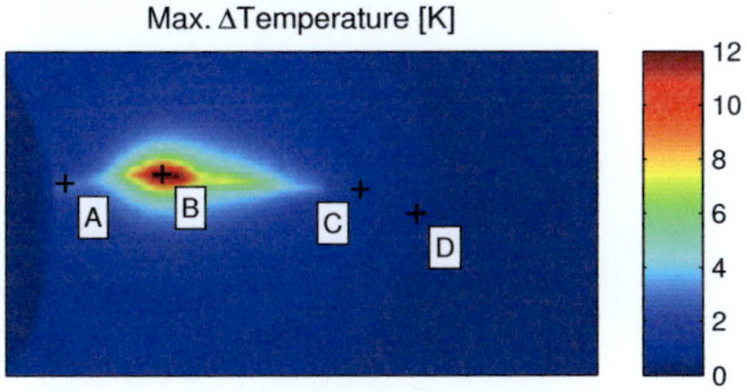

(e) Force: −18 kN

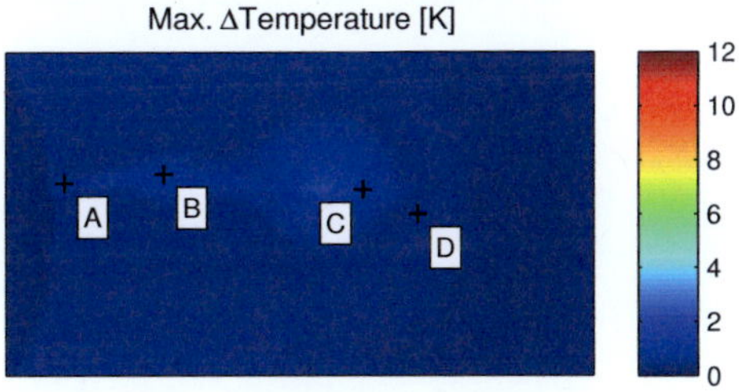

(f) Force: −12 kN

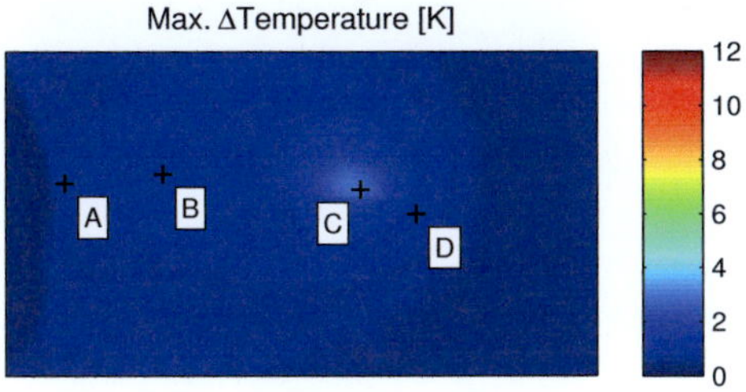

(g) Force: −6 kN

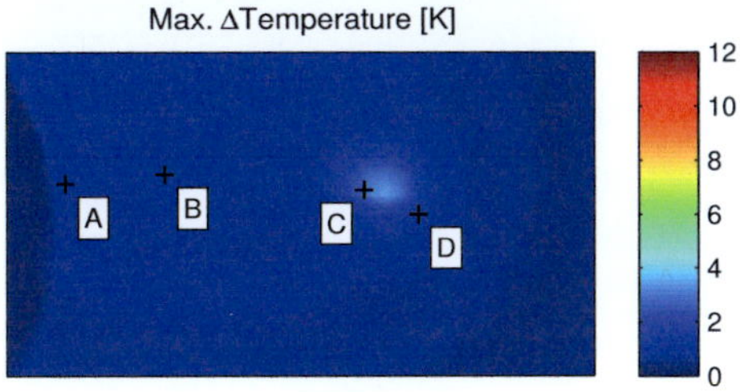

(h) Force: 0 kN

Figure A.19: Distribution of max. ΔT achieved at different preloading states; specimen 10-12-III-B.

(a) Force: −80 kN

(b) Force: −60 kN

(c) Force: −40 kN

(d) Force: −20 kN

(e) Force: 0 kN

(f) Force: 20 kN

Figure A.20: Crack mouth opening displacement at varying machine forces; specimen 10-12-V-B.

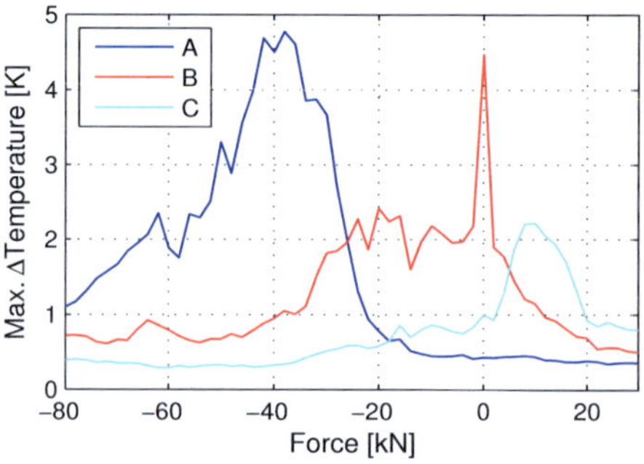

Figure A.21: Maximum temperature increase at points A, B and C achieved at varying machine forces; specimen 10-12-V-B.

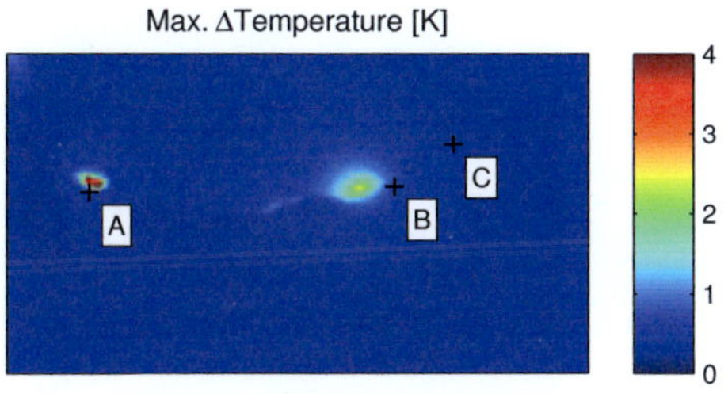

(a) Force: −80 kN

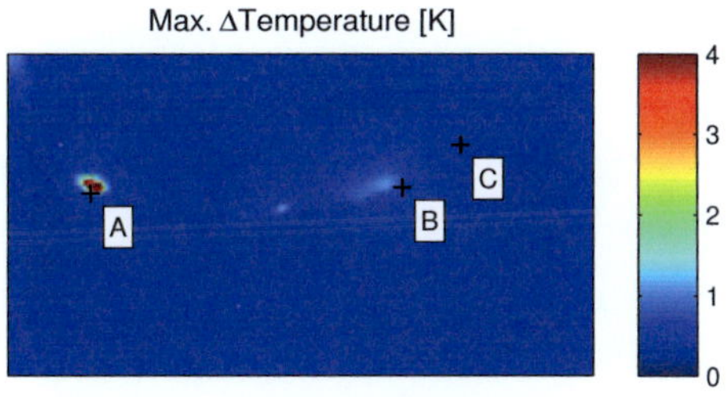

(b) Force: −68 kN

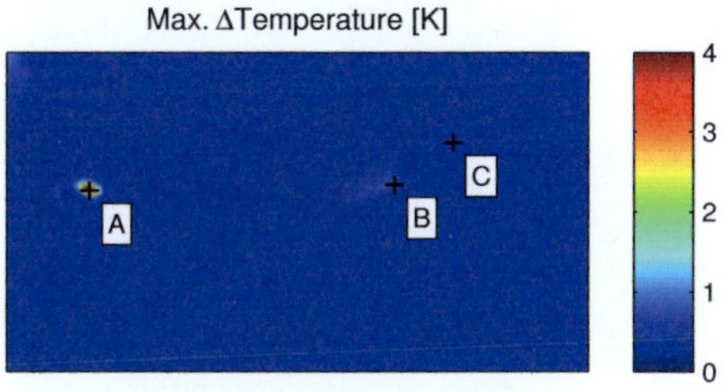

(c) Force: −56 kN

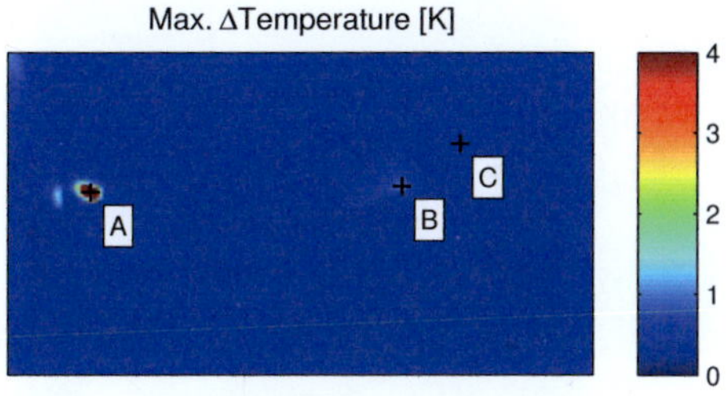

(d) Force: −44 kN

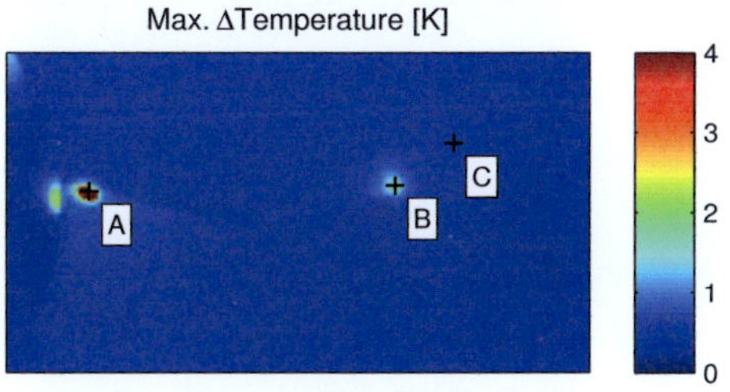

(e) Force: −32 kN

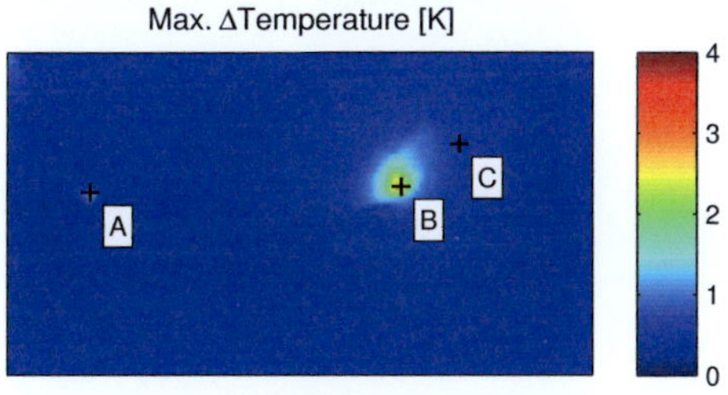

(f) Force: −20 kN

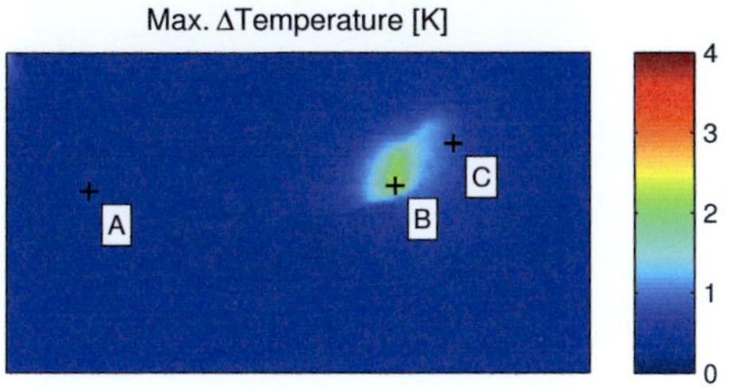

(g) Force: −8 kN

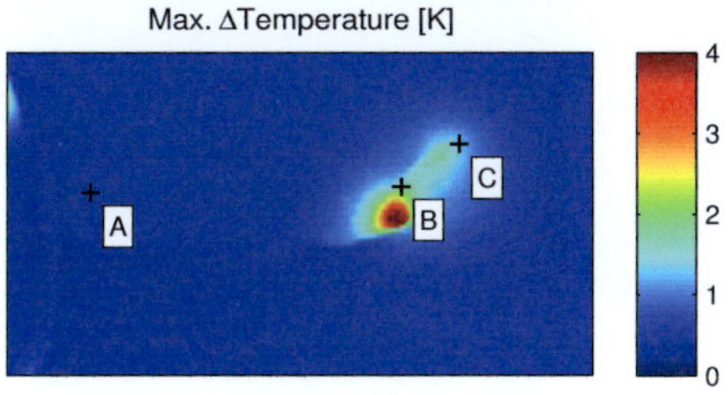

(h) Force: 4 kN

Figure A.22: Distribution of max. ΔT achieved at different preloading states; specimen 10-12-V-B.

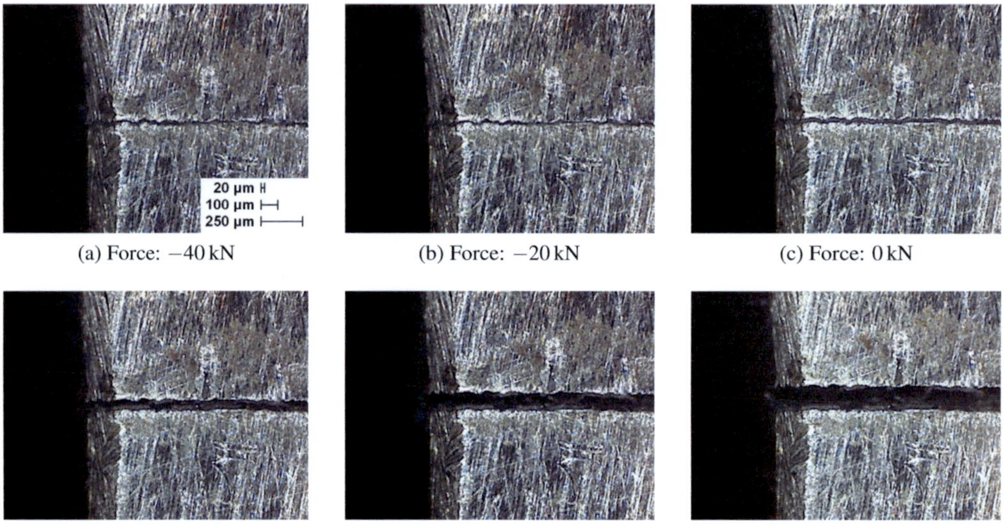

(a) Force: −40 kN (b) Force: −20 kN (c) Force: 0 kN

(d) Force: 20 kN (e) Force: 40 kN (f) Force: 60 kN

Figure A.23: Crack mouth opening displacement at varying machine forces; specimen 10-24-II-B.

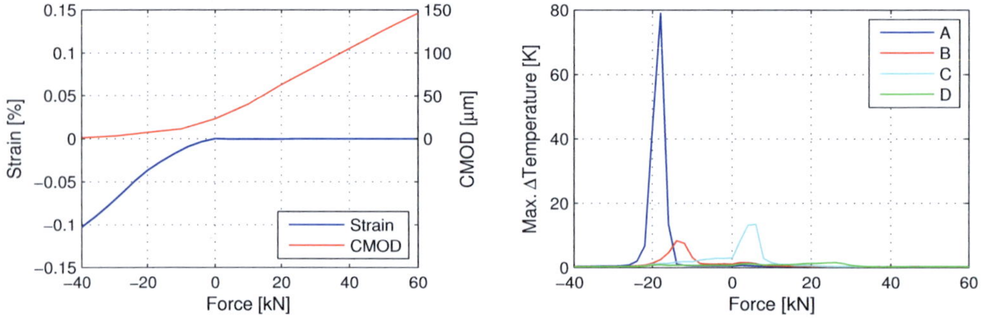

(a) Strain and CMOD (b) Max. ΔT over force

Figure A.24: Axial strain near the fillet edge, crack mouth opening displacement and max. ΔT at points A, B, C and D achieved at varying machine forces; specimen 10-24-II-B.

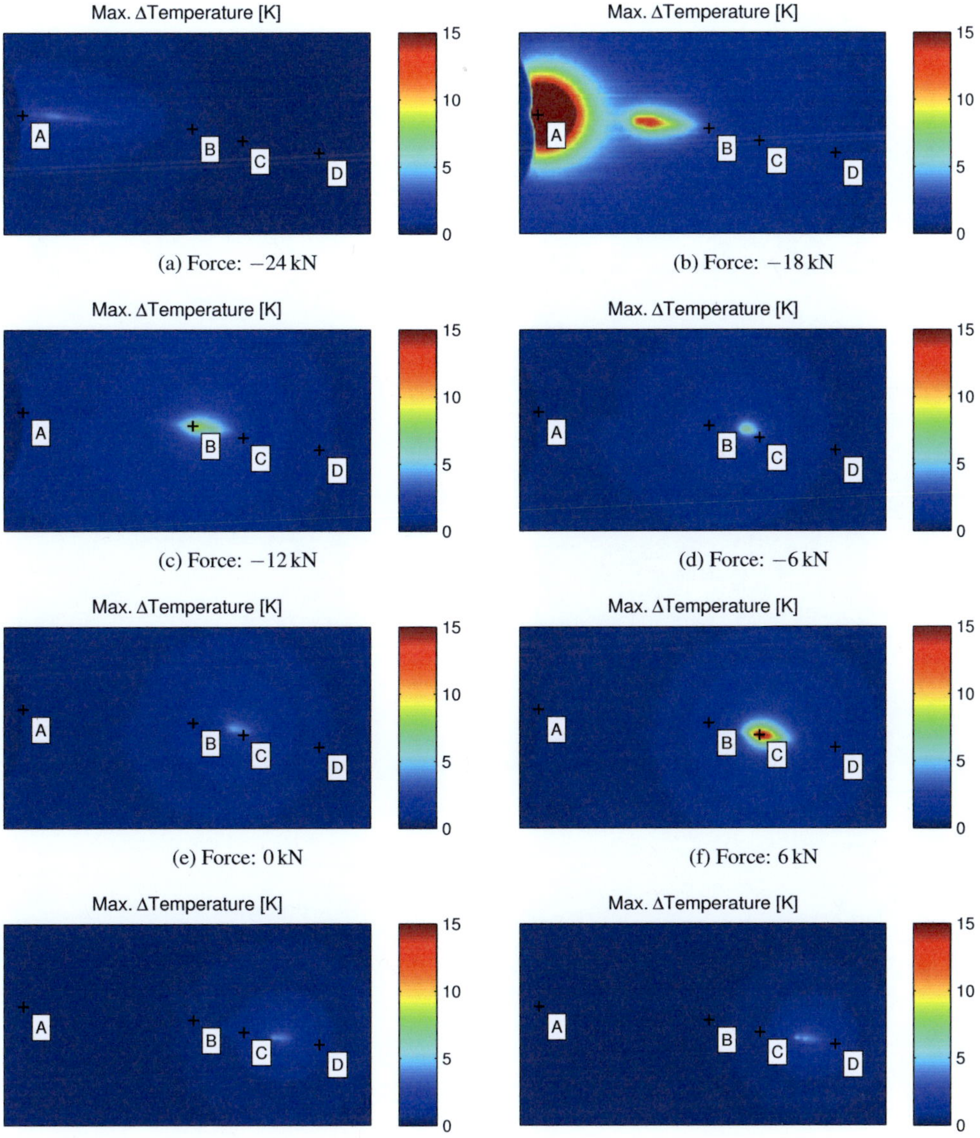

(a) Force: -24 kN
(b) Force: -18 kN
(c) Force: -12 kN
(d) Force: -6 kN
(e) Force: 0 kN
(f) Force: 6 kN
(g) Force: 12 kN
(h) Force: 18 kN

Figure A.25: Distribution of max. ΔT achieved at different preloading states; specimen 10-24-II-B.

A.5 Roughness characterisation of fracture surfaces

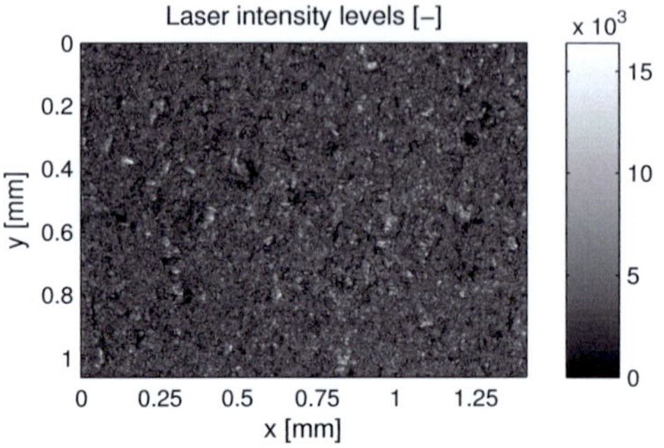

(a) Laser intensity image (10× magnification)

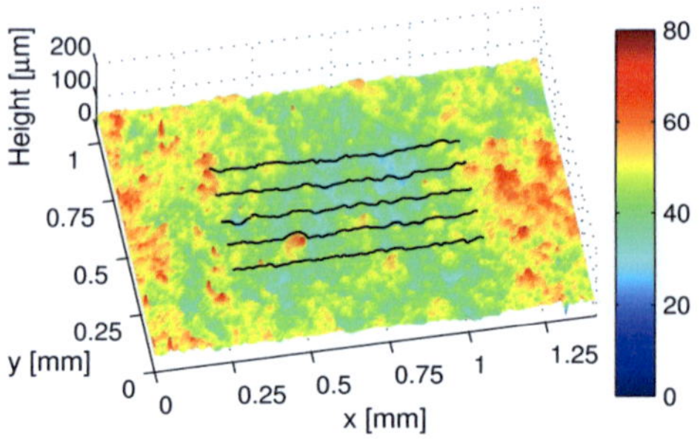

(b) Filtered map of surface height values

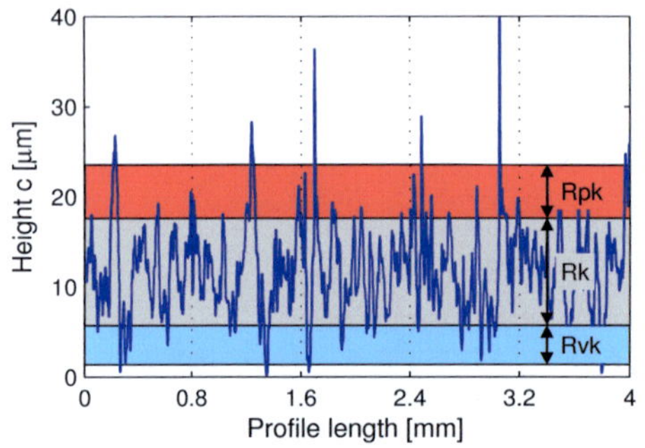

(c) Roughness profile

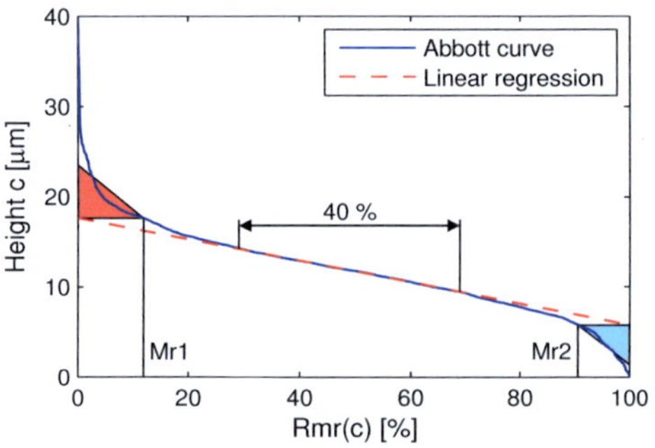

(d) Abbott-Firestone (BAC) curve

Figure A.26: Microscopic roughness measurement and evaluation; specimen M-I, position 2 (regular fracture surface).

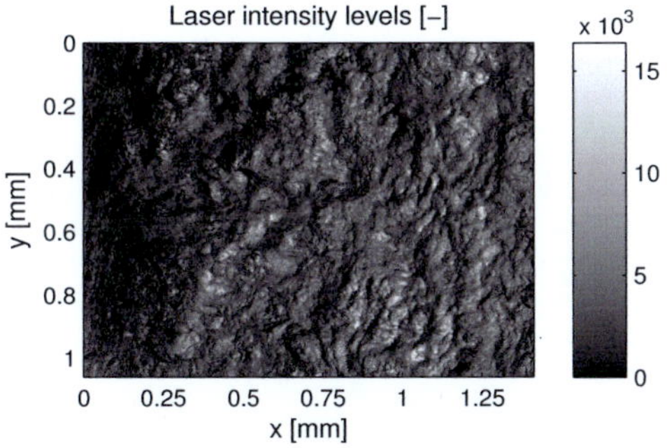

(a) Laser intensity image (10× magnification)

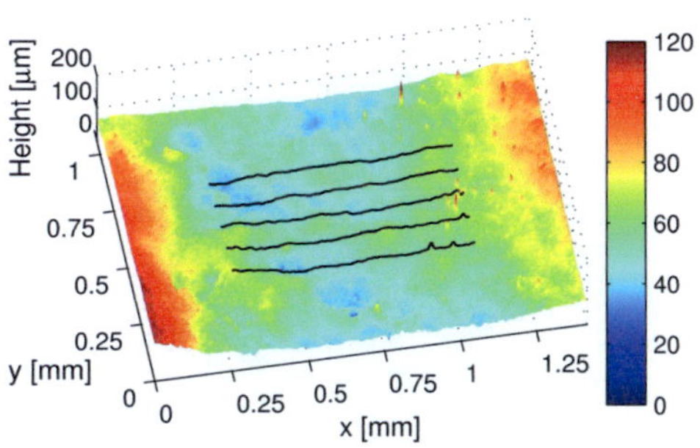

(b) Filtered map of surface height values

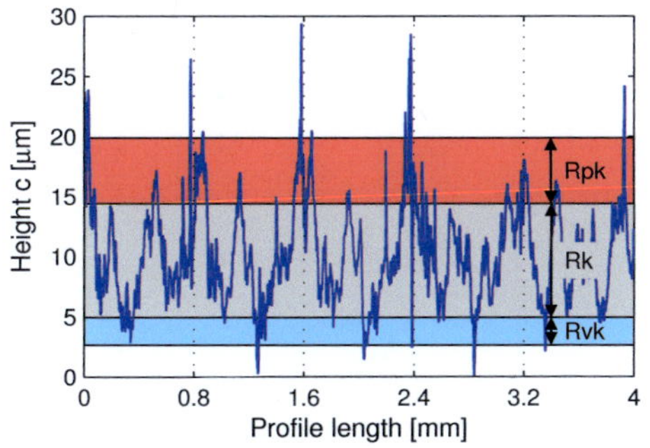

(c) Roughness profile

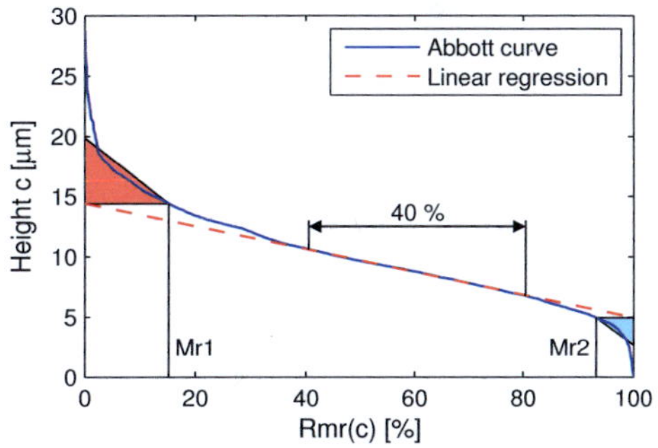

(d) Abbott-Firestone (BAC) curve

Figure A.27: Microscopic roughness measurement and evaluation; specimen M-I, position 1 (metallic/oxidised region).

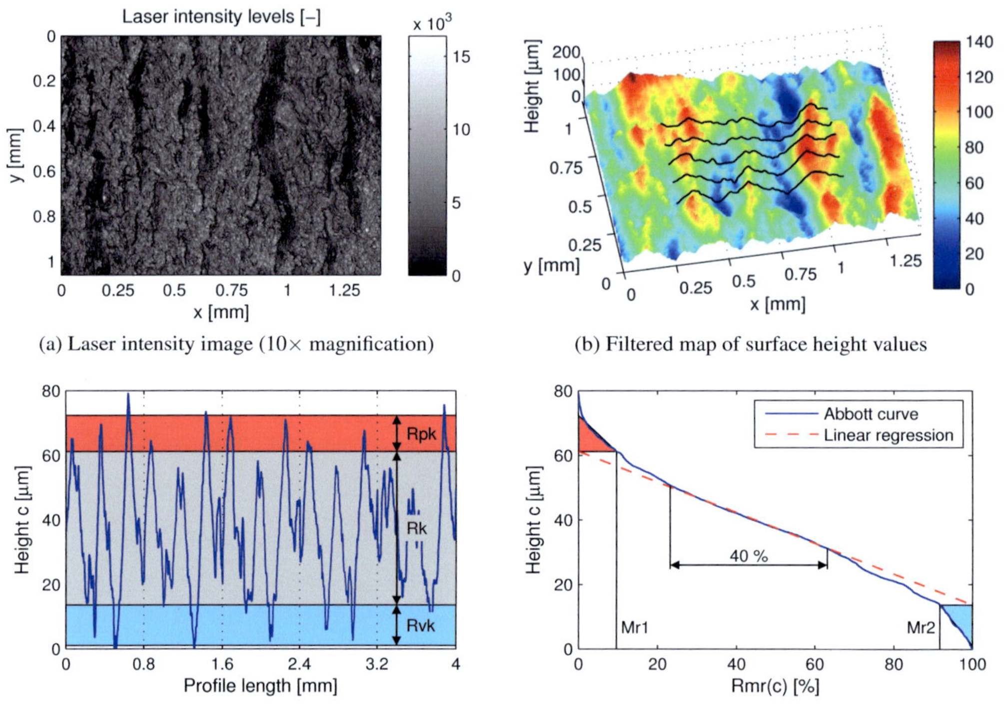

(a) Laser intensity image (10× magnification)

(b) Filtered map of surface height values

(c) Roughness profile

(d) Abbott-Firestone (BAC) curve

Figure A.28: Microscopic roughness measurement and evaluation; specimen 10-24-I, position 2 (regular fracture surface).

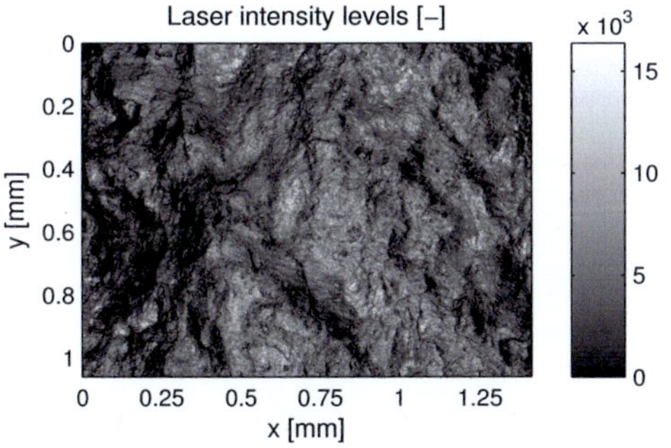

(a) Laser intensity image (10× magnification)

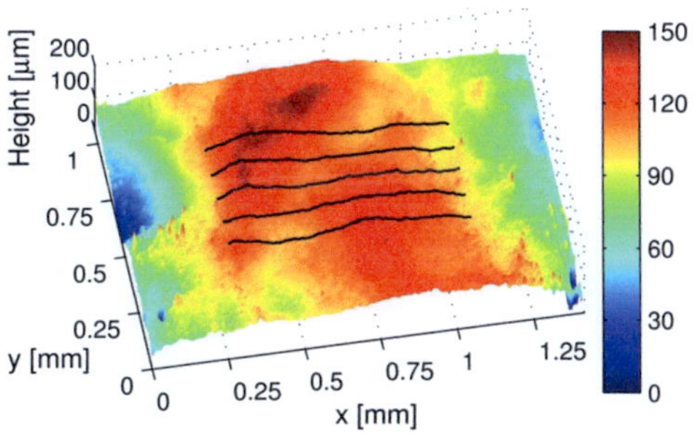

(b) Filtered map of surface height values

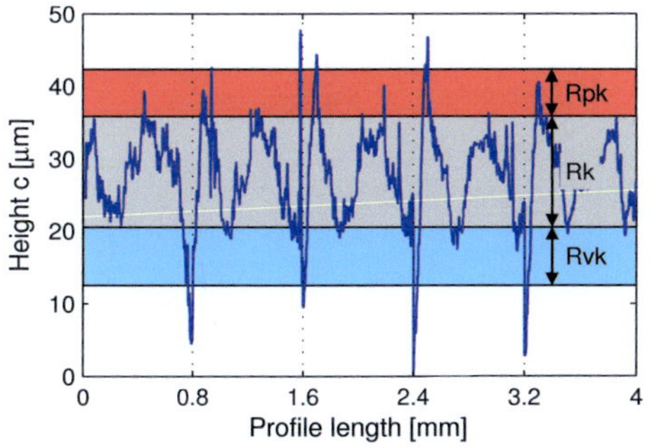

(c) Roughness profile

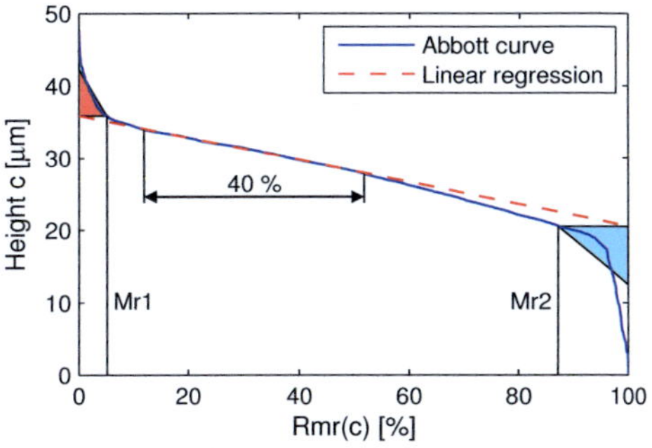

(d) Abbott-Firestone (BAC) curve

Figure A.29: Microscopic roughness measurement and evaluation; specimen 10-24-I, position 1 (metallic/oxidised region).

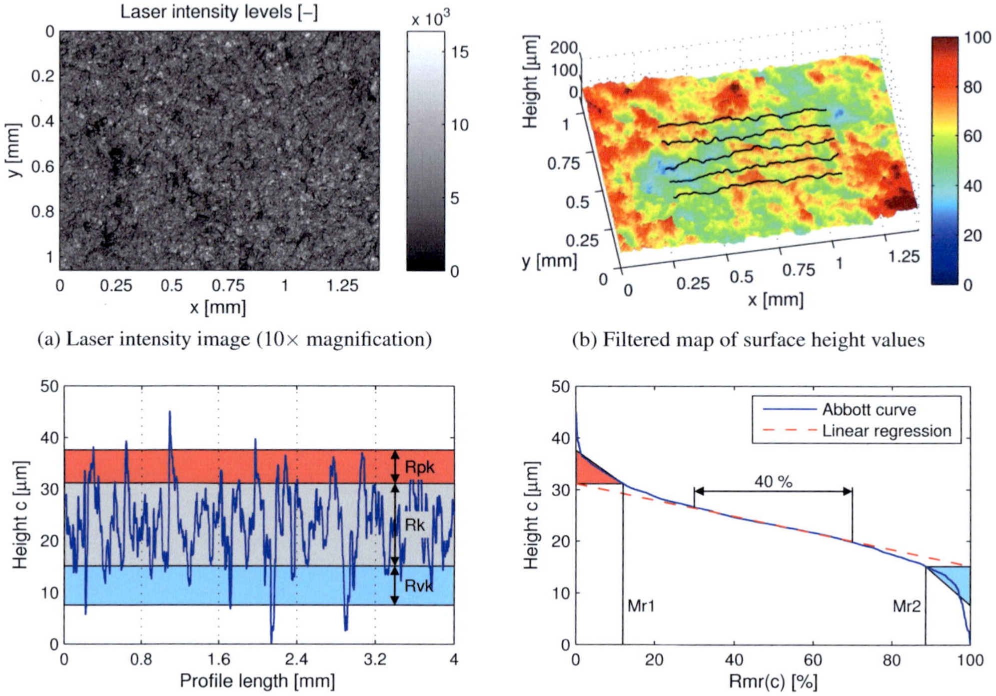

(a) Laser intensity image (10× magnification)

(b) Filtered map of surface height values

(c) Roughness profile

(d) Abbott-Firestone (BAC) curve

Figure A.30: Microscopic roughness measurement and evaluation; specimen M-IV, position 2 (regular fracture surface).

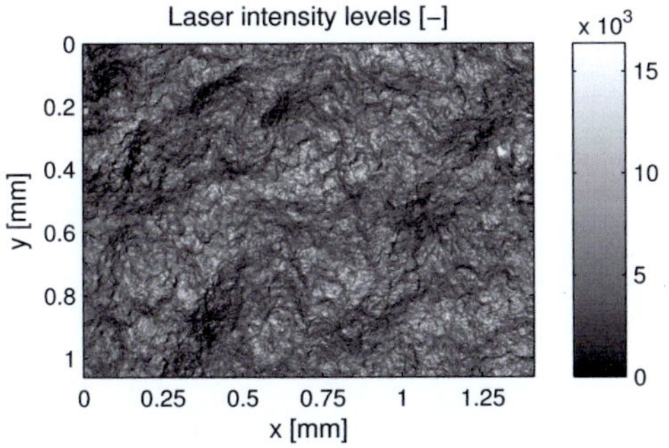

(a) Laser intensity image (10× magnification)

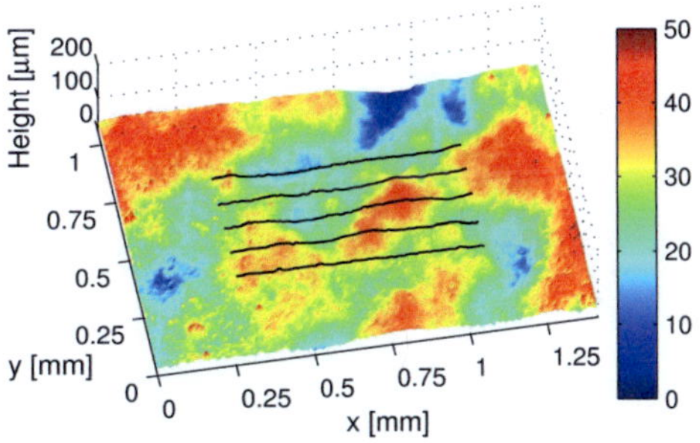

(b) Filtered map of surface height values

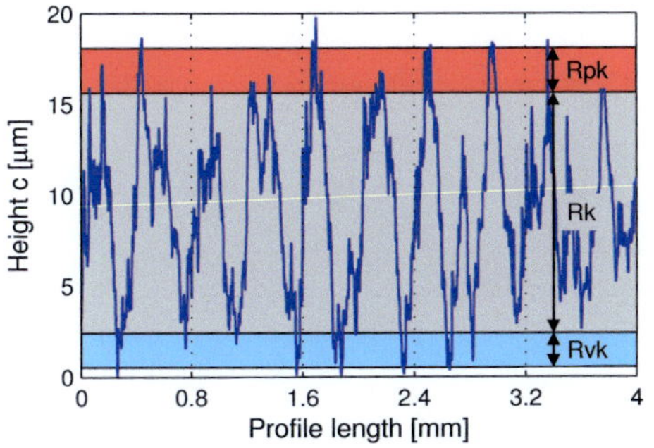

(c) Roughness profile

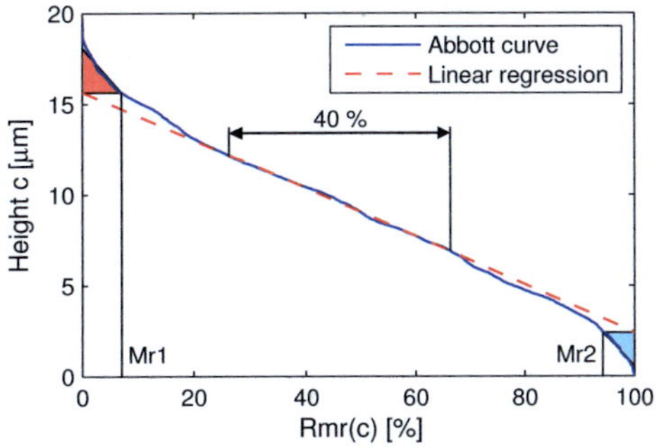

(d) Abbott-Firestone (BAC) curve

Figure A.31: Microscopic roughness measurement and evaluation; specimen M-IV, position 1 (metallic/oxidised region).

A.6 Harmonic distortion of ultrasound transducers

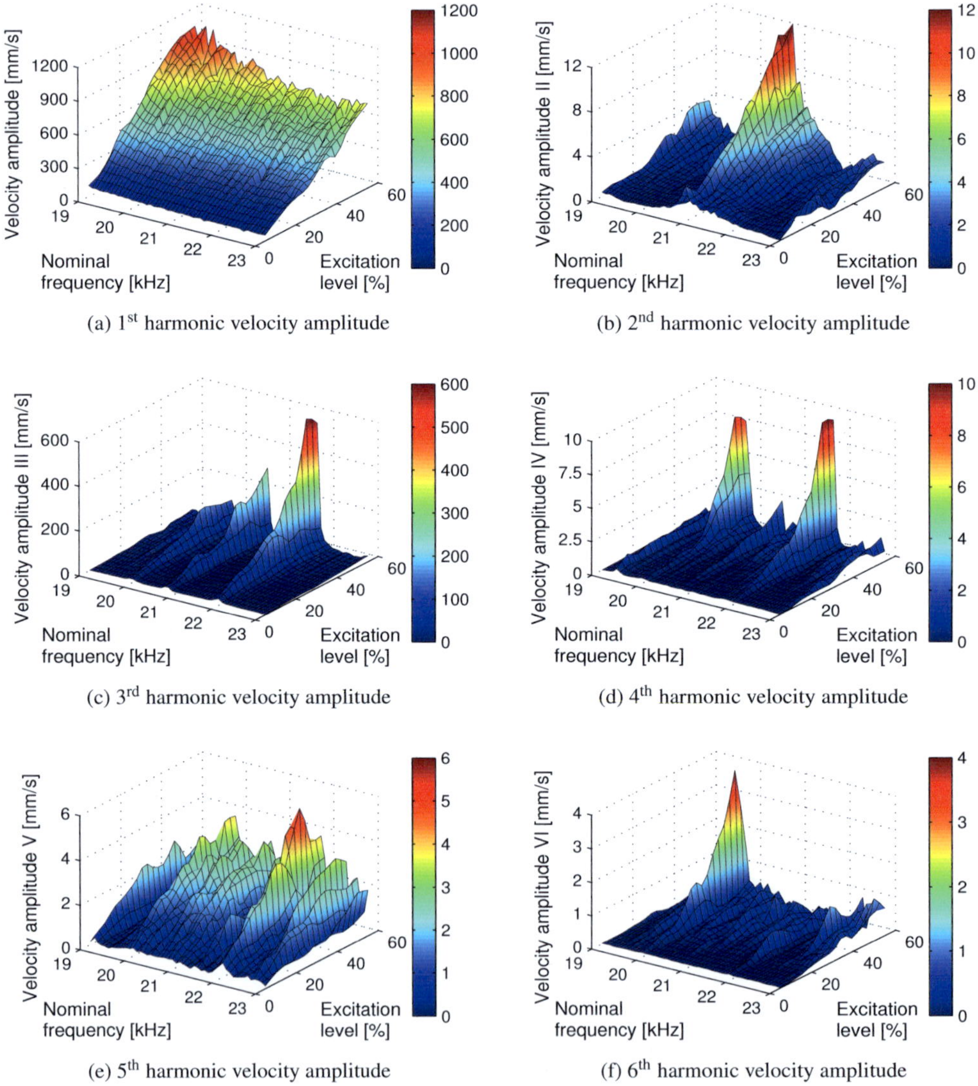

(a) 1st harmonic velocity amplitude

(b) 2nd harmonic velocity amplitude

(c) 3rd harmonic velocity amplitude

(d) 4th harmonic velocity amplitude

(e) 5th harmonic velocity amplitude

(f) 6th harmonic velocity amplitude

Figure A.32: Velocity amplitudes of the Branson transducer according to the 1st to 6th harmonic (fundamental tone and 2nd to 5th overtone).

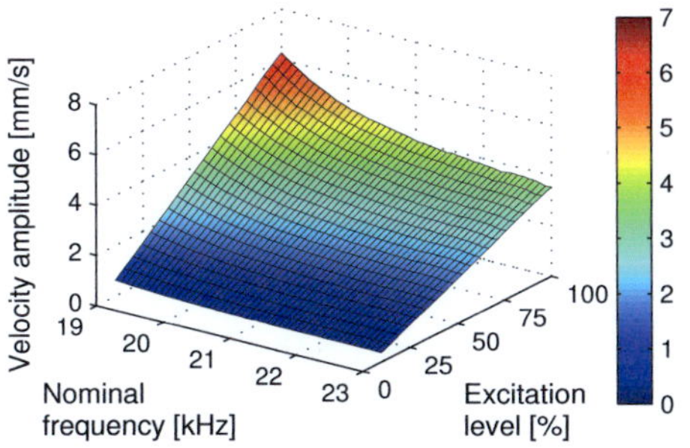

(a) 1st harmonic velocity amplitude

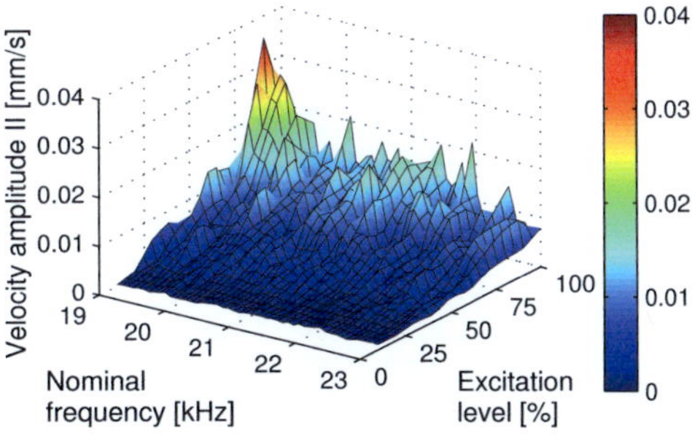

(b) 2nd harmonic velocity amplitude

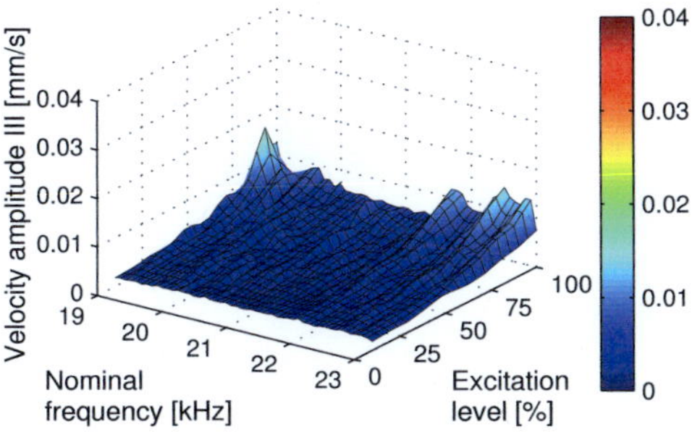

(c) 3rd harmonic velocity amplitude

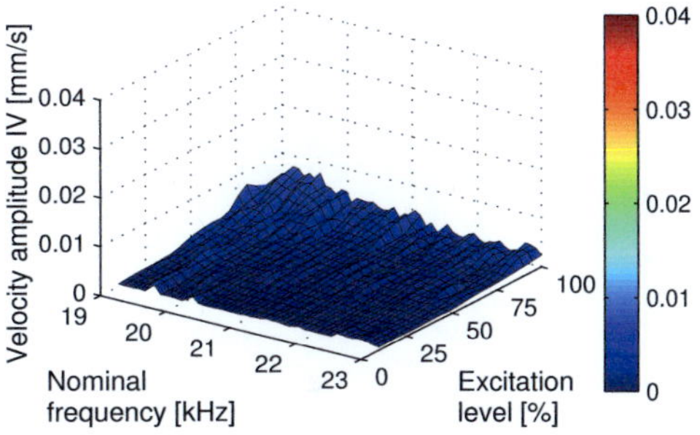

(d) 4th harmonic velocity amplitude

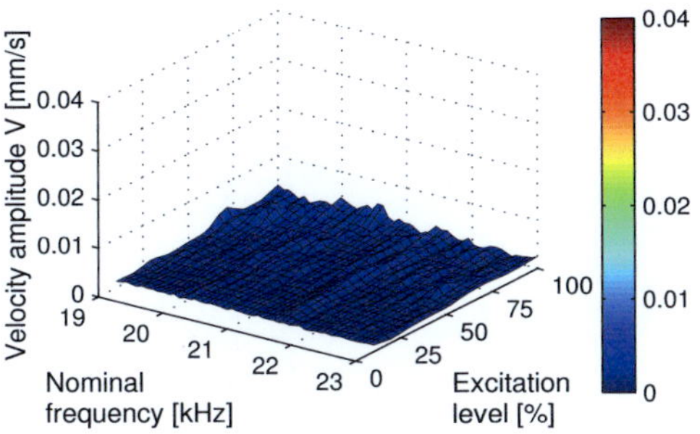

(e) 5th harmonic velocity amplitude

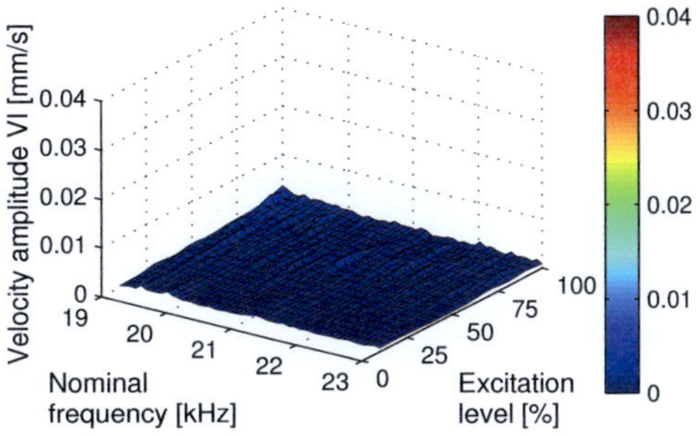

(f) 6th harmonic velocity amplitude

Figure A.33: Velocity amplitudes of the TIB transducer according to the 1st to 6th harmonic (fundamental tone and 2nd to 5th overtone).

A.7 Correlation between crack heating and displacement

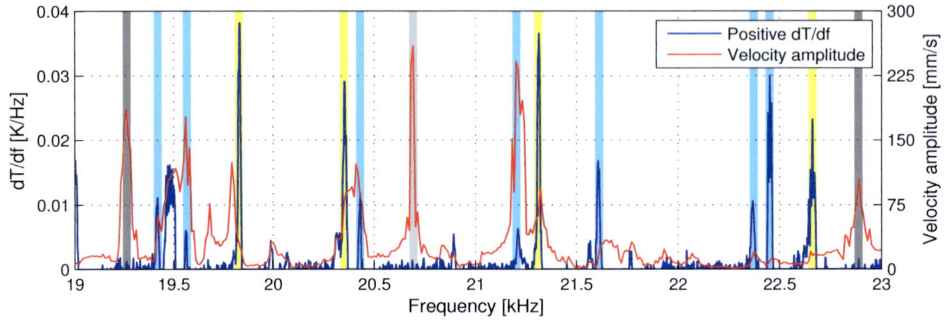

(a) Histories of $\mathrm{d}T/\mathrm{d}f$ and mode II velocity amplitude of crack side 1

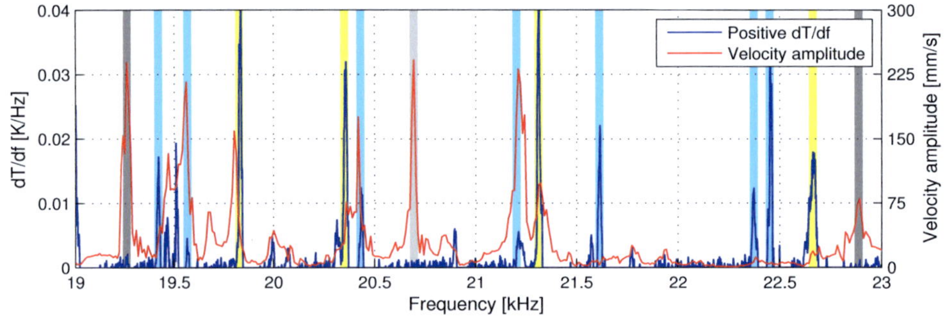

(b) Histories of $\mathrm{d}T/\mathrm{d}f$ and mode II velocity amplitude of crack side 2

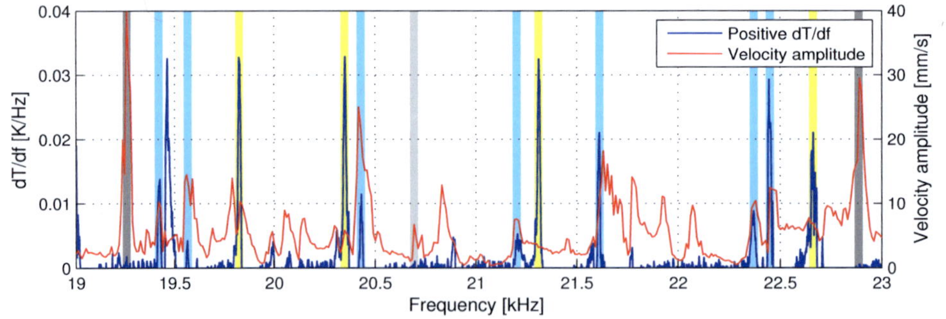

(c) Histories of $\mathrm{d}T/\mathrm{d}f$ and mode II differential velocity amplitude

Figure A.34: Histories of $\mathrm{d}T/\mathrm{d}f$ and mode II velocity amplitudes; specimen 06-06-B; blue: large diff. amplitude and heating, crack sides in-phase; yellow: mode III is determinative of the diff. amplitude; light grey: low diff. amplitude, no heating; dark grey: large diff. amplitude, no heating.

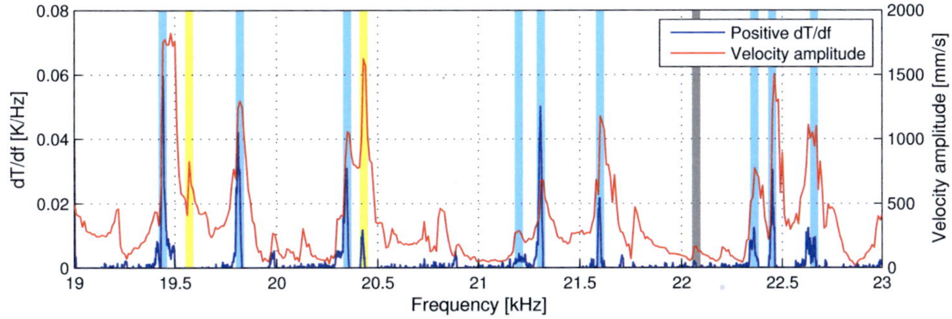

(a) Histories of $\mathrm{d}T/\mathrm{d}f$ and mode III velocity amplitude of crack side 1

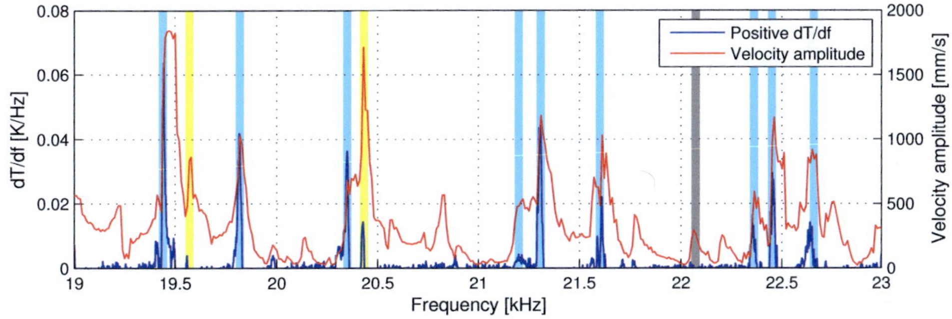

(b) Histories of $\mathrm{d}T/\mathrm{d}f$ and mode III velocity amplitude of crack side 2

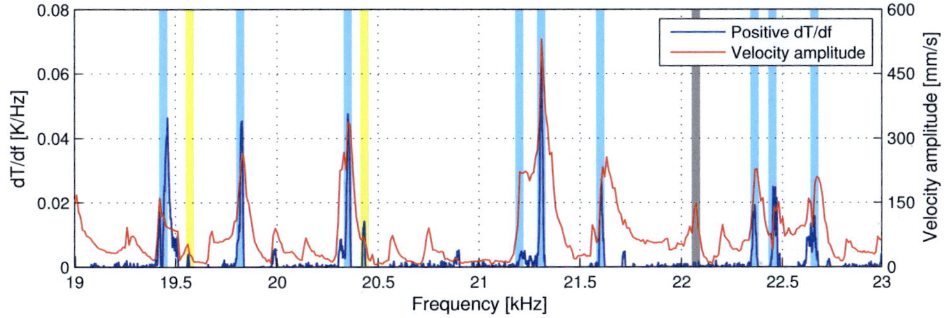

(c) Histories of $\mathrm{d}T/\mathrm{d}f$ and mode III differential velocity amplitude

Figure A.35: Histories of $\mathrm{d}T/\mathrm{d}f$ and mode III velocity amplitudes; specimen 06-06-B; blue: large diff. amplitude and heating, crack sides in-phase; yellow: mode II is determinative of the diff. amplitude; dark grey: large diff. amplitude, no heating.

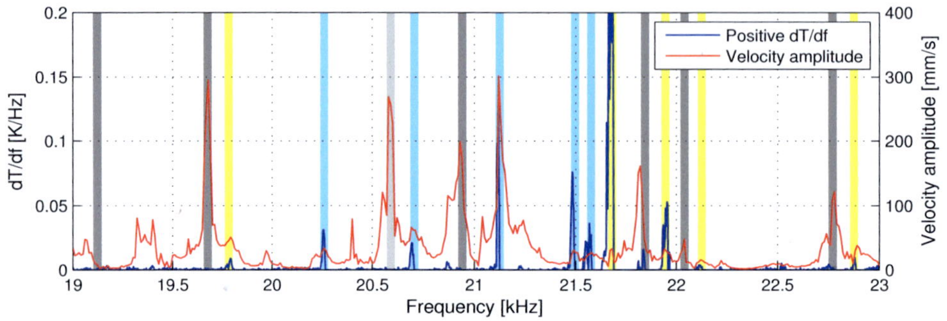

(a) Histories of $\mathrm{d}T/\mathrm{d}f$ and mode II velocity amplitude of crack side 1

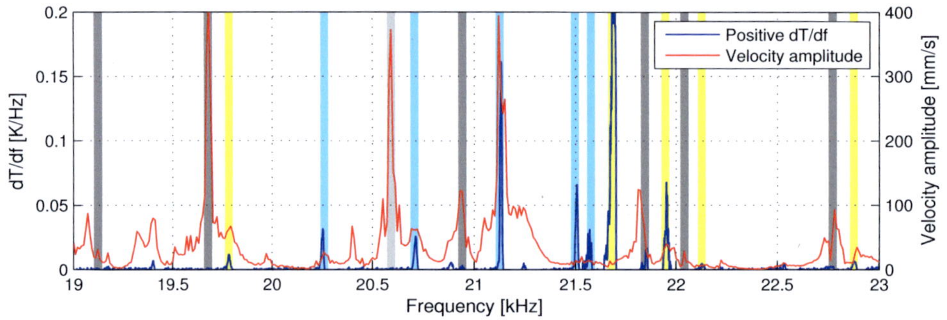

(b) Histories of $\mathrm{d}T/\mathrm{d}f$ and mode II velocity amplitude of crack side 2

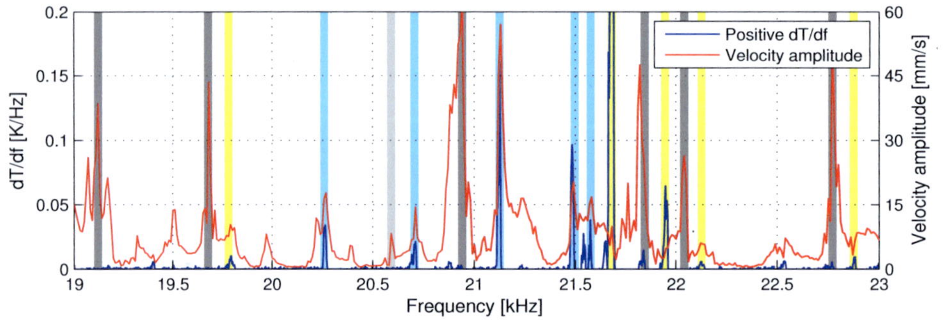

(c) Histories of $\mathrm{d}T/\mathrm{d}f$ and mode II differential velocity amplitude

Figure A.36: Histories of $\mathrm{d}T/\mathrm{d}f$ and mode II velocity amplitudes; specimen 10-12-I-B; blue: large diff. amplitude and heating, crack sides in-phase; yellow: mode III is determinative of the diff. amplitude; light grey: low diff. amplitude, no heating; dark grey: large diff. amplitude, no heating.

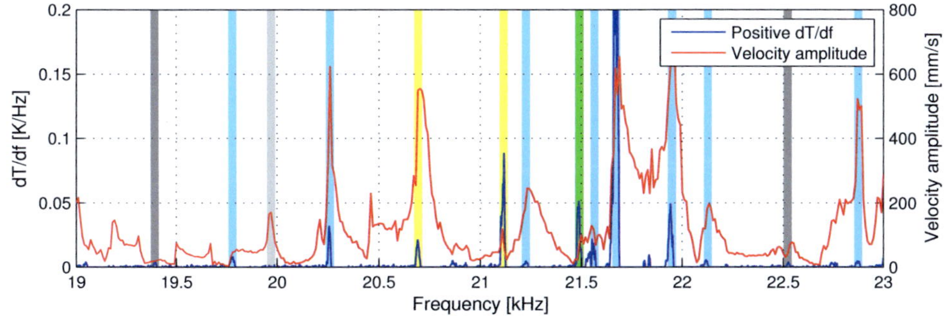

(a) Histories of $\mathrm{d}T/\mathrm{d}f$ and mode III velocity amplitude of crack side 1

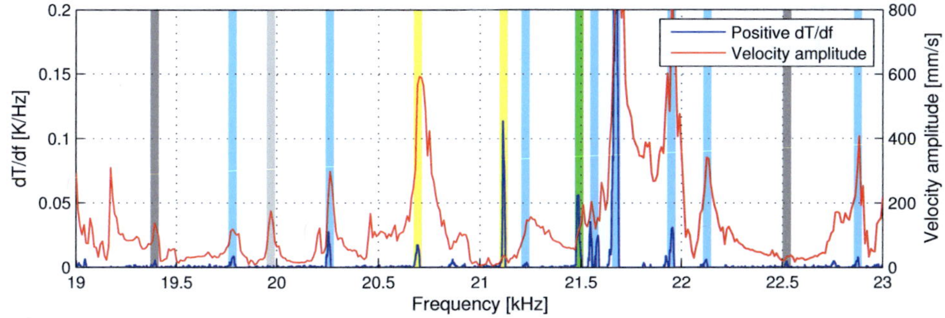

(b) Histories of $\mathrm{d}T/\mathrm{d}f$ and mode III velocity amplitude of crack side 2

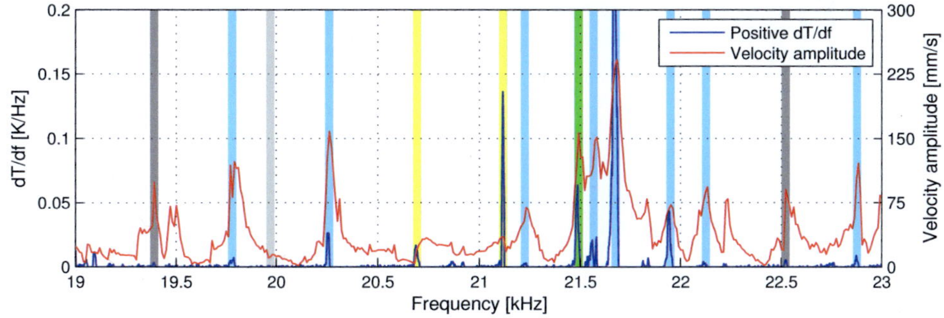

(c) Histories of $\mathrm{d}T/\mathrm{d}f$ and mode III differential velocity amplitude

Figure A.37: Histories of $\mathrm{d}T/\mathrm{d}f$ and mode III velocity amplitudes; specimen 10-12-I-B; blue: large diff. amplitude and heating, crack sides in-phase; green: large diff. amplitude and heating, crack sides out-of-phase; light grey: low diff. amplitude, no heating.

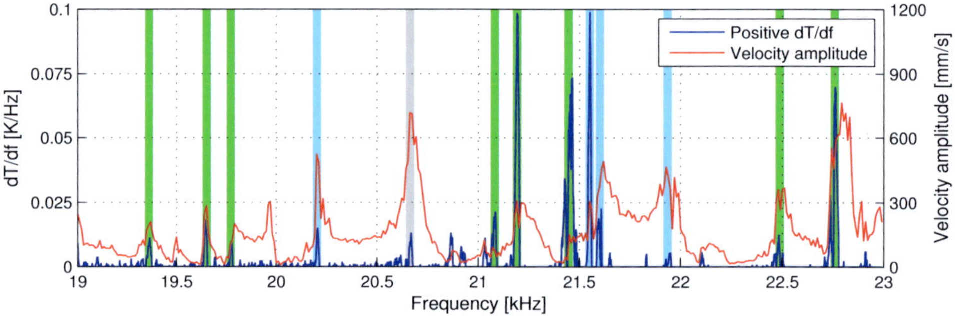

(a) Histories of $\mathrm{d}T/\mathrm{d}f$ and mode III velocity amplitude of crack side 1

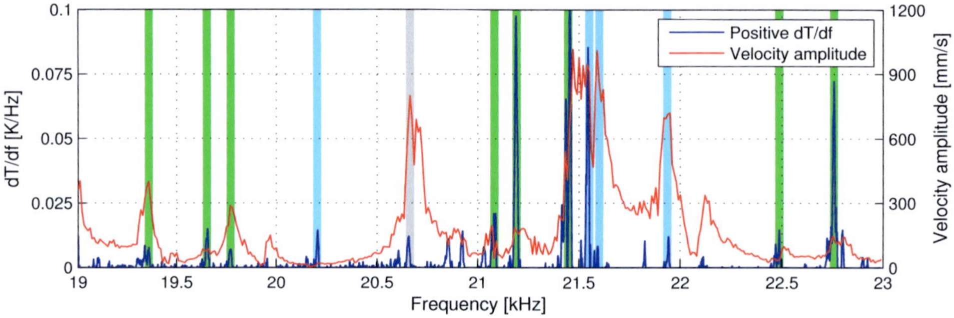

(b) Histories of $\mathrm{d}T/\mathrm{d}f$ and mode III velocity amplitude of crack side 2

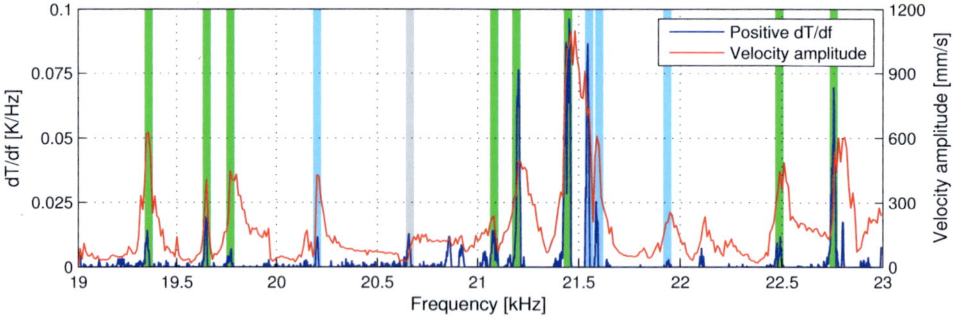

(c) Histories of $\mathrm{d}T/\mathrm{d}f$ and mode III differential velocity amplitude

Figure A.38: Histories of $\mathrm{d}T/\mathrm{d}f$ and mode III velocity amplitudes; specimen 10-12-III-B; blue: large diff. amplitude and heating, crack sides in-phase; green: large diff. amplitude and heating, crack sides out-of-phase; light grey: low diff. amplitude, no heating.

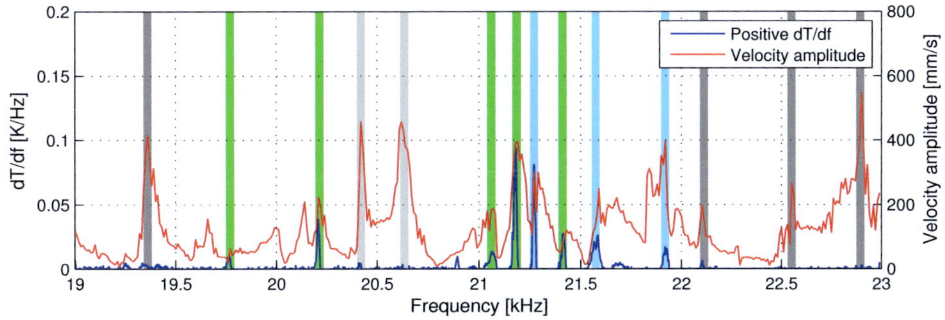

(a) Histories of dT/df and mode III velocity amplitude of crack side 1

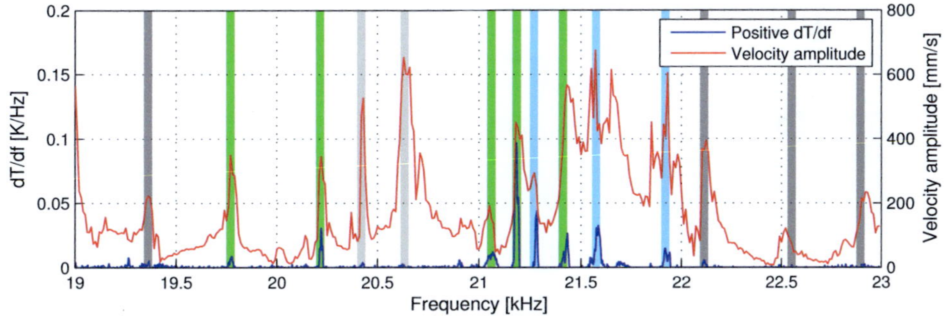

(b) Histories of dT/df and mode III velocity amplitude of crack side 2

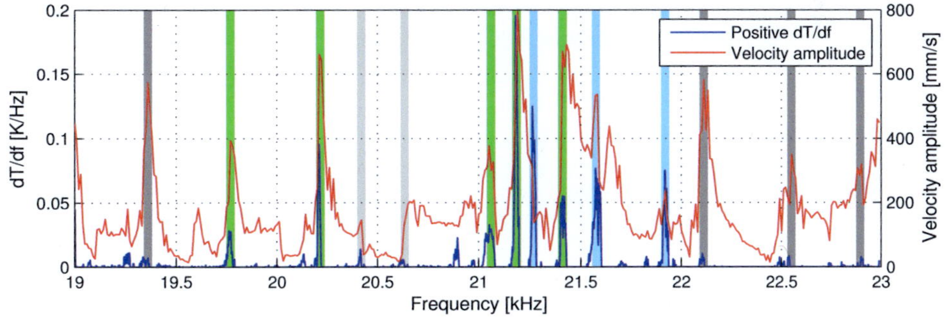

(c) Histories of dT/df and mode III differential velocity amplitude

Figure A.39: Histories of dT/df and mode III velocity amplitudes; specimen 10-24-III-B; blue: large diff. amplitude and heating, crack sides in-phase; green: large diff. amplitude and heating, crack sides out-of-phase; light grey: low diff. amplitude, no heating; dark grey: large diff. amplitude, no heating.

A.8 Three-dimensional vibration mode shapes

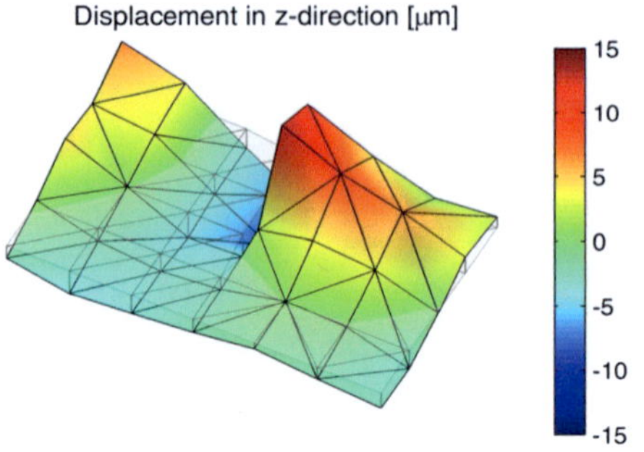

(a) Out-of-plane component, mode III (Branson)

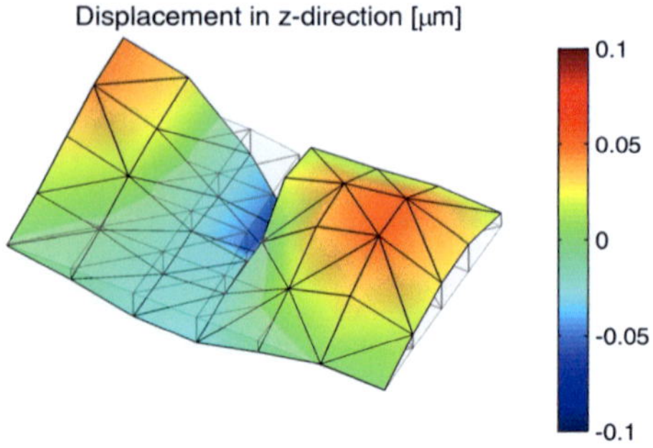

(b) Out-of-plane component, mode III (TIB)

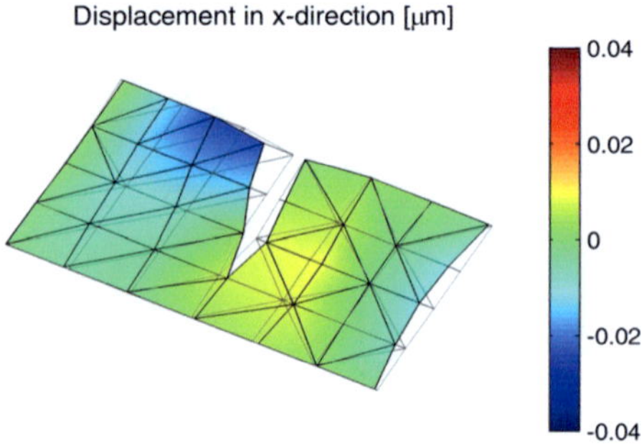

(c) In-plane component, mode I (TIB)

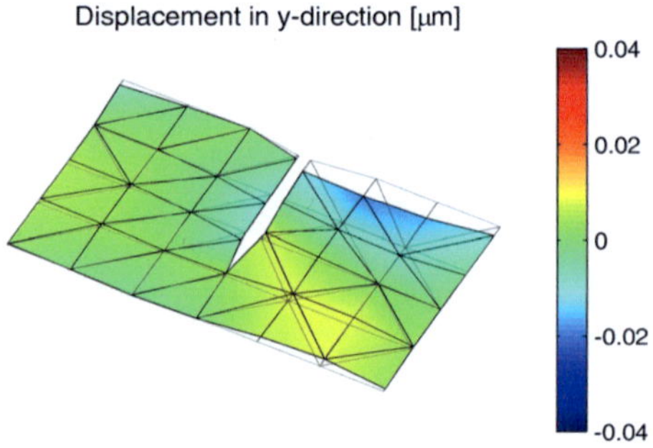

(d) In-plane component, mode II (TIB)

Figure A.40: Displacement components of the mode shape of specimen 06-24-B at 19.42 kHz.

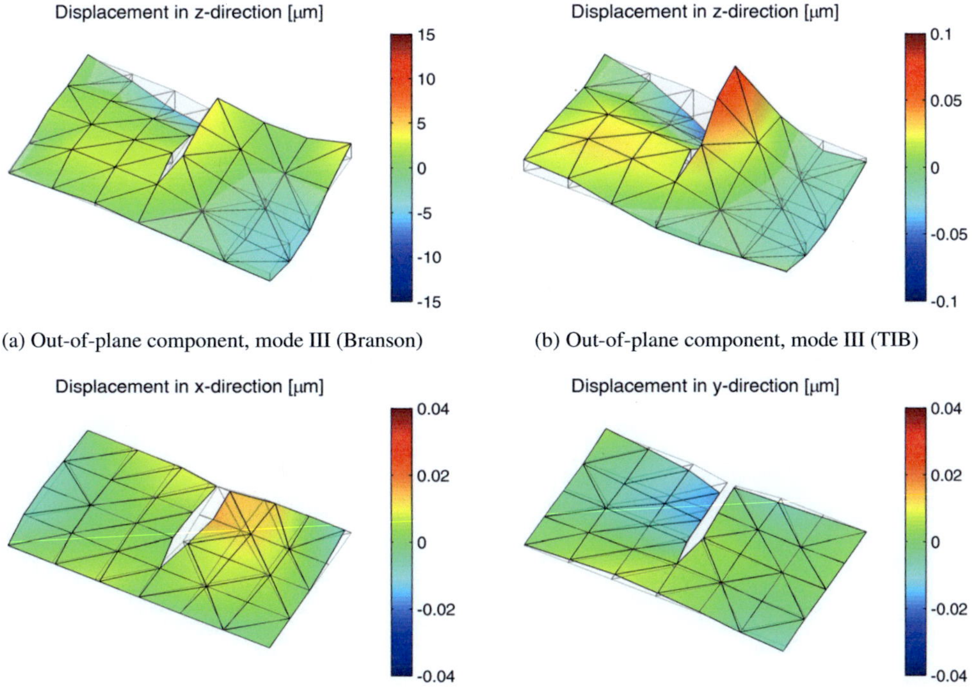

(a) Out-of-plane component, mode III (Branson)

(b) Out-of-plane component, mode III (TIB)

(c) In-plane component, mode I (TIB)

(d) In-plane component, mode II (TIB)

Figure A.41: Displacement components of the mode shape of specimen 06-24-B at 19.965 kHz.

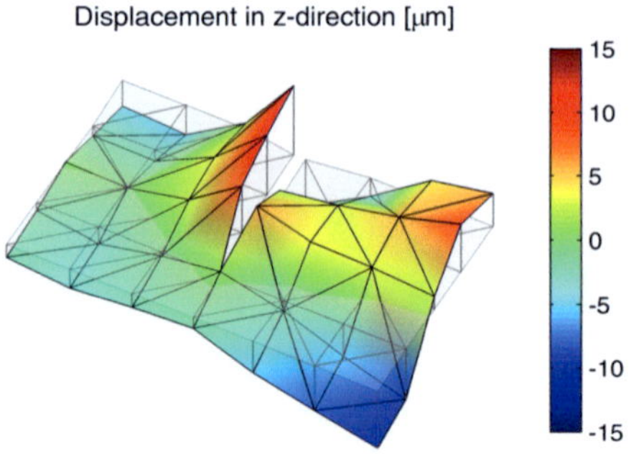

(a) Out-of-plane component, mode III (Branson)

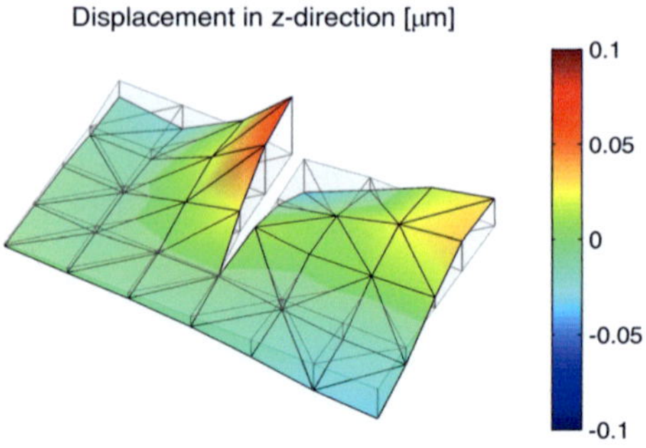

(b) Out-of-plane component, mode III (TIB)

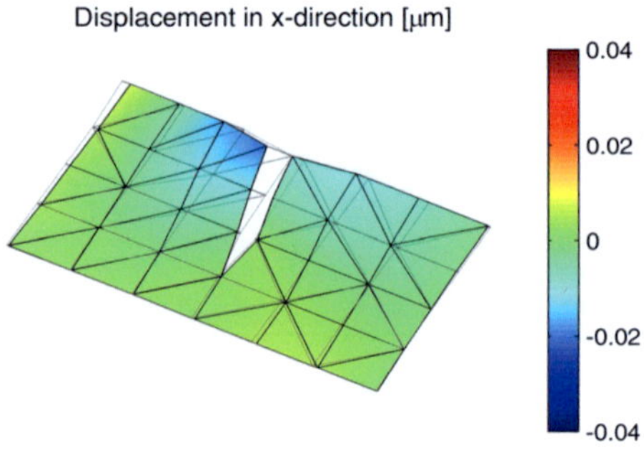

(c) In-plane component, mode I (TIB)

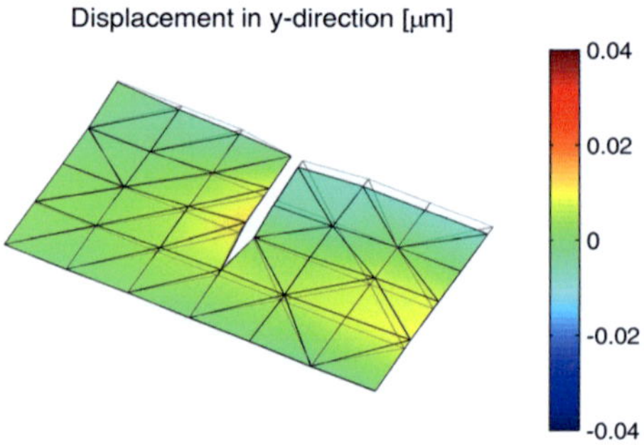

(d) In-plane component, mode II (TIB)

Figure A.42: Displacement components of the mode shape of specimen 06-24-B at 21.76 kHz.

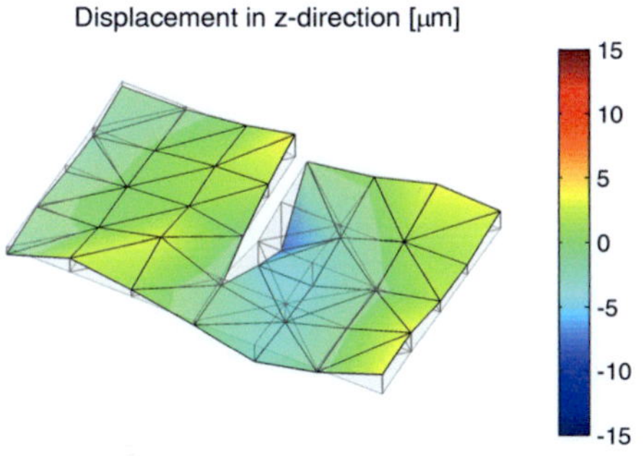

(a) Out-of-plane component, mode III (Branson)

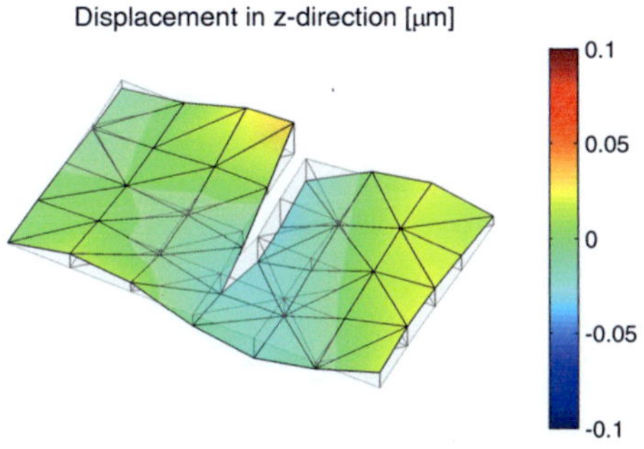

(b) Out-of-plane component, mode III (TIB)

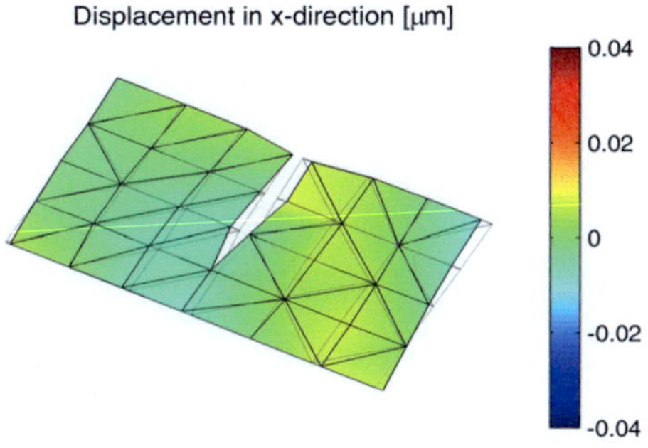

(c) In-plane component, mode I (TIB)

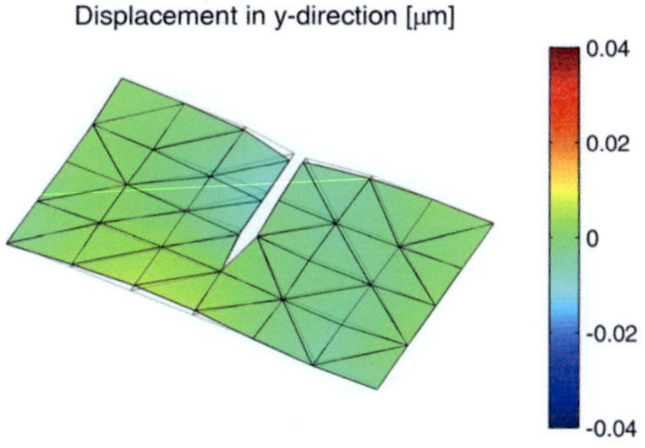

(d) In-plane component, mode II (TIB)

Figure A.43: Displacement components of the mode shape of specimen 06-24-B at 22.42 kHz.

A.9 Thermophysical properties of air and heat transfer coefficients

Table A.4: Thermophysical properties of air under atmospheric pressure [vB06].

T	k	C_p	ρ	α^1	η	υ^1
[°C]	[10^{-3} W/(m K)]	[J/(kg K)]	[kg/m^3]	[10^{-6} m^2/s]	[10^{-6} kg/(m s)]	[10^{-6} m^2/s]
10	24.94	1007	1.230	20.14	17.74	14.42
20	25.69	1007	1.188	21.47	18.24	15.35
30	26.43	1007	1.149	22.84	18.72	16.29
40	27.16	1007	1.112	24.25	19.20	17.27
50^2	27.88	1008	1.079	25.65	19.67	18.24
60	28.60	1009	1.045	27.12	20.14	19.27
70^2	29.31	1010	1.015	28.59	20.60	20.28

[1]Calculated values based on (3.101) and (3.133): $\alpha = k/(\rho C_p)$; $\upsilon = \eta/\rho$
[2]Values of k, C_p, ρ and η interpolated

Table A.5: Dimensionless parameters and heat transfer coefficients.

ΔT^1	Pr	Gr	Ra	Nu [−]			α_c [W/(m^2 K)]			$\alpha_r{}^2$
[K]	[−]	[10^3]	[10^3]	(6.6)	(6.7)	(6.8)	(6.6)	(6.7)	(6.8)	[W/(m^2 K)]
20	0.7132	315.2	224.8	11.26	11.75	11.54	5.95	6.21	6.10	6.33
40	0.7119	561.2	399.5	13.03	13.56	13.32	7.08	7.36	7.24	6.99
60	0.7112	754.5	536.6	14.06	14.60	14.34	7.84	8.14	8.00	7.72
80	0.7105	900.9	640.1	14.72	15.25	14.99	8.42	8.73	8.57	8.51
100	0.7095	1016.9	721.4	15.19	15.71	15.45	8.90	9.21	9.05	9.36

[1]Ambient temperature $T_{\mathrm{env}} = 20\,°\mathrm{C}$
[2]Emissivity $\epsilon = 1.0$

List of Figures

List of Tables